T0396732

The World IT Project

Global Issues in
Information Technology

World Scientific–Now Publishers Series in Business

ISSN: 2251-3442

The World Scientific–Now Publishers Series in Business publishes advanced text-books, research monographs, and edited volumes on a variety of topics in business studies including accounting, entrepreneurship, finance, management, marketing, operations, and strategy. The Series includes both applied and theoretical topics that present current research and represent the state-of-the-art work in their respective fields. Contributed by academic scholars from academic and research institutions worldwide, books published under this Series will be of interest to researchers, doctoral students, and technical professionals.

Published:

The complete list of titles in the series can be found at
https://www.worldscientific.com/series/ws-npsb

(Continued at the end of the book)

World Scientific – Now Publishers Series in Business: **Vol. 17**

The World IT Project

Global Issues in Information Technology

Editors

Prashant PALVIA
The University of North Carolina at Greensboro, USA

Jaideep GHOSH
Shiv Nadar University, India

Tim JACKS
Southern Illinois University Edwardsville, USA

Alexander SERENKO
Ontario Tech University, Canada
University of Toronto, Canada

Aykut Hamit TURAN
University of Nizwa, Oman
Sakarya University, Turkey

Published by

World Scientific Publishing Co. Pte. Ltd.

5 Toh Tuck Link, Singapore 596224

USA office: 27 Warren Street, Suite 401-402, Hackensack, NJ 07601

UK office: 57 Shelton Street, Covent Garden, London WC2H 9HE

and

now publishers Inc.
PO Box 1024
Hanover, MA 02339
USA

Library of Congress Cataloging-in-Publication Data
Names: Palvia, Prashant, editor. | Ghosh, Jaideep, editor. | Jacks, Tim, editor. |
    Serenko, Alexander, editor. | Turan, Aykut Hamit, editor.
Title: The World IT Project : global issues in information technology / Prashant Palvia
    (The University of North Carolina at Greensboro, USA), Jaideep Ghosh (Shiv Nadar University,
    India), Tim Jacks (Southern Illinois University at Edwardsville, USA), Alexander Serenko
    (University of Ontario Institute of Technology, Canada and University of Toronto, Canada)
    and Aykut Hamit Turan (Sakarya University, Turkey).
Description: USA : World Scientific, 2019 | World scientific-now publishers series in business;
    volume 17 | Includes bibliographical references.
Identifiers: LCCN 2019947676 | ISBN 978-981-120-863-8 (Hardcover) |
    ISBN 978-981-120-864-5 (ebook) | ISBN 978-981-120-865-2 (ebook other)
LC record available at https://lccn.loc.gov/2019947676

British Library Cataloguing-in-Publication Data
A catalogue record for this book is available from the British Library.

For any available supplementary material, please visit
https://www.worldscientific.com/worldscibooks/10.1142/11508#t=suppl

Desk Editors: Ramya Gangadharan/Sylvia Koh

Typeset by Stallion Press
Email: enquiries@stallionpress.com

Preface

It was June 2003

Some attendees at the Annual Conference of the Global Information Technology Management Association were at the foothills of Banff near Calgary, Canada, and were just having a good time. Then the conversation turned somewhat serious. Being part of an international Information Systems (IS) conference, we asked ourselves whether IS research was truly international. Unfortunately, the answer was a clear and unequivocal no. Then we asked ourselves what we could do to change the landscape. Thus, the seeds of the World IT Project were planted.

Then in December 2012 and 2013

After years of deliberation and consultation with like-minded IS researchers and a week-long retreat in December 2012, the World IT Project was officially launched in January 2013. We set an expansive and ambitious goal for ourselves. The project would examine various IT employee issues, such as organizational IT issues, technology issues, and individual issues in the context of different countries, representing different cultures, levels of economic growth, societal and religious beliefs, and political systems. We took our inspiration from Geert Hofstede's seminal work on national culture and the GLOBE project by Robert House and his colleagues on culture and leadership, and sought to examine important IS/IT issues in many parts of the world. Ultimately, the project would provide a methodologically grounded representation of the entire world and include 37 countries which exhibit diversity in terms of cultural, economic, political, religious, and societal backgrounds.

The long journey from 2013 to now

After finalizing and testing an instrument grounded in the literature, we sought to obtain data from countries that represented every major region of the world, and included different cultures, levels of economic growth, religious beliefs, and political systems. We recruited and selected local country investigators after a careful screening process. Local country investigators were needed because they understood the local culture and how to best approach IT employees in local businesses to participate. In the end, 37 countries participated in the project and there were about 90 country investigators. The data collection was completed at the end of 2017 and it was subjected to rigorous manual inspection as well as statistical testing to ensure construct reliability. Major challenges were encountered in managing the mega-project and the work of the country investigators, but each challenge was successfully overcome with perseverance and extensive communication.

The Book

The World IT Project has many deliverables. This book "Information Systems Management and Technology: The World Landscape" is an important and major contribution to the IS field in that it describes and analyzes the organizational, technological, and individual issues of IT employees in the 37 countries included in the project. The book starts with an introductory chapter which describes the goals and objectives of the World IT Project, its general framework and major research questions, the relevant literature and theoretical background, methodological details, expected outcomes and publications, and important contributions. This chapter is a *must read* for every reader in order to set the stage and for better understanding of the rest of the book. Then there is one chapter for each country. Each chapter, in alphabetical order of country, briefly describes the country's background, its history, and its information technology developments. Results are then presented for about three types of IT employee issues and interpreted in the country's context. Most chapters have co-authors from the country of focus (who were assisted by the central core team) and they had the requisite experience and expertise to be able to offer nuanced explanations of the findings.

Contributions and Who Should Read the Book

A good understanding of the critical IT issues facing firms and their employees within their surrounding contexts is important from the firm, national, and international points of view.

At the firm level, our results help management and staff in formulating business and IT related policies and strategies. At the national level, it allows stakeholders, such as policymakers, governments and vendors, to address important issues of the times. In international business, it helps firms and governments to respond to the needs of partners and stakeholders in other countries. For academic research, the World IT Project offers current and future scholars a grounded understanding of the international IT environment and provides a validated framework to launch many international IT studies.

The audience for the book is truly international and readers from every country would find it useful. Even if your own country is not among the 37 countries, you will find many countries that are similar to your country and many who you wish to aspire to or compare and contrast with.

More specifically, we see several groups who would benefit from the book. One such group is educators and students. Educators who are either conducting international IS research or teach courses would enhance their knowledge of the global IT environment. If you are teaching a course in International Business or International IS or even an advanced research course, the book will be an excellent resource for your students. For the same reasons, college libraries and institutional libraries may want to include it in their collection. In addition, multinational companies, international agencies, and governmental and non-governmental organizations would find it valuable for pursuing their own goals. Senior managers, IT managers, and Human Resource (HR) managers of virtually all organizations in the 37 countries that were included will benefit significantly from the analysis and discussions reported in the book. These findings may help them refine their IT and HR related initiatives and policies.

Enjoy the book and please provide your feedback.

Prashant Palvia, USA
Jaideep Ghosh, India
Tim Jacks, USA
Alexander Serenko, Canada
Aykut Hamit Turan, Turkey

Foreword

There is a popular belief that technology has nothing to do with people. The contrary is true. The Internet, once pioneer country, is becoming a "digital twin" of the real world. It is becoming increasingly obvious that preexisting structures and functions in society get reproduced, in new guises, in the electronic world. Cybersecurity and e-government differ from country to country. Information sharing systems in companies follow those countries' cultures. Communication technologies and social media are used differently in different cultures. International software development projects face cross-cultural misunderstandings. Institutional and political practices, themselves products of their respective cultures, get reproduced in the e-world.

On the surface, the world becomes unified through Information Technology (IT). IT tools enable the slaying of frontiers. Below the surface, however, there are cultural undercurrents. We thus have a global marketplace, to some extent, and a global economic battlefield. At the same time, the Web has not become a unified world. At the level of firms and countries, we defend our various "cultural villages" with passion. The "Global Village" of the 1990s is not happening. This creates tensions that are felt, among others, by all those working in or with IT.

It is therefore an important step to attempt an international comparative study, such as the World IT project. The project's ambition is to "provide a world view of IT issues that will be relevant to stakeholders at the firm, national, and international levels". I like that aim. This is not about finally resolving the issue; it is more about squarely putting it on the agenda. As the electronic world becomes smaller, while geopolitical weights shift, cultural differences take on more urgency at all these levels.

In the balance between relevance and rigor, this study heavily leans toward the former. That is good for a pioneering study. There will be more to come. It will be possible to sharpen the methodology and increase sample sizes. New IT will raise new questions. The great merit of this study is to create a bridge between IT issues and people issues, and to get so many countries on board. It provides a useful overview and will allow further investigations. After all, we are just getting started on the age of IT.

I warmly recommend this study.

Gert Jan Hofstede

Gert Jan Hofstede is currently a Personal Professor at Wageningen University, the Netherlands. He obtained a Master's degree in Population Biology, joined the software industry, and then returned to Academia for a Doctorate in planning systems called "Modesty in Modelling" (1992). He developed simulation games about IT in organizations, and worked on transparency and trust across international supply chains. He wrote on Synthetic Cultures as a simulation gaming scripting device, and joined his father Geert as an author of the best-selling Cultures & Organizations: software of the mind (2005, 2010; the book has been widely translated and is still appearing in new languages). More recently, he turned to Artificial Sociality, the modeling of human social behavior in agent-based models, often of socio-technical or socio-ecological systems.

About the Editors

Prashant Palvia, Ph.D., is a Joe Rosenthal Excellence Professor in the Bryan School of Business & Economics at the University of North Carolina at Greensboro, USA. Dr. Palvia received his Ph.D., MBA and MS from the University of Minnesota. He has worked extensively in the field of Global IT Management and chairs the conferences of the Global IT Management Association. Professor Palvia is the Editor-in-Chief of the Journal of Global Information Technology Management and Associate Editor for Information & Management. His research interests include global IT management, societal issues of IT, healthcare IT, and security and privacy. He has published 116 journal articles, 5 books, 21 book chapters, and numerous conference proceedings. His articles have appeared in such journals as the *MIS Quarterly, Decision Sciences, Communications of the ACM, Communications of the AIS, Information & Management*, and *Decision Support Systems*. He has co-edited five books on Global IT Management. In 2013, he formed an international research team and launched *The World IT Project*, which looks at important IT issues in 37 countries across the world.

Jaideep Ghosh is Professor of Decision Sciences, Operations Management, and Information Systems at the School of Management and Entrepreneurship, Shiv Nadar University, India. His research interests include applications of social networks, big data analytics, system dynamics, financial modeling and econometric methods in IS, operations, and decision sciences. He is an editorial board member of several peer-reviewed journals. His research articles have appeared in the *Journal of the Association for Information Science and Technology, Sociological Methods & Research, IEEE Transactions on Engineering Management, International Journal of*

Production Research, Communications of the Association for Information Systems, the *Journal of Global Information Technology Management, Social Indicators Research, and Scientometrics*, among others. His awards include several best paper awards, the best track chair award, and the Ramanujan Fellowship awarded by the Science and Engineering Research Board of the Department of Science & Technology, Government of India.

Tim Jacks is an Associate Professor at Southern Illinois University Edwardsville in the USA. He has been an active member of the World IT Project core research team since its inception in 2013. His research interests include culture (at the country, organization, and occupation levels), business/IT strategic alignment, and healthcare informatics. He is a pioneer in the area of IT Occupational Culture and its impact on organizations. He has published in a variety of academic journals including Communications of the Association of Information Systems, the DATA BASE for Advances in Information Systems, Journal of Global Information Technology Management, Information Technology and People, Business Process Management Journal, and Decision Support Systems.

Alexander Serenko is an Associate Professor of Management Information Systems in the Faculty of Business and IT, University of Ontario Institute of Technology and a Lecturer in the Faculty of Information, University of Toronto. Dr. Serenko holds a Ph.D. in Management Information Systems from McMaster University. His research interests pertain to scientometrics, knowledge management, and technology addiction. Alexander has published more than 80 articles in refereed journals, including *MIS Quarterly, European Journal of Information Systems, Information & Management, Communications of the ACM*, and *Journal of Knowledge Management*. He has also won six Best Paper awards at Canadian and international conferences.

Aykut Hamit Turan works as a Professor at the College of Economics, Management and Information System at the University of Nizwa, Oman and in the School of Management, Department of Management Information Systems at the Sakarya University, Turkey. Dr. Turan has done research in the field of Management Information Systems. His research interests include global IT management, healthcare IT, IT acceptance and adoption, and IT diffusion in SMEs. He has published a number of journal articles in such outlets as the *European Journal of Information Systems*, the *Journal of Global Information Technology Management, Information & Management*, the *Journal of Theoretical and Applied Electronic Commerce Research*, and *Communication of the ACM*. He has also published papers in the proceedings of more than 25 international conferences.

About the Contributors

Dolphy M. Abraham is a Professor of Systems and Operations Management, and Program Director of the Doctoral Program at the Alliance School of Business, Alliance University till 2016. Prior to joining Alliance University, Dr. Abraham was the Dean (Academics) at St. Joseph's College of Business Administration, Bangalore. He was also a Professor of Computer Information Systems at the College of Business Administration in Loyola Marymount University, Los Angeles, USA, for 15 years. Dr. Abraham earned his Ph.D. from the Joseph M. Katz Graduate School of Business, University of Pittsburgh. He is interested in studying how information technology impacts managerial decision-making and drives organizational change.

Ulla-Riitta Ahlfors has a Doctoral degree in Economics from the University of Jyväskylä, Finland, where she has worked for 17 years as an Assistant Professor in Marketing. She also has a history of business experience. In academia, Dr. Ahlfors has cooperated both internationally and nationally with entities in information management. Today, Dr. Ahlfors is a retired, independent researcher. Her research interests are strategic electronic business innovations, media business, and international business.

Chadi Aoun is an academic in the Information Systems program at Carnegie Mellon University in Qatar. He holds a Ph.D. in Information Systems from the University of New South Wales, along with qualifications in business, education, and environmental science. He is an expert in green information systems, sustainability management, geographic information systems, and e-collaboration. Dr. Aoun's research applies trans-disciplinary perspectives towards studying the pivotal role of information systems in complex socio-material contexts pertaining to sustainability, collaboration, and climate change. His academic experience includes appointments at universities in the US, Australia, UK, and Thailand. Dr. Aoun received teaching awards in 2010 and 2018 for his teaching excellence. He also served as a president of SIGGreen at the Association for Information Systems.

Ricardo Luis Parés Arce, is the cofounder of Nearshore Delivery Solutions and NDS Cognitive Labs, Mexican cognitive computing companies, and a board member of diverse companies attending the needs of thousands of clients in Mexico and the region. He has participated in international forums as a speaker and panel invitee, sharing his knowledge of the IT ecosystem in countries such as South Korea, Italy, and the United States. He holds a Bachelor's Degree in Business from Universidad Iberoamericana in Mexico City.

Carlo Gabriel Porto Bellini earned a BSc degree in Computer Science and a PhD degree in Information Systems & Decision Support from UFRGS, Brazil. He started his career in IT as a developer of database and Web applications in sectors such as air transportation, banks and telecom. He is currently an associate professor of information systems at the Department of Management, UFPB, Brazil. He is also the editor-in-chief of Brazilian Administration Review and senior editor of IT & People. In 2016–2017, he was visiting scholar at the Department of Information Systems & Supply Chain Management, Bryan School of Business & Economics, University of North Carolina Greensboro, USA, working under the supervision of Dr. Prashant Palvia. He published articles in Journal of Global Information Technology Management, Communications of the ACM, IT & People, Computers in Human Behavior, Team Performance Management, Personnel Review, and CyberPsychology & Behavior.

Gokul Bhandari is an Associate Professor of Information Systems at the University of Windsor, Odette School of Business. He got his Ph.D. from McMaster University, Canada. His research interests include data analytics, health informatics, and emerging technologies. He has published articles in various journals such as *Decision Support Systems, Behavior and Information Technology,* and *JCIS,* among others.

Nick Bontis is a Chair of Strategic Management at the DeGroote School of Business at McMaster University. He received his Ph.D. from the Ivey Business School at Western University. He is the first McMaster professor to win Outstanding Teacher of the Year and Faculty Researcher of the Year simultaneously. He is a 3M National Teaching Fellow; an exclusive honor only bestowed upon the top university professors in Canada. He is recognized the world over as a leading professional speaker and consultant and has amassed over 30,000 Google scholar citations. His most recent book is titled *Information Bombardment: Rising above the Digital Onslaught.* Visit his website at www.NickBontis.com.

João Álvaro Carvalho is a Full Professor at the Department of Information Systems, School of Engineering, University of Minho and a collaborator at the United Nations University Operating Unit on Policy-Driven Electronic Governance. His academic interests focus on the fundamentals of information systems and on enterprise development interventions that involve the adoption, use, and exploitation of information technology. He is also interested in research approaches and methods and on information systems curricula and education. He was one of the founders of the APSI — the Portuguese association in IS (currently an AIS national chapter) and of its annual conference — CAPSI. He also served as a national representative in IFIP TC8 for several years until 2009, was a Co-programme Chair of ECIS 2017, and was a member of the MSIS 2016 taskforce.

Wachara Chantatub is the Director of the Ph.D. and M.Sc. Programs in IT in Business at the Chulalongkorn Business School, Chulalongkorn University, Bangkok, Thailand. She graduated from the University of Sheffield, U.K., with a Ph.D. in Computer Science. Her teaching areas are in Database Management, Advanced Data Analytics, IT Governance, and Digital Transformation. Her research areas are in IT Governance and Digital Transformation. In addition to scholarly engagement, she also works closely with IT vendors. She consults many client organizations in both private and public sectors to share her knowledge and experiences to help them get the most out of their IT investment.

Alicia Cortagerena is a Chair Professor (Profesor Titular) of Information Systems at the Business School at Universidad Maimonides and an Associated Professor (Profesor Adjunto) of Administrative Systems at Facultad de Ciencias Económicas, Universidad de Buenos Aires (FCE-UBA). She is a Ph.D. Candidate (UBA). Alicia also holds a bachelor degree in Accountancy and a postgraduate degree in Pedagogy for Business and Economics. She was director of the Administration Program in an educational institution.

She has authored four books published by Pearson Education on the topics of business administration and ICT. Her professional experience focuses on advising and implementing curricular changes for various educational institutions.

Jocelyn Cranefield is a Senior Lecturer at New Zealand's Victoria University of Wellington. Her research focuses on the intersection of human behavior, management, and information and technology. Areas of special interest are ethical digital leadership, managing IT and data in transformative settings, the impact of technological trends on work, using online communities and networks for driving and embedding change, and managing sustainable smart cities. Jocelyn's research has been published in journals including *JAIS*, *CAIS*, and *Knowledge and Process Management*. She regularly presents at international information systems conferences such as ICIS, PACIS, ECIS and HICSS, is co-editor of a book on social knowledge management, and the author of a number of book chapters.

Olayinka David-West is an Information Systems Professional with over two decades of experience in the Nigerian IT industry. Olayinka is a senior fellow in the Operations, Information Systems and Marketing Division of Lagos Business School where she currently leads the Sustainable and Inclusive Digital Financial Services (SIDFS) Initiative. She holds a doctorate in Business Administration (DBA) from Manchester Business School; an M.Sc. in Business Systems Analysis and Design from City University, London; and a B.Sc. in Computer Science from the University of Lagos, Nigeria. Dr. David-West is also a Certified Information Systems Auditor (CISA), Certified in the Governance of Enterprise IT (CGEIT), and a qualified practitioner of the Skills Framework for the Information Age (SFIA).

Jerry Godwin Diabor is a lecturer at the Faculty of Architecture, Computing & Humanities at University of Greenwich, UK and NCC, UK at IPMC College of Technology, Ghana. Jerry holds a master in Management Information Systems from Sikkim Manipal University, India. His research interests pertain to business and data analytics, knowledge management, and operations research modeling and application. Jerry has co-authored 18 articles. He has also won the Best Student Award in Management Information Systems and the Best Teacher in the 2018 academic year, which is the top award given by IPMC College of Technology. In 2017, Jerry was ranked among the best five influential and productive lecturers in the business and information technology discipline by NCC, UK and University of Greenwich.

Eder Espinos Diaz holds a bachelor's degree in Electronics and Communications from the Technology University of Mexico. He has wide experience in the IT and Telecommunication industries, managing several large-size projects. He joined NDS Cognitive Labs and Nearshore Delivery Solutions, Mexican Cognitive Computing Companies in 2014 and has been participating in the IT ecosystem, collaborating with International and Mexican IT firms, helping them to expand capabilities in the LATAM Region.

Luis Alfredo Martins do Amaral is an Associate Professor at the Department of Information Systems in the School of Engineering of the University of Minho where he has been teaching since 1984. His other affiliations include: a researcher at the Algoritmi Center; a Scientific Coordinator of GÁVEA — Laboratory for Study and Development of the Information Society; a Founding member of APSI and APDSI; a Coordinator of the Network of Houses of Knowledge since 2013; a Chairman of the Board of the CCG — Computer Graphics Center since September 2005; a Coordinator of Portuguese Universities Foundation for the affairs of East Timor since 2011; a Pró-Rector of the University of Minho between July 2006 and October 2009; the President of the National College of Informatics (Order of Engineers) since 2010; and an Elected member of the General Council of the University of Minho since October 2013.

Octavian Dospinescu graduated from the Faculty of Economics and Business Administration in 2000 and the Faculty of Informatics in 2001. He received a Ph.D. in 2009 and he has published more than 30 articles. He is the author and co-author of 10 books, and he works as an associate professor in the Department of Information Systems in the Faculty of Economics and Business Administration, University Alexandru Ioan Cuza, Lasi. Since 2010, he has been a Microsoft Certified Professional, Dynamics Navision, Trade & Inventory Module. In 2014, he successfully completed the course "Programming Mobile Applications for Android Handheld Systems" authorized by Maryland University, USA. He is interested in mobile devices' software, computer programming and decision support systems.

Doina Fotache is a Professor of Business Information Systems at Alexandru Ioan Cuza University of Iaşi, Romania, in the Faculty of

Economics and Business Administration. She received her Ph.D. in Accounting Information Systems in 2000, and she is the author and co-author of 17 books, 15 articles in ISI Thomson Conference Proceedings and more than 100 articles published in journals and in Romanian and international conferences proceedings. Her research interests include information technology organizational impact, enterprise resource planning, groupware and collaborative technologies. Her teaching interests include business information systems, enterprise resource planning, customer relationship management and groupware. In her free time, she enjoys working in the garden, reading books, learning English, and traveling.

Claudio Freijedo is Chair Professor on various courses about Information Systems and Strategy at Universidad de Buenos Aires. He is Secretary of Technology Transfer at FCE-UBA, where he has held various academic and administrative positions. He is an invited professor in the Universities Nacional de Rosario (UNR), Católica Argentina (UCA), San Andres (UDESA) and Nacional De Asunción (UNA). He received a Ph.D. degree in Information Systems from the Universidad de Buenos Aires (UBA), an MBA and a BS in Accountancy. He has several books published by Pearson Education. He has participated as an organizer, coordinator, and lecturer in several professional conferences. He has experience in the industry as a programmer, systems analyst, project manager, manager and vice president of Operations and Systems in financial sector organizations.

Chiara Frigerio is an Assistant Professor of Organization Science and Information Systems at Università Cattolica of Milan. She holds a Ph.D. in Information Systems. She is a deputy director of CeTIF — a research center on innovation for financial services at Università Cattolica. Her research interests pertain to IT innovation and IT management in complex organizations, such as banks and insurance companies. Dr. Frigerio is currently doing research on blockchain and IoT technologies in order to understand the ethical and organizational impacts of designing and managing such innovations. She is a deputy director of the master degree in Fintech and Digital Innovation for Financial Services, hosted by Università Cattolica.

Damian Gajda, Ph.D., works as a researcher at the Department of Statistics at the University of Gdansk, Poland. His primary research interests include statistics, information society, and e-business. Dr. Gajda is an expert at designing and conducting statistical surveys and building structural equation models using SPSS and AMOS. He is an experienced lecturer for degree courses in Financial Mathematics, Descriptive and Mathematical Statistics, and Probability. Damian Gajda has co-authored numerous journal publications, conference papers, and book chapters.

Rimantas Gatautis (1974–2018) was a principal investigator leading the Digitalization Research Group at Kaunas University of Technology. He obtained his Ph.D. degree in Economics in 2002. Prof. Gatautis had wide-ranging experience of implementing European Commission funded projects in the Leonardo da Vinci, EUREKA, COST, Interreg, FP6, and H2020 programs. He was also a member of several scientific and advisory committees responsible for evaluating national government programs in the domains of electronic business and electronic government. His latest research focused on IST influence on enterprise transformation in the transition economies, e-business models in transition economies, socio-economic aspects of IST adoption, IST influence on processes of governance and democracy, knowledge-driven innovations, social networks, and gamification.

Rolando A. Gonzales is a researcher from ESAN University in Lima-Peru. He got a Ph.D. from ESADE Business School in Spain and an MBA from Pennsylvania State University in the USA. His primary research topics are business intelligence, analytics, and culture related to IT. He teaches several courses in the Graduate School of Business and in the undergraduate program, mainly related to analytics. He has published several articles in prestigious academic journals and presented them at major conferences. He is fluent in Spanish, English, and a little Portuguese. He likes to travel around the world and knows several different cultures. He is also impressed by the role that artificial intelligence plays in modern life.

Mary Ellen Gordon is a Senior Lecturer in Information Management and Director of the Masters of Information Management Programme at Victoria University of Wellington. Her academic research focuses on organizational data capabilities and marketing technology. Mary Ellen has extensive experience working in applied research and data-related roles, including: founding Apple's Market Research Analytics team, which supports marketing of Apple products worldwide; heading research at Flurry, including mining app data collected from more than a billion smartphones and tablets; and founding and leading organizations providing research and analytics consulting (Market Truths), data-driven content marketing (Story in the Data), and data training services (Analytics for Us). Mary Ellen's Ph.D.

is from the University of Massachusetts, and her bachelor's and master's degrees are from Babson College.

Alexandre Reis Graeml holds a Master's and Doctor's Degrees in Business Administration (Information Systems Management and Operations Management, respectively) from Fundação Getulio Vargas (FGV-EAESP), Brazil. His B.Sc. Degree is in Electrical Engineering from the Federal University of Technology — Paraná (UTFPR), also in Brazil. He is a professor at the Graduate School of Applied Computing (PPGCA) and the Graduate School of Business (PPGA) at UTFPR, in Curitiba. His international academic experience includes being a research associate (2002) and, more recently, a visiting scholar (2018) at the University of California — Berkeley. He also regularly teaches in the Business Information Systems graduate program at Ecole Supérieure d'Ingénieurs (Esigelec) in France. He is currently the President of the Association for Information Systems' LACAIS chapter (Latin America and the Caribbean).

Yue Guo is a Professor and Executive Dean at the Aurora International Research Institute, a joint research institute involving CODA research center at King's Business School, London, the Chinese Academy of Sciences, and Hohai Business School. Professor Guo's research interests are in data analytics, online platforms, online social media and social networks. His work has appeared in the *Harvard Business Review, Information System Journal, Information & Management, Journal of Information Technology, Tourism Management, Energy Policy, Computers in Human Behavior, Psychology & Marketing, Electronic Commerce and Research and Applications*, among other academic journals. Professor Guo maintains an active international research agenda. He has received multiple research grants from National Science Foundation of China and industries.

Penny Hart is a Senior Lecturer in Information Systems at the University of Portsmouth, UK. Her research interests include: knowledge sharing, organizational learning, virtual teams and communities of practice, qualitative methods (Soft Systems and Grounded Theory) and action research. She served as the co-chair of the organizing committee for the European Conference on Information Systems (2018) and has over 12 years of experience in the IT industry.

Md. Rakibul Hoque is an Associate Professor of Management Information Systems at University of Dhaka, Bangladesh. He is also a Visiting

Scholar, School of Business, Emporia State University, USA. His research interests include technology adoption, big data, e-health and ICT4D. Dr. Hoque has published a number of research articles in peer-reviewed academic journals and has presented papers at international conferences. He had the opportunity to work on a number of research projects in Bangladesh, USA, Australia, Japan, China, and Saudi Arabia. Dr. Hoque is the member of Association for Information Systems (AIS), UNESCO Open Educational Resources Community, Asia e-Health Information Network, Information Systems Audit and Control Association (ISACA), IEEE, and Internet Society.

Luminiţa Hurbean is an Associate Professor in the Business Information Systems Department of the Faculty of Economics and Business Administration at West University of Timişoara, Romania. She has a Ph.D. in Economic Informatics (1999) and has 28 years of teaching experience in the field of business information systems. She is the author or co-author of 17 books and more than 90 papers published in journals or conference proceedings. She has an extensive experience in teaching and researching integrated information systems (enterprise resource planning). Her research interests also include IT-driven change management or e-government systems.

Moazzam Hussain graduated from Mitchell College of Business, the University of South Alabama, USA, with an MBA degree. Moazzam also holds a B.S. Hons in Computer Science from Beacon House Informatics and Punjab University. He is an experienced business professional with 9 years of diverse finance, marketing, and IT experience in hospitality, education, telecommunications and public sector entities in the US and Pakistan. He has worked in academia for the last 5 years as an Assistant Professor in the University of Sialkot and has his research interests in MIS, m-learning, cloud computing, and strategic management domains. He has supervised several theses of M.Phil. students. Currently, he is working on his Ph.D. proposal in m-learning and e-Health for developing countries.

Mario Pezzillo Iacono is an Associate Professor of Organization Studies, Department of Economics — Vanvitelli University — where he currently teaches Human Resource Management and Organizational Design. Mario holds a Ph.D. in Organization Design and Human Resource Management and is a Visiting Researcher at Cardiff Business School, Cardiff

University. His research interests are focused on knowledge management and innovation, human resource management and organizational diversity/identity, organizational control in private and public sectors, and inter-organizational coordination mechanisms in private and public sectors.

A.K.M. Najmul Islam works as a University Research Fellow in the Department of Future Technologies, the University of Turku. Dr. Islam holds a Ph.D. (Information Systems Science) from the University of Turku, Finland, and an M.Sc. (Eng.) from Tampere University of Technology, Finland. He has more than 50 publications. His research has been published in outlets such as *Computers & Education, Information Technology & People, Telematics & Informatics, Communications of the AIS, Journal of Information Systems Education, AIS Transaction on Human–Computer Interaction, Computers in Human Behavior,* and *Behavior & Information Technology.*?

Noor Ismawati Jaafar is an Associate Professor and Deputy Dean (Research & Development) at the Faculty of Business and Accountancy, the University of Malaya. She obtained Doctor of Business Administration (DBA) qualification from Macquarie Graduate School of Management (MGSM), Macquarie University in Sydney, Australia. She holds a professional qualification as a Certified Financial Accountant (CFIA) from the Malaysian Institute of Certified Public Accountants (MICPA). She is a member of the Association for Information Systems (AIS). She has published in many refereed journals such as *Government Information Quarterly, Computer in Human Behavior, Behavior and Information Technology, Industrial Data and Management Systems, Telematics and Information International Journal of Mobile Communication,* and *Internet Research.* Her research areas are IT management, IT governance, and human–computer interaction.

Michel Kalika is an emeritus Professor at the University Jean Moulin, Iaelyon School of Management. He has published 25 books and approximately 100 other publications in strategy and information technology in premier journals. His research focuses on the use of Information and Communication Technologies by managers (the "Millefeuille" theory) and the impact of Business Schools. He is the co-Director of the Business School Impact System (BSIS) methodology he developed for FNEGE & EFMD.

This process has so far been used in about 30 international Business Schools. Michel Kalika is also the President-Scientific Advisor of the Business Science Institute (international DBA programme).

Adam Katona is a Ph.D. candidate at the Institute of Marketing and Media at Corvinus University of Budapest, Hungary. He is also working for a multinational technology company as a Marketing and Sales Director.

Nikolay Kazantsev is a Research Associate/Post-Graduate Researcher at the Management Sciences and Marketing Division and at the Alliance Manchester Business School, UK. Nikolay is also a Lecturer of Business Informatics at the Faculty of Business and Management at National Research University "Higher School of Economics," the Russian Federation. Nikolay holds an MBA from National Research University "Higher School of Economics," Russian Federation and an M.Sc. in Information Systems from University of Munster, Germany. His research interests include digital economy, collaboration, digital/smart manufacturing, industry 4.0, and knowledge management. Nikolay has published more than 30 articles and international conference papers, including two papers that received the Best Paper Award. In 2018, Nikolay received an Earlier Career Researcher Placement Award and a mini-sabbatical at the University of Queensland.

Hajer Kefi is a Professor of Management Science in PSB Paris School of Business. She is invited by the National University of Singapore. Her research interests include social media usage, big data analytics and digital marketing, and Information Technology ethics. She received a Ph.D. degree from the University of Paris Dauphine and a post-doctoral degree in Research Supervision (HDR) from the University of Paris Sud, France. She has won several awards, such as the Best Information Dissertation in France and the Best Papers at international conferences. Her work has appeared in premier scientific journals, such as the *Journal of Business Research,* the *Journal of Strategic Information Systems, Information Technology and People, Management International, Systèmes d'information et Management, Information Resources Management Journal,* and *Communication of the AIS.*

Tamara Keszey is an Associate Professor at the Corvinus University of Budapest, Hungary. Prior to her academic career, she had been working as an IT consultant for KPMG and participated in various IT system implementations at firms, such as Lufthansa in Germany and Lyreco in France. Her research interests include the organizational aspects of firm-level IT system adoption and technology acceptance by the individual. She is the member of several research collaborations with scholars from the University of Groningen and Norwegian University of Life Sciences. Tamara is the author of more than 80 academic publications. Her studies have been published in the Journal of Business Research, Information Systems Management, Journal of Knowledge Management, International Business Review, etc.

Youngkyun Kim is a Professor in the College of Business at Incheon National University, Korea. He received his Ph.D. from Inha University in Korea, and Master's degree from Indiana State University. Prior to his employment in academia, he worked at Asiana Airlines, Kumho Tires, Statefarm Insurance, and other firms for more than 16 years. His research has been published in academic journals such as Technology Forecasting and Social Change, Asian Journal of Technology Innovation, Journal of Targeting, Measurement, and Analysis for Marketing among others. He serves as an associate editor in the Journal of Korean Society of Industrial Information Systems. His research interests include high-tech marketing and strategic marketing of technology firms.

Zlatko Kovačić is director of a statistical consultancy agency. He has over 30 years of experience as a University professor, international researcher and government consultant, and was formerly Associate Professor in Statistics at The Open Polytechnic of New Zealand. He has published widely in the areas of time series, forecasting, multivariate analysis, general applied statistics, information systems, e-commerce, e-government, learning styles, online education, and financial time series.

Kyootai Lee is a Professor at the Graduate School of Management of Technology at Sogang University, Korea. He received his Ph.D. from the University of Missouri-St.Louis. His research has been published in *Information Systems Journal, Journal of Organizational Behavior, Leadership Quarterly, Technology Forecasting and Social Change, Journal of Business Research, European Management Journal, DataBase*, and *Journal of Global Information Technology Management*, among others. His recent research has been focused on decisionmaking in innovation implementation processes, individual and organizational innovativeness, status quo biases, and the management of innovation. Recently, he has become interested in applying statistical learning (a.k.a. machine learning) techniques to his research areas.

Nikola Levkov is an Associate Professor of Management Information Systems at the Faculty of Economics — Skopje, Ss. Cyril and Methodius University in Skopje, Republic of Macedonia. He holds a Ph.D. and an MBA from the Faculty of Economics — Skopje, and an M.Sc. in information management from KU Leuven, Belgium. He has completed the JFDP (Junior Faculty Development Program) in the field of business administration at George Washington University, Washington, USA. His research interests are related to social business–IT alignment and IT professional business communication. His research has been published in many journals and international conference proceedings.

Marcello Martinez is a Full Professor of Organization Studies at Università della Campania Luigi Vanvitelli. He is also the Coordinator of the Faculty Board of the Ph.D. course in Entrepreneurship and Innovation. He is the President of ASSIOA — Association of Italian Organization Studies Academics and a member of AIDP — Italian Association of Human Resources Management and AIDEA — Accademia Italiana di Economia Aziendale. Marcello is the Editorial Chief of the journal "Prospettive in Organizzazione" ISSN: 2465-1753, and of the journal "Studi Organizzativi" FrancoAngeli, ISSN 0391-8769". He is a member of the scientific committee of the journal "Sviluppo e organizzazione" Este editore, ISSN: 0391-7045 and of the journal RU Pubblica amministrazione, Maggioli ISSN: 1723-9877. He is a member of the editorial board of the journal "Law and Economics Yearly Review" ISSN 2050-9014.

Nicholas Mavengere is a Lecturer in Business Information technology at Bournemouth University, UK. He completed his Ph.D. in Information Systems at Tampere University, Finland in 2013. His research was selected as the best Ph.D. at a top European conference and received a university award. His research interests include technology adoption and application by SMEs, digitalization of public services, technology application in higher education, business and IT alignment, strategic agility in business and ICT4D. He has published in various journals, such as the Journal of Education and Technologies, the Journal of Information Technology Cases & Applications Research, Electronic Journal for IS Evaluation, and International Journal of Agile Systems and Management. He has reviewed for several journals and conferences. He is a vice chair of IFIP WG 3.4.

Valter Moreno Jr. has a Ph.D. degree in Business Administration from the University of Michigan. He holds a joint appointment as a Professor of Management Information Systems in Rio de Janeiro State University (UERJ) and Faculdades Ibmec, Brazil, as well as a visiting position at the German Graduate School of Management and Law (GGS). He has published more than 70 papers in the proceedings of international conferences and in peer-reviewed journals. He has also been involved in consulting and executive education activities for organizations such as EY, Petrobras, Bradesco Seguros, SHV, Endesa, and the United Nations Development Programme. His research interests are centered on the influence of organizational and IT resources, the generation of business value, and firm performance and competitiveness.

Eddy A. Morris is Ph D candidate in Information Technology and Communications in La Salle Ramon Llull University, Barcelona, Spain. He also is the Director of the MBA and Master of Science in Information Technology in ESAN University in Peru, and a renowned consultant in Information Technology and Business, for the more important companies in Peru.

Elisabeth F. Mueller is an Assistant Professor in Management at the University of Passau (Germany). She holds a Ph.D. in Management from the University of Passau. She was a Visiting Scholar at the University of Pennsylvania (USA) and taught at the Corvinus University (Budapest, Hungary) and the Turkish-German University (Istanbul, Turkey). She recently received research grants from the German Research Foundation (Deutsche Forschungsgemeinschaft) and the German Academic Exchange Service (Deutscher Akademischer Austauschdienst). Her current research interests lie at the intersection of strategy, international business, and entrepreneurship. Specifically, she focuses on research questions in the areas of cooperative strategy, governance, and business models of small and medium-sized companies and family firms as well as on governance of inter-firm cooperation.

Hamid Nemati is full Professor of Information Systems at the University of North Carolina at Greensboro. He received his PhD from the University of Georgia in Information Technology and Management Science and his MBA from The University of Massachusetts. He is internationally recognized for his research in various aspects of Information Technology, including data analytics, big data, information security and privacy, organizational and behavioral aspects of Information Technology development and use. He has extensive professional experience as a developer, an analyst and project leader and has been a consultant for numerous major corporations. He has published nine books and over 120 peer reviewed academic publications in various premier scholarly and professional journals and conference proceedings.

Kennedy Njenga, Ph.D., is an Associate Professor in the Department of Applied Information Systems, the University of Johannesburg in South Africa. He has published and presented his research both nationally and internationally in the field of information systems and technology. His research focuses on methodological, philosophical and behavioral issues related to the protection of technology systems such as deviant computer behavior. He also has a special research interest in protecting technology around the Internet of Things. His work has been published in outlets such as the *European Journal of information Systems*, the *African Journal of Information Systems*, and conference proceedings such as the *Pacific Asian Conference on Information Systems*.

Vicente Cubells Nonell holds a Ph.D. in Computer Science and a Master's degree in Telematics and his professional career in Engineering in Computer Science. He has more than 20 years of experience in computer systems, both in the productive and academic fields. He has collaborated in various national and international projects and has taught professional and postgraduate levels in various educational institutions. Since 1995, he develops software applications to areas such as architecture, civil engineering, manufacturing, education, telecommunications, retail, and manufacturing. Among his areas of interest are programming languages, algorithms, parallel and concurrent programming, data processing, machine learning, and cognitive computing.

Eugene Ohu is a Senior Lecturer and the Head of the Department of Organizational Behavior/Human Resource Management, Lagos Business School, Nigeria. His current research focuses on the psychology of human–computer interaction in the workplace, to promote individual wellbeing and organizational productivity. Eugene is intrigued by how people navigate multiple roles across work and non-work domains. As technology introduces flexibility in the nature and design of work, Dr. Ohu studies the implications for work–life balance that this permeability introduces. He also uses Big Data to provide insights useful for human resources and marketing. His research has been published in outlets such as the *Journal of Occupational Health*

Psychology. He is an executive director of the Nigeria Internet Registration Association (NiRA), which manages the .ng domain.

Gillian Oliver is an Associate Professor of Information Management at Monash University in Australia and a Director of the Center of Organizational and Social Informatics. Previously, she led teaching and research into archives and records at Victoria University of Wellington and the Open Polytechnic of New Zealand. Her research interests focus on the information cultures of workplaces and issues relating to the continuity of digital information. She is co-editor in chief of *Archival Science*, co-author (with Fiorella Foscarini) of *Records Management and Information Culture: Tackling the People Problem* (Facet, 2014) and co-author (with Ross Harvey) of the 2nd edition of *Digital Curation* (Facet, 2017).

Gustavo Parés, born in Mexico City, holds a bachelor's degree in Management Information Systems and a master's degree in Business Administration from the Monterrey Institute of Technology and Higher Education. He also received an Executive Certificate in Management and Leadership at the Massachusetts Institute of Technology and attended executive education courses at Harvard, Singularity University, and Stanford. Gustavo is a digital entrepreneur and a partner of Nearshore Delivery Solutions and NDS Cognitive Labs, Mexican cognitive computing companies. He is an experienced technology advisor, and he is known for helping, optimizing, and supporting customers as they develop strategic solutions for the future of their businesses.

Vasile-Daniel Pavaloaia obtained his Ph.D. in Accounting Information Systems in 2008 and he is an Associate Professor in the Department of Accounting, Business Informatics and Statistics at *Alexandru Ioan Cuza University* of Iaşi, the Faculty of Economics and Business Administration. His teaching subjects and research interests include enterprise resource planning, information technologies for business administration and public sector, business process modeling, and artificial intelligence for business. Over the last 10 years, he has published more than 40 individual and co-authored articles, authored 10 books, and presented his work at more than 20 international conferences.

Luis Humberto Rojas Pineda holds a bachelor's degree in International Relations from the Monterrey Institute of Technology and Higher

Education. He has a keen interest in the confluence of technology and society, specifically in the role technology companies play in the development of countries and their economies. He joined Nearshore Delivery Solutions, a Mexican cognitive computing company, after a season working for Mexico's Ministry of Economy in New York. He has participated actively in the research of Mexico's IT ecosystem and the development of the national cognitive computing hub.

Anne Powell is Professor and Chair in the Computer Management and Information Systems department at Southern Illinois University Edwardsville. She received both her MBA and Ph.D. in the Management of Information Systems from Indiana University. She has ten years corporate experience as a systems analyst and business analyst. Her primary teaching responsibilities are in Database and Systems Analysis and Design courses. Her research specializations include virtual teams, the impact of new technology on individuals, teams, and organizations, and technology-enhanced education. Her work has been presented at numerous conferences and has been published in numerous Information Systems academic journals.

Ijaz A. Qureshi is the Vice Chancellor and the Dean of the Faculty of Business and Administrative Sciences at the University of Sialkot. Dr. Qureshi holds a Ph.D. from Argosy University, San Francisco, California, an MBA in e-commerce from John F. Kennedy University, California, and an M.S. in Strategic Management from American International University, UK. His areas of interest include business process management, management information systems, disruptive technologies and modern businesses, RFID and its application in the drivers licenses and ID cards, radio frequency identification technology and its application in consumer goods, intellectual property rights and innovation, retail industry and customer services for customer retention, FMCGs and disruptive technologies, and strategy and digitization. He has numerous international journal publications in top tier journals and has attended conferences with renowned academies of management.

Federico Rajola is a Full Professor at the Management Faculty of Università Cattolica of Milan where he teaches Organization Design. He is the Rector Delegate for Innovation and Information Systems. He is a scientific director of the CeTIF research Center on Technology, Innovation, and Finance. He has authored several papers and books on Information

Systems, Business Intelligence, CRM, and Innovation in Organizations. He has chaired several international academic conferences and workshops. Federico has coordinated and managed several innovative European research projects under the Esprit and IST programs. He is a member of the Editorial Board of *Information Systems and e-Business Management Journal*, published by Springer. He was a Visiting Scholar of Innovation Trends at INSEAD, and a Visiting Professor of Artificial Intelligence and Innovative Systems at the University of Darmstadt.

Myriam Raymond holds a Ph.D. from Université de Nantes in banking information systems. She held a Postdoc position on the subject of Big Data & Risk Analytics in the banking sector. She now holds a professor position at Université d'Angers in France and is an invited lecturer at Audencia Business School as well as at ESLSCA MBA program. As an active member of several academic counsels and workgroups, she is actively involved in curricula development and pedagogy. She is an international Faculty Visiting Staff for a number of European universities; she equally publishes original research articles in scientific journals and daily newspapers. With devotion and energy, she also serves as a consultant onboard several foundations and social entrepreneurship projects.

Hassan Raza graduated from Cardiff Business School (UK, M.Sc.) and is doing a Ph.D. at the University of Malaya, Malaysia. His research interests are social media, e-learning, m-learning, consumer engagement, social identity, and identity construction.

Pablo Rota is a Professor Titular (Chair Professor) of Information Technology at Facultad de Ciencias Económicas, Universidad de Buenos Aires (FCE-UBA) and a former professor at Comahue's National University and Patagonia's National University. He is heading as Academic Secretary at the FCE-UBA, where he was a Board Member, a member of the Advisory Commission of the Degree in Information Systems, and a former Secretary at the Graduate Specialization in Information Systems. He is a Ph.D. candidate (UBA) and has a degree in Accountancy. Pablo participated in several conferences and professional symposiums. He is a co-author of *Managerial Information Systems: Technology to Add Value to Organizations* published by Pearson (2011). His professional experience mainly focused on the development and application of information system's projects in both public and private organizations. Pablo is an external auditor of several organizations.

Mikko Ruohonen is a Professor of Business and Information Systems at the Tampere University, Finland. He has worked for more than 30 years in information and computer sciences, business development, and information society research. His interests are in information strategies, electronic business, knowledge management, inter-organizational learning, ICT for development, mass customization and strategic agility. He has published more than 150 articles or reports, authored or co-authored 10 textbooks and a number of large industry reports on technology, innovations and organizational development. International Federation of Information Processing (IFIP) granted him the Silver Core Award 2007 under the Technical Committee 3, IT and education.

Mijalche Santa, Ph.D., is an Associate Professor at the Faculty of Economics — Skopje, Ss. Cyril and Methodius University in Skopje, Republic of Macedonia, and is a postdoctoral fellow at Gent University in Belgium. He received his Ph.D. from University Paris 1, Pantheon, Sorbonne in 2014. His research focuses on information systems and digital innovation. He has published in journals such as Knowledge Management Research and Practice and the Learning Organization and presented at various conferences, such as ECIS, ICEIS, and others.

Hiroshi Sasaki is a Professor of IT systems and marketing research in the college of business at Rikkyo University, Japan. He has been serving as the Dean of Career Center for five years. Dr. Sasaki holds a Ph.D. from Osaka University. Prior to joining Rikkyo, he was a Professor at Momoyama Gakuin University. Before that, he worked as a systems analyst at Fujitsu, and a business consultant at the Sanwa Research Institute (currently Mitsubishi UFJ Research & Consulting). He was also a Visiting Professor at Michigan State University in 1999 and at the Turku School of Economics in Finland in 2010. Dr. Sasaki is a board member of several academic societies in Japan. His recent interests include the co-creation paradigm and the business ecosystem perspective.

Osam Sato received his B.Sc., MA, and Ph.D. degrees from Hitotsubashi University in Tokyo, Japan. He joined the Faculty of Business Administration, Tokyo Keizai University (TKU) as a lecturer in 1985. He is a Full Professor at the Faculty of Business Administration, TKU. His research interest includes information systems, supply chain management, and e-learning. His research has been published in many scholarly journals, including the *Journal of Transnational Management Development*, *ACM SIGSAC Review*, *Information & Management*, and *Pacific Asia Journal of AIS*. He has taught courses in diverse topics including programming, information systems, and quantitative analysis. He is an academic member of the AIS and the ACM.

Brenda Scholtz, Ph.D., is an Associate Professor at the Nelson Mandela University in Port Elizabeth. She was awarded her Ph.D. in ERP skills in 2012. Her research interests include business process management, enterprise systems, business intelligence, data science, and business analytics. She has more than 70 accredited research publications and has supervised many masters and Ph.D. students. She is also currently serving as a project manager for several research collaboration projects between South Africa and Germany.

Shaista Shahid has her MBA in Management Sciences from McGrath Institute of Business Administration, KSA, and BBA from British International Business School, KSA. She is an active researcher, lecturer, and facilitator at the University of Sialkot. She has done multiple research projects on Saudi Arabia's organizations that included strategic management, international relations, human resources, and information systems management. She is currently working on her Ph.D. proposal in information systems in education.

Rui Dinis Sousa holds a PhD in Management Information Systems from the University of Georgia, USA. He is an Assistant Professor in the Department of Information Systems and the vice-coordinator of the Information Systems and Technologies for the Transformation of Organizations and Society research group. His current research includes the implantation of enterprise systems, IT governance, business-IT alignment, and enterprise engineering making use of business process management and enterprise architecture approaches for digital transformation. He is on the board of the Portuguese Association for Information Systems and also a co-founder and vice-president of the Portuguese Association for Organizational Innovation. Having been the Deputy Director of the Department of Information Systems for two 2-years terms, he is now the coordinator of the SAP Next-Gen Lab at UMinho.

Ainin Sulaiman obtained her MBA from Stirling University and her Ph.D. from the University of Birmingham. Currently, she is attached to the Halal

Research Center, the University of Malaya, Malaysia, as a Senior Professor. In addition, she is serving as the Dean of the Social Advancement and Happiness Research Cluster, the University of Malaya. Her research interests include organizational performance, information, computer and communication technology, social networks, big data, and Halal Tourism.

Naciye Güliz Uğur is a faculty member of the Department of Management Information Systems at Sakarya University. Uğur received her Ph.D. in the field of MIS. She has more than five years of industry experience in management and teaches courses on information systems and system analysis and design. Her research interests include technology acceptance and behavioral aspects of emerging technologies. Uğur has also published articles in the proceedings of more than 25 international conferences.

Eglė Vaičiukynaitė is a Researcher at the Digitalization Research Group, Kaunas University of Technology (KTU). She has been working for more than six years on national and international research projects related to customer emotions and information and communication technologies (ICT). Her research interests include customers' emotions, customer engagement in the context of ICT, neuroscience, and neuromarketing. Eglė Vaičiukynaitė is a co-founder of two university organizations, such as KTU Ph.D. Students' Association and KTU Marketing Lab for students. Eglė Vaičiukynaitė has received a distinguished award for her research, business, and social work activities, titled Petras Vileišis Nomination from Lithuanian Confederation of Industrialists.

Jean-Paul van Belle is a professor in the Department of Information Systems at the University of Cape Town (UCT) and Director of the Center for IT and National Development in Africa (CITANDA). His research areas are the adoption and use of emerging technologies in developing world contexts including mobile, cloud computing, open and big data. His passions are ICT for development (ICT4D) — with a focus on emerging technologies as well as data for development (D4D) in an SDG context — and adoption of ICTs by small organizations. He has over 200 peer-reviewed publications including 25 chapters in books and about 40 refereed journal articles. Jean-Paul has active collaborations with researchers in India, UK, Ethiopia, Kenya and Mauritius.

Savanid Vatanasakdakul is a visiting associate professor of information system at Carnegie Mellon University, Qatar campus. She was a senior lecturer at Macquarie University, Australia, for 10 years. She holds a Ph.D. in Information Systems from the University of New South Wales, where she received a prestigious Australian Postgraduate Scholarship Award. Her research interests include strategic fit and cultural aspects of IT transfer globally, particularly to Asia, IT utilization and success, and IT risk management and governance. She is the author of nearly 50 international peer-review journals, conference and book publications. She served in the Association for Information Systems including a founding committee member, a secretary and treasurer of SIGGreen at the Association of Information Systems, and a chair of over 20 tracks at the leading international IS conferences.

Victor Veriansky is Profesor Asociado (Associated Professor) of Languages and Algorithms Theory and Information Technology at the Facultad de Ciencias Económicas of the Universidad de Buenos Aires (FCE-UBA). He participates as a Professor in the University Education Project in Prison. He is a deputy director of the Department of Information Systems of the FCE-UBA. One of the primary functions is to organize an annual academic conference of national scope. He was Lecturer in several professional congresses. He received a degree in Information Systems from the Universidad de Buenos Aires (UBA) and is an MS candidate (UBA). He has experience as a project manager in development projects, in both public and private organizations.

Elena Vitkauskaitė is a Lecturer of Marketing at the School of Economics and Business, Kaunas University of Technology, Lithuania. She delivers courses on Digital Marketing, IT Management, e-Business, Social Media Marketing, and Marketing in Virtual Environments. With student teams in following courses, multiple times she became a Google Online Marketing Challenge winner in Google+ Social Media Marketing category. Ms. Vitkauskaitė is a Researcher of Digitalization Research Group, Kaunas University of Technology, Lithuania. She works on various research projects funded by the Lithuanian government and European Commission (e.g., H2020, 6th Framework, Interreg IVC) related to e-Business and e-Government. Elena' current research interests include modeling of business processes, cross-cultural issues on the web, social network sites, gamification, and the impact of Internet of Things on business models.

Stanislaw Wrycza is a Professor of Business Informatics at University of Gdansk. His research interests include: business informatics, IS modeling, technology acceptance models and digital transformation. His research has been published in IS journals including among others, *Communications of the AIS* and *Information Systems*. Prof. Wrycza is the member of Editorial or Advisory Boards of several IS journals, such as *Information Systems Journal, Information Systems Management* and others. He is serving as the President of the Polish Chapter of AIS, awarded four times as the AIS Outstanding Chapter. He has been the General Chair of ECIS' 2002, held in Gdansk, and the series of PLAIS EuroSymposia. He developed master studies in Business Informatics in English at University of Gdańsk.

Hao Wu is a Ph.D. student and Graduate Assistant at the University of North Carolina at Greensboro, USA. He holds a Bachelor's degree in Information Management and Information Systems, and a Master's degree in Library and Information Science. He is passionate and enthusiastic in the information technology area with research interests including Information Behavior, Online Identity, etc. Besides the ability to do research, he thinks that the most critical characteristics for people in this field are the ability to communicate with others, deliver and manage information, and handle multiple tasks simultaneously. Caven also has work experience in an executive search company, a big data company, a language school, an international office in a university setting, etc. He is working towards obtaining a position at a higher education institution.

ShiKui Wu is an Assistant Professor in the Faculty of Business Administration at Lakehead University. He holds a Ph.D. degree in Management Information Systems from Concordia University. His research interests include e-commerce, design science research, business innovations, and supply chain management. His work has been funded by the Canadian federal and provincial government agencies. Dr. Wu's research has been published in leading journals and presented at prestigious international conferences in management information systems. He also wrote a book on a data-driven approach in designing and implementing e-negotiation systems and received a patent from the US Patent and Trademark Office for multi-attribute auction models and systems.

Benjamin Yeo is an Assistant Professor in the Albers School of Business and Economics at Seattle University, teaching courses in MIS and data mining. His research interests include innovation- and technology-driven economic growth and social informatics, using heavily quantitative and qualitative methods. His recent projects include studies on the impact of ICTs on industry performance, the impact of CSR on firm performance, and the drivers of job satisfaction among IT workers, among several others. Prior to academia, Benjamin Yeo worked in the industry as a Senior Research Analyst in Regional Economics. He received a Ph.D. in Information Science from the College of Information Sciences and Technology at the Pennsylvania State University.

Jie Yu (Joseph) is currently an Assistant Professor in Information Systems at the University of Nottingham Business School, China. He obtained his Ph.D. in Information Systems from the National University of Singapore in 2010. Prior to joining Nottingham University, he worked at the National University of Singapore as a teaching assistant in Information Systems (2007– 2009). His research interests include social media, healthcare information systems, and business analytics. His research work has been presented at a number of international conferences (PACIS, AMCIS, EurOMA, SIGCHI) and research seminars. He has published in top international journals including Journal of Management Information Systems, Journal of the Associations for Information Systems, International Journal of Production Economics, and IEEE Transactions on Professional Communication.

Contents

Chapter 1

The World IT Project: A Long Journey in the Making

Prashant Palvia[*,**], Jaideep Ghosh[†,††], Tim Jacks[‡,‡‡],
Alexander Serenko[§,¶,§§], and Aykut Hamit Turan[‖,¶¶]

[*]University of North Carolina at Greensboro,
Greensboro, North Carolina, USA
[†]Shiv Nadar University, Tehsil Dadri,
Greater Noida, Uttar Pradesh, India
[‡]Southern Illinois University at Edwardsville,
Edwardsville, Illinois, USA
[§]University of Toronto, Toronto, Canada
[¶]University of Ontario Institute of Technology, Oshawa, Canada
[‖]Sakarya University, Sakarya, Turkey
[**]pcpalvia@uncg.edu
[††]jghosh20770@gmail.com

‡‡ *tjacks@siue.edu*
§§ *a.serenko@utoronto.ca*
¶¶ *aykut.turan@gmail.com*

A journey of a thousand miles begins with a single step
— Lao Tzu

Summary

The World IT Project, the largest study of its kind in the IS academic field, was conceptualized more than a decade ago. Long time in the making, the project was motivated by the dominant and pervasive bias in IS research towards American and Western views. In very broad terms, the World IT Project captures the organizational, technological, and individual issues of IT employees across the world and relates them to cultural and organizational variables. The project was officially launched in 2013 and is now in the publication phase. Because of the enormous global scale of the project, a single or even two or three publications cannot fully describe our findings. This book is a major publication that describes and analyzes the organizational, technological, and individual issues of IT employees in the 37 countries that were included in the project.

This first chapter provides the necessary background for the remaining country chapters, which are co-authored with specific country teams. In here, we describe the goals and objectives of the World IT Project, its general framework and major research questions, the relevant literature and theoretical background, methodological details, expected outcomes and publications, and important contributions. The goal is to provide a world view of IT issues that will be relevant to stakeholders at the firm, national, and international levels.

1.1   Introduction

While the journey begins with a single step, we believe the World IT Project is a single giant step. In fact, it is a journey within a journey.

It is widely recognized that Information Systems (IS) academic research is dominated by American and Western views. This is not due to any grand design but more so because much of IS research began in the US (in the seventies and eighties) and later in Western Europe. However, in this age and times, information technology (IT) has permeated the entire world, and yet the situation has persisted. This pervasive bias does not do justice to the rest of the world as other countries and regions do not find their topics investigated or have to rely on Western results, which may not be necessarily applicable to their contexts. Many scholars find themselves using the models and results from the Western world directly and unscrupulously applying them to other countries. The consequences of such actions can be misguided, misleading, or even harmful.

An international team of researchers recognized this deep divide in IS research and initiated the World IT Project in 2013 — in order to investigate the major IS/IT[1] issues in different parts of the world. The World IT Project, although huge in size and scope, is yet a modest but important step to move IS research to incorporate views from major parts of the world. In our pursuit, we developed a standard instrument to track important issues in different countries of the world. The instrument focuses on organizational, technological, and individual issues of IT employees across the world and relates them to contextual factors such as the organization itself, organizational culture, environment, IT occupational culture, and national culture.

While it is not feasible to investigate each country in the world, we wished to maximize the usefulness of our results. So, we looked for countries exhibiting diversity in terms of their cultural, economic, political, religious, and societal backgrounds. Over a period of over 3 years, we were able to collect the same data from 37 countries. These countries provide a good representation of the diversity in the world. The data we collected is huge and enormous; our results will be captured in a series of articles and books. However, it is worth noting that the project's history, goals, governance, challenges, and lessons were reported in Palvia *et al.* (2017) and the major research questions were presented in Palvia *et al.* (2018).

This book was conceived to share with our audience the organizational, technological, and individual issues of IT employees in each of the 37 countries. There are 37 chapters that follow. Each chapter, in alphabetical order of country, briefly describes the country's background, its history, and its IT developments. Results are then presented about the three types of IT employee issues and interpreted in the country's context. Most chapters have co-authors from the country of focus, who were assisted by the central research team. Readers may read all chapters or focus on the countries that are of most interest to them.

In this beginning chapter, we describe the project goals and objectives, relevant literature and theoretical background, overall methodology and the research journey, outcomes and planned publications, and finally our contributions and limitations.

---

[1] As is often the case, we use the terms IS and IT interchangeably.

1.2 Project Goals and Objectives

As early as 2003, at the GITMA World Conference (www.gitma.org) in Calgary, Canada, and later in 2004 in San Diego, California, several IS colleagues started expressing concerns about IS research in general and especially the key IS management issue studies published in the top IS journals. The published research was clearly Western-centric. For example, two types of IS issues have been tracked in the US on an annual basis: organizational IT issues and technology issues (e.g., Kappelman *et al.*, 2018). While there have been fragmented and sporadic attempts to address issues in specific countries or regions (e.g., Ifinedo, 2006; Luftman *et al.*, 2012; Watson *et al.*, 1997), no one had examined much of the world's IT issues in any systematic or comprehensive manner. It took us almost 10 years to crystallize our ideas and put a team together to launch the project in 2013. After much deliberation, the project's core team (the authors of this chapter) settled on the following charter and goals for the World IT Project.

The World IT Project is designed to examine important issues confronting IT employees, both staff and management, in many countries of the world. The proposed project requires survey data collection from different countries, representing different cultures, levels of economic growth, societal and religious beliefs, and political systems. More than forty countries from all parts of the globe will be targeted for this research. In terms of scope, the project is akin to Hofstede's research on national culture and the GLOBE project on culture and leadership, and builds on their research.

Specifically, the project will examine various IT employee issues, such as organizational IT issues, technology issues, and individual issues. Among organizational IT issues are the roles of IT strategic planning, IT-business alignment, business process reengineering, security and privacy, and IT reliability and efficiency, to name a few, in the nature and experience of IT employment. Technology issues include how cloud computing, social media, mobility, ERP systems, business intelligence, and big data, again to name just a few, are perceived by IT workers to influence their jobs. Some of the factors concerning individuals include job satisfaction, efficacy, and role ambiguity. For a deeper understanding of these, the context is important. Also examined will be contextual factors such as organizational variables (including structure and strategy), organizational culture, IT occupational culture, and national culture.

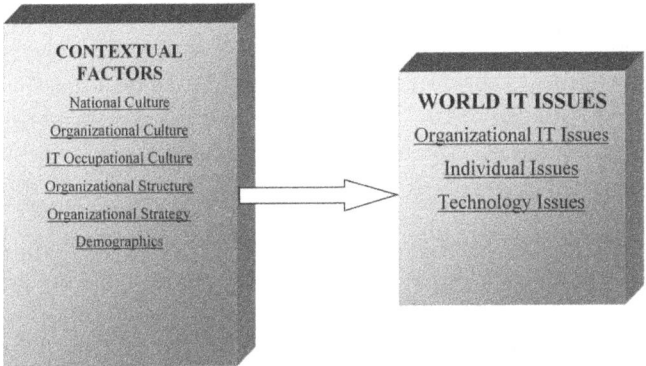

Fig. 1.1: General Framework for the World IT Project

Figure 1.1 provides the general framework for the project and captures the goals and objectives of the project in a succinct manner, albeit at a very high level. It is worth noting that many research questions and research models can be derived from the general framework. The more important research questions were reported in Palvia *et al.* (2018). Since then, we have added one more question. These questions are summarized as follows:

- **Research Question 1:** What are the important organizational IS issues in different countries and regions of the world?
- **Research Question 2:** What are the important technology issues in different countries and regions of the world?
- **Research Question 3:** What are the important individual issues of IT employees in different countries and regions of the world?
- **Research Question 4:** How do the national cultural values of IT employees compare with the national culture values of the general population in each country?
- **Research Question 5:** Do the national culture values of IT employees exhibit similarities across countries?
- **Research Question 6:** Do IT occupational culture values differ by country and/or region of the world? And, if so, how?
- **Research Question 7:** What are the antecedents of job satisfaction among IT employees, and how do they differ from country to country?
- **Research Question 8:** What are the antecedents of turnover and turnaway among IT employees, and how do they differ from country to country?

- **Research Question 9:** What are the differences due to gender in the individual variables and the relationships between the antecedents and the consequents? How do gender effects vary from country to country?
- **Research Question 10:** How do IT employees use social capital and draw from friendship circles when contending with dynamic elements of the organization? What are the differences from country to country?

The focus of this book is on research questions 1, 2, and 3. In subsequent chapters, we provide the results of organizational IS issues, technology issues, and individual issues for each of the 37 countries. Results related to the remaining questions will be included in forthcoming journal articles and conference papers.

1.3   Literature and Theoretical Background

Given the scope of the book, we describe only the literature related to organizational IS issues, technology issues, and individual issues. Much of the following description is adopted from Palvia *et al.* (2017), where it was first reported.

Academic studies on organizational IT issues have typically been conducted only in the US and have been tracked annually for more than a decade. Past research has generally tracked two types of issues: organizational IT issues and technology issues, from the perspective of IT managers. These results are published annually in *MIS Quarterly Executive*; the last results are from 2017 (Kappelman *et al.*, 2018). We evaluate organizational IT issues and technology issues worldwide from the perspective of IT employees. We were interested in IT employees' views and deliberately chose them, as they are closer to the IT profession and thus not unduly influenced by general management or other stakeholders. There are other studies that have examined specific IT personnel issues, such as job satisfaction (McMurtrey *et al.*, 2002), technostress (Ayyagari *et al.*, 2011), and turnover intention (Shih *et al.*, 2013). We grouped these and similar issues under "individual issues" and included them in the World IT Project.

The original source of organizational IT issues was the key issue studies published annually in *MISQ Executive*, such as Luftman and Ben-Zvi (2010) and Luftman *et al.* (2012). The various issues listed in these studies were examined for their relevance in the global context. Items were added,

Table 1.1: Organizational IT Issues

Business productivity and cost reduction
Alignment between IT and business
Business agility and speed to market
Revenue-generating IT innovations
IT cost reduction
IT strategic planning
Business process reengineering
Enterprise architecture
Security and privacy
IT reliability and efficiency
IT service management (e.g., ITIL)
Globalization
Outsourcing
Attracting and retaining IT professionals
Bring your own computing device (BYOD)
Continuity planning and disaster recovery
Project management
Knowledge management

deleted, and modified based on various international studies and existing literature. The final list included eighteen issues, as shown in Table 1.1.

A primary and important part of IT employees is to interface with the technology itself and deal with related issues. In many ways, technology (e.g., computer hardware, telecommunications, software, and services) defines the entire IT occupation (Guzman *et al.*, 2008; Jacks, 2012). There are a myriad technologies and related issues: old, new, and emerging; so, we included broad technologies based on their appearance in the IS literature (e.g., the annual key issue studies cited above) and our knowledge of the IT industry. A total of 16 technology issues were included, as shown in Table 1.2.

Individual issues for IT employees describe the attitudes, beliefs, and behaviors that prior research has shown to be critically important in the workplace. These issues were measured by several constructs, namely: job satisfaction (Moore, 1997), perceived work overload (Kirmeyer and Dougherty, 1988), perceived work/home conflict (Kreiner, 2006), strain (Moore, 2000), professional self-efficacy (Moore, 1997), job insecurity (Ashford *et al.*, 1989), turnover intention in the organization (Moore, 1997), and turnover intention in the IS profession (Moore, 1997). Each construct was operationalized by multiple items, as shown in Table 1.3.

Table 1.2:   IT Related Issues

Business intelligence/analytics
Cloud computing
Enterprise resource planning (ERP) systems
Collaborative and workflow tools
Customer relationship management (CRM) systems
Mobile and wireless applications
Enterprise application integration
Business process management systems
Big data systems
Mobile apps development
Networks/telecommunications
Social networking/media
Virtualization (desktop or server)
Software as a service
Data mining
Service-oriented architecture (SOA)

1.4   Methodology and the Journey

Much of the methodology details described here are documented in Palvia *et al.* (2017).

The first critical step of the World IT Project was the preparation of the standard survey instrument. As much as possible, we used previously validated items for the constructs so that the instrument had good psychometric properties. Pilot tests also helped us refine the instrument. Ultimately, the instrument contained 160 items. Although the instrument appeared to be long, all questions required the respondent to select one answer from a menu of options, and it could be completed in about 25 min. For example, majority of the questions required a selection on 1–5 scale, where 1 represented most importance and 5 represented no importance, or 1 represented strongly agree and 5 represented strongly disagree. The instrument was frozen at the end of 2013. As an exception and in special cases, country investigators (CIs) could add a few of their own questions. This was necessary to assure uniform data across all countries. The Institutional Review Board at the University of North Carolina at Greensboro approved and exempted the instrument from further review. The core team members and CIs also received approvals from their home Institutional Review Boards if necessary.

Table 1.3:   Individual Issues

Job Satisfaction	In general, I like working here. All in all, I am satisfied with my current job. In general, I don't like my current job.
Perceived Workload (Work Pressure)	I feel that the number of requests, problems or complaints that I deal with at work is more than expected. I feel that the amount of work I do interferes with how well it is done. I feel busy or rushed at work. I feel pressured at work.
Work/Home Conflict (Work–Life Balance)	There is a blurring of boundaries between my job and my home life. My work-related responsibilities create conflicts with my home responsibilities. I do not get everything done at home because I find myself completing job-related work.
Strain (Workload and Burnout)	I feel drained from activities at work. I feel tired from my work activities. Working all day is a strain for me. I feel burned out from my work activities.
Professional Self-Efficacy (Sense of Accomplishment)	I feel I'm making an effective contribution to what this organization does. In my opinion, I do a good job. I have accomplished many worthwhile things in this job. At my work, I feel confident that I am effective at getting things done.
Job Insecurity (Threats to One's Job)	I am worried that future technology advancements may pose a threat to my job. I believe that other people may be able to perform my work activities. I am concerned that my job may be eliminated soon. I am concerned that my job may be outsourced soon.
Career Plans: Turnover Intention — Organization	I will be with this organization 1 year from now. I will take steps during the next year to secure a job at a different organization. I will be with this organization 5 years from now.
Career Plans: Turnover Intention — IS Profession	I will be working in the IT field 1 year from now. I will take steps during the next year to secure a job outside the IT field. I will be working in the IT field 5 years from now.

The core team members guided the overall project and served as liaisons to individual CIs and provided them with support. Each country had its own team. Each country team generally had one to three CIs. In some cases, we allowed four members per team. We solicited CIs through professional contacts and conferences, requests on the AISWorld listserv, and direct emails to faculty listed on the AIS faculty directory. We also organized information sessions at conferences, such as the GITMA and AMCIS conferences in 2013 and 2014, in order to attract CIs. In each communication, we described the benefits to the CIs and their roles and responsibilities.

One of the primary goals of the project was to obtain data from major regions of the world and include countries that represent different cultures, economic status, religious beliefs, and political systems. We needed local CIs because they would understand the local culture and how to best approach local companies to participate. The CIs were also responsible for the translation/back-translation of the instrument to the local language, if necessary. As a result, the instrument has been translated into the following languages: Chinese, French, Italian, Japanese, Korean, Malay, Polish, Portuguese, Russian, Spanish, Thai and Turkish.

The CIs were given general directions and guidelines for collecting the data from IT employees in organizations. As they knew their situation best, they were given discretion in what specific method(s) they used to collect the data and how to approach organizations in their country. The general guidelines included:

- All respondents needed to be in the IT profession.
- Try to get a minimum usable sample size of 300 IT employees.
- Collect responses from 10 to 15 IT employees from 20 to 30 companies.
- Try to include a variety of industries.
- Assure complete anonymity in data collection.

Most countries exceeded the target sample size of 300. There are a few who fell short because of a variety of reasons; nevertheless, we decided to include them in our analysis.

Recommendations were made to the CIs on the actual data collection, e.g., mail surveys, face-to-face surveys, email surveys, Web-administered surveys, and going through the CEO/CIO or another senior executive to recruit multiple IT employees from the same organization. The CIs used the above methods, but also came up with their own innovative solutions to collect the data, e.g., hire a consulting company specializing in conducting surveys, have the surveys completed at industry conferences, and organize

Table 1.4:   Countries in the World IT Project

Argentina	Iran	Portugal
Bangladesh	Italy	Romania
Brazil	Japan	Russia
Canada	Jordan	South Africa
China	Lithuania	South Korea
Egypt	Macedonia	Taiwan
Finland	Malaysia	Thailand
France	Mexico	Turkey
Germany	New Zealand	UK
Ghana	Nigeria	USA
Greece	Pakistan	Vietnam
Hungary	Peru	
India	Poland	

special events for the purpose. Many CIs used multiple methods to collect the data.

The initial cutoff date for data collection was December 2016. However, given the enormity of the projects and challenges encountered in data collection, the date was extended to December 2017. Our initial target for the number of participating countries was 30, which we later increased to 40. The final count was 37 when we finished the data collection effort. Table 1.4 shows the list of countries participating in the World IT Project.

Once the data collection was complete, the CIs were asked to code the data in an Excel file and send it to the core team. They were sent a coding template and instructions. The core team examined each country data both visually as well as applied statistical tests. The visual examinations looked for things like: missing data, same codes for all fields, out of range data, and similar codes across multiple records. Any anomalous records were deleted. For example, if the missing data was excessive, the entire record was deleted. If the row with missing data was retained, the missing data were replaced by the number 9. IBM SPSS 23 was used for statistical analysis. Statistical tests included construct reliability assessment. This was done both at country level and the entire data set level. Problems were flagged and will be accounted for in our future analysis. It is worthy of note that at least three country datasets were completely rejected, and the CIs were given the opportunity to collect the data all over again. Two of the countries complied. One did not and is not part of the 37 countries.

The complete database of all the 37 countries is enormous; it has more than 11,000 records and more than 1.7 million data items.

1.5   Outcomes and Publications

Subsequent to the goals and research questions presented in an earlier section, we describe here some specific studies that will result from the analysis of the massive and rich database of the World IT Project. The types of studies conducted can be viewed along two dimensions: scope and epistemology, as shown in Table 1.5. On the scope dimension, we plan to conduct three types of studies: single country studies, multiple country studies, and global studies. The multiple country and global studies will provide a comparative examination as well as an integrated view of findings across countries and regions. On epistemology, each study can be either descriptive, i.e., providing analysis of stand-alone issues and findings, or theoretically grounded explanation of various phenomena and relating them to a variety of antecedents and contextual factors.

This book is a compilation of single country studies, and it reports descriptive analysis of the organizational, technological, and individual issues of IT employees in each of the 37 countries (i.e., the top left cell of the above matrix). We plan to also prepare descriptive analyses of IT occupational culture and national culture values of IT employees.

Studies in the other cells of the above matrix are underway and will be published in other outlets. As an example of theory-based study in a single country, we recently completed a study in Brazil (Porto Bellini *et al.*, 2019), which examines the following research question: What are the antecedents of turnover and turnaway among IT workers in the context of a national crisis, and how are they affected by age differences? Note that turnover refers to changing jobs within the IT profession and turnaway refers to moving to another profession altogether. We had a fortuitous situation, from a research point of view, that a major national crisis developed in Brazil during the time data were collected in the country. We were able to develop a theoretically grounded model employing the constructs of professional self-efficacy, job satisfaction, and job insecurity along with turnover and turnaway, with national crisis and age as control variables. The model was evaluated by using structural equation modeling (SEM) techniques.

Table 1.5:   Types of Studies from the World IT Project

Scope/Epistemology	Single Country	Multiple Country	Global
Descriptive			
Theory-based			

Another example of a theory-based study using the entire global dataset, currently underway, is to examine the relationship between IT occupational culture and national culture. Possible research questions to examine are: Do the national culture values of IT employees exhibit similarities across countries and do they differ from the general population? Do the IT occupational culture values of IT employees exhibit similarities across countries? Can the countries be clustered based on IT occupational culture and if so, how many clusters? Can links be found between national culture and IT occupational culture?

Another important area of investigation would be to examine the impact of national culture and IT occupational culture on important issues and dependent variables, e.g., the need for security and privacy, IT strategic planning, and technostress. Along these lines, the role of gender on these issues would be an important one to examine.

As can be seen, numerous research questions can be addressed using the World IT Project's rich dataset. Along with this book, we have had three journal publications and multiple conference papers. At this time, we have plans to complete eight to 10 mores studies.

1.6 Contributions and Limitations

The World IT Project started with an ambitious and expansive goal. Fortunately, we have been able to accomplish our objectives and the project is coming to fruition. The implications for both research and practice are enormous. As we said in our project charter, a good understanding of the critical IT issues facing firms and their employees within their surrounding contexts will be important from the firm, national, and international points of view.

Specifically, at the firm level, our results would help management and staff in formulating business and IT related policies and strategies. At the national level, it would allow stakeholders, such as policymakers, governments and vendors, to address the pressing issues of the times. In international business, it would help firms and governments respond to the needs of partners and stakeholders in other countries. A comparative examination across countries and world regions would help facilitate global understanding, cooperation, and knowledge transfer among many nationalities.

For academic research, the World IT Project offers current and future scholars a grounded understanding of the international IT environment and provides a validated framework to launch many international IT studies.

Truly global IT studies are acutely needed in IS research; occasional and sporadic forays do not lead to cumulative knowledge and a good understanding of global phenomena. Other publications from the project will provide theoretically grounded models which researchers would find useful to build and extend their work. We even offer the possibility, on a selective basis, of sharing our database or part of it with other researchers as long as their goals coincide with ours and help the mission of greater global understanding.

As for the contributions from this book itself, readers, whether from academia or industry, may find value in several ways. The readers may focus only on one or a few chapters, based on their countries of interest. Each chapter is independently written and can be read without the knowledge of other chapters. More likely, readers would be more interested in multiple chapters so that they can compare, contrast, and integrate issues from several countries. The more avid reader is encouraged to read the entire book to develop a more comprehensive understanding of the global issues in IT.

As for the limitations of the World IT Project, it is worth mentioning a few. First, our survey instrument was long with 160 questions; yet it may not have been able to capture all the dimensions of global IT. But we do believe that we have captured most of the important dimensions within the constraints of what was achievable in one single project, albeit a very large one. By the same token, many of the lists and items we provided to the respondents may not have captured all of the relevant issues; although again, we did our homework in combing through the literature, pilot-testing and validating the various constructs and items. Our research also suffers from the usual limitations of survey research, e.g., sample size, randomness, and representativeness. Although it was almost impossible to draw our respondents completely randomly from representative sub-populations, we did our best to draw respondents from various industries and organizations of different sizes. Finally, while using a standard instrument offered benefits in terms of comparability of results, there were instances when some terms in the instrument had different semantic interpretations in other countries and cultures. Again, we exercised caution and relied on local CIs to handle such situations.

1.7 Conclusion

In this chapter, we have provided the necessary background for the World IT Project as a precursor to the remaining 37 chapters, one for each country that was investigated by the project. We described the goals and objectives of the World IT Project, its general framework and major research questions, the relevant literature and theoretical background, methodological details, expected outcomes and publications, and important contributions. The project will have several deliverables that should significantly enhance our understanding of the global IT environment, and this book is one of its first deliverables. Each chapter that follows is a systematic assessment of the needs and issues of IT employees in organizations across the globe. As pointed out earlier, the problems and opportunities associated with these needs have profound implications for researchers and practitioners in all parts of the world. Finally, this research effort should serve as an exemplar of global IT research and encourage diversity in research and the use of multiple paradigms beyond the current American and Western-centric views. Just as we were able to accomplish our goals, we encourage research collaboration among scholars across the world, thus leading to higher synergy and relevance.

References

Ashford, S. J., Lee, C., & Bobko, P. (1989). Content, cause, and consequences of job insecurity: A theory-based measure and substantive test. *Academy of Management Journal*, 32(4), 803–829.

Ayyagari, R., Grover, V., & Purvis, R. (2011). Technostress: Technological antecedents and implications. *MIS Quarterly*, 35(4), 831–858.

Guzman, I. R., Stam, K. R., & Stanton, J. M. (2008). The occupational culture of IS/IT personnel within organizations. *ACM SIGMIS Database*, 39(1), 33-50.

Ifinedo, P. (2006). Key information systems management issues in Estonia for the 2000s and a comparative analysis. *Journal of Global Information Technology Management*, 9(2), 22–44.

Jacks, T. (2012). *An Examination of IT Occupational Culture: Interpretation, Measurement, and Impact* (doctoral dissertation). The University of North Carolina at Greensboro.

Kappelman, L., Johnson, V., McLean, E., & Maurer, C. (2018). The 2017 SIM IT Issues and Trends Study. *MISQ Exec*, 17(1), 53–88.

Kirmeyer, S. L., & Dougherty, T. W. (1988). Work load, tension, and coping: Moderating effects of supervisor support. *Personnel Psychology*, 41(1), 125–139.

Kreiner, G. E. (2006). Consequences of work-home segmentation or integration: A person-environment fit perspective. *Journal of Organizational Behavior*, 27(4), 485–507.

Luftman, J., & Ben-Zvi, T. (2010). Key issues for IT executives 2009: Difficult economy's impact on IT. *MIS Quarterly Executive*, 9(1), 203–213.

Luftman, J., Zadeh, H. S., Derksen, B., Santana, M., Rigoni, E. H., & Huang, Z. D. (2012). Key information technology and management issues 2011–2012: An international study. *Journal of Information Technology*, 27(3), 198–212.

McMurtrey, M. E., Grover, V., Teng, J. T., & Lightner, N. J. (2002). Job satisfaction of information technology workers: The impact of career orientation and task automation in a CASE environment. *Journal of Management Information Systems*, 19(2), 273–302.

Moore, J. E. (2000). One road to turnover: An examination of work exhaustion in technology professionals. *MIS Quarterly*, 24(1), 141–168.

Moore, J. E. (1997). *A Causal Attribution Approach to Work Exhaustion: The Relationship of Causal Locus, Controllability, and Stability to Job-Related Attitudes and Turnover Intention of the Work-Exhausted Employee* (doctoral dissertation). Indiana University.

Palvia, P., Jacks, T., Ghosh, J., Licker, P., Romm-Livermore, C., Serenko, A., & Turan, A. H. (2017). The World IT Project: History, trials, tribulations, lessons, and recommendations. *Communications of the Association for Information Systems*, 41(18), 389–413.

Palvia, P., Ghosh, J., Jacks, T., Serenko, A., & Turan, A. (2018). Trekking the globe with the World IT Project. *Journal of Information Technology Case and Application Research*, 20(1), 1–6.

Porto Bellini, Carlo, G., Prashant, P., Valter, M., Tim, J., and Graeml, A. (2019). "Should I stay or should I go? A study of IT professionals during a national crisis." Information Technology & People (2019).

Shih, S. P., Jiang, J. J., Klein, G., & Wang, E. (2013). Job burnout of the information technology worker: Work exhaustion, depersonalization, and personal accomplishment. *Information & Management*, 50(7), 582–589.

Watson, R. T., Kelly, G. G., Galliers, R. D., & Brancheau, J. C. (1997). Key issues in information systems management: An international perspective. *Journal of Management Information Systems*, 13(4), 91–115.

Chapter 2

Information Technology Issues in Argentina

Claudio Freijedo*, Victor Veriansky*, Alicia Cortagerena*, Pablo Rota*, and Tim Jacks†

*Universidad de Buenos Aires,
CABA, Argentina
†Southern Illinois University Edwardsville,
Edwardsville, IL, USA

Summary

The IT industry in Argentina has demonstrated strong growth in recent years, despite economic turmoil. The software and service segment has shown more dynamic growth than the hardware and supply segment. From 2008 to 2017, the software and service sector saw employment growth at an annual average of 4%. This growth drives a strong demand for IT personnel, but can also cause high employee turnover. The top IT organizational issues for IT workers in Argentina were alignment between IT and business, IT reliability and efficiency, IT strategic planning, and security and privacy. Top technology issues included enterprise application integration, business intelligence/analytics, and networks/telecommunications. The typical IT employee exhibits moderately high levels of job satisfaction. The overall assessment is that the IT sector in Argentina seems to exist as an island that is somewhat insulated from the larger economic challenges of the country.

2.1 Introduction

In this chapter, we present a brief description of the history, geography, and economy of the Republic of Argentina, followed by a brief description of the evolution of the IT industry in recent years, detailing its rate of growth and employment. In the methodology section, we describe the way the survey was prepared and administered, as well as the way the information was collected and analyzed. In addition to demographic information, the results of the analysis include the top IT organizational concerns, the highest ranked technology issues, and the most frequent individual issues and attitudes of IT workers in Argentina.

2.2 Country Background and History

Argentina is located in the southern half of South America. The country is bordered by Chile (5,308 km) to the west, Bolivia (742 km) to the northwest, Paraguay (1,699 km) to the northeast, and by Brazil (1,132 km) and Uruguay (887 km) to the east. With a mainland area of 2,791,810 km^2, Argentina is the largest Spanish-speaking country in the world. Its 6.5 million km^2 continental platform expands all the way to Antarctica, where the country has six permanent bases, reaching the South Pole. This large geographical reach from north to south and the differences in altitude in the country (from sea level on the Atlantic coast to 6,692 m at Mount

Aconcagua, the highest in the Western Hemisphere) result in a broad variety of different climates, with tropical and subtropical regions in the north, vast hills and wet plains in the center and extremely cold temperatures with glaciers in the south (IGN, 2018; Argentina, 2018). Argentina's geography contains five natural and six cultural locations that are included in The United Nations Organization for Education, Science and Culture's World Heritage List (UNESCO, 2018).

At the time of the country's independence in 1816, its population comprised an indigenous base, a small number of Spanish colonists — creating a large mixed-race population together — and immigrant African slaves. Between 1850 and 1940, large waves of European immigrants (mainly Spanish and Italian) were integrated into the considerably smaller local population. Starting around the middle of the 20th Century, waves of immigrants from other South American countries, principally Bolivia and Paraguay, arrived in Argentina.

Today, the Argentine Republic is a Federal State comprising 23 provinces and the Autonomous City of Buenos Aires, where the seat of the Executive, Legislative, and Judicial powers is located. With 40 million residents according to the last national census (2010) and current estimates reaching more than 44 million residents, Argentina's population is ranked 32nd worldwide and third in the Americas, after the USA and Brazil (The World Bank, 2018). Argentina has demonstrated significant scientific and cultural development, as it is the Latin American country with the most Nobel Prizes (5) (Nobel Prize Organization, 2018). It also has shown important developments in the biotechnological, nuclear and space activity fields.

Argentina's economy is ranked 5th in the Americas (after the United States, Brazil, Canada and Mexico) and 21st in the world in terms of gross domestic product (GDP) (The World Bank, 2018). The country was one of the founders of the Mercosur Agreement in 1991 and is part of the G20 since its creation in 1999. Argentina's economy has had a very turbulent history, including extended periods of hyperinflation in 1989 and 1990 and sovereign default in 1827, 1890, 1982, and 2002 where the country's government defaulted on its debts.

2.3 Information Technology in Argentina

The IT industry has shown strong growth in recent years. From 2008 to 2017, the IT sector increased its sales in dollar terms at an annual accumulated average rate of 2.2%; including in that average are some years of contraction caused by economic instability (CESSI, 2018; CICOMRA, 2018).

In the Information and Communication Technology (ICT) market, communication accounts for close to two thirds of receipts and IT the other third. The growth of each subsector has been similar. In the IT industry, the software and service segment has demonstrated more dynamic growth than the hardware and supplies segment, rising from 49% in 2006 to around 63% of the market in 2015. Similarly, the software and information services industry has shown a sustained period of growth, with 50% growth in the number of businesses in the sector from 2006 to 2015.

Employment in ICT grew 42.2% from 2008 to 2017, at a 4% annual rate, accounting for more than 96,000 reported jobs. Software and service sales to foreign clients grew at an 8.5% annual rate in the last 10 years, 50% of which is to the United States, followed by Uruguay with 10% and then smaller percentages for other countries. In 2017, exports rose to almost US$1,700 million, reaching 29% of the sector's total receipts. 56% of the income comes from custom software, while 25% is from the sale of ICT products and associated services.

Companies that do not participate in the ICT industry show the same tendency regarding employment growth in their internal IT departments. This growth drives a strong demand for IT personnel, but can also cause high employee turnover. The companies in this sector experience an average annual turnover of more than a quarter of their employees, which is fairly high.

2.4   Methodology

Our first step was the translation of the English instrument for the World IT Project (Palvia et al., 2017; Palvia et al., 2018) survey in order for it to be administered in Spanish. A different member of our team then back-translated from Spanish to English. Both versions were compared and needed adjustments were made in order to achieve the final instrument to be used.

The determination of the sample was made using non-probabilistic methods. The "Judgement Sampling" framework (Adams et al., 2007) was applied, searching for a representative distribution of small, medium and large companies within the area of the Autonomous City of Buenos Aires and Buenos Aires Province — the geographical area to which we limited our study.

The survey was done as a self-administered questionnaire. A thousand copies of the instrument were delivered, which were distributed by the researchers among teachers and students of Buenos Aires University who administered it to IT workers in different work places. 336 surveys were received resulting in a 34% response rate which we consider reasonable taking into account the distribution procedure used in this case. About 27 surveys were discarded for being significantly incomplete, giving a final number of 309 answered surveys. The compilation of the information was done by one of the researchers. The analysis of the results was done as a team.

In our sample, most of the IT workers (65%) are not older than 29 years of age and 28% are between 30 and 39. Around 6% of the workers are between 40 and 59 years old and 1% are over 60. The most common roles were analysis & design (11%), programming (9%), consulting (9%), and project management (9%) (see Table 2.1).

Table 2.1: Descriptive Statistics

Characteristics	N	%	Characteristics	N	%
Education:			Years of work experience:		
High school or less	32	10	0–4 years	59	19
Associate degree (2 year	22	7	5–9 years	127	41
degree) or some college			10–19 years	94	30
Bachelor's degree	241	78	20–29 years	25	8
Master's degree	14	5	30+ years	4	1
Ph.D.	0	0			
			Organizational location:		
Years of IT experience:			IT department	160	52
0–4 years	74	24	employee		
5–9 years	152	49	IT worker in a non-IT	46	15
10–19 years	68	22	department		
20–29 years	11	4	Contract employee	28	9
30+ years	4	1	Consultant	64	21
			Vendor employee	11	4
Work as:					
Mostly full time	254	82	Are You?		
Mostly part time	45	15	Not part of	245	79
Mostly over time	10	3	management		
			In lower management	43	14
Been laid off from IT job:			In middle management	17	6
Yes	5	2	In Senior Management	4	1
No	304	98			

2.5   Organizational IT Issues

The survey gathered the participants' opinions about the level of importance of 18 IT organizational topics, using a scale from 1 to 5 (1 very important, 5 not important). Table 2.2 shows all of the IT organizational issues ranked according to their average score.

The highest importance among the issues evaluated is accorded to alignment between business and IT. This result reflects growth in the average level of organizational maturity level, with regard to the cultural adoption of the goal of alignment between business and IT. The Society for Information Management in its "IT Issues and Trends Study" (Kappelman *et al.*, 2018), puts this topic in second place for organizational relevance. This seems to indicate that what happens in Argentina matches what is observed in the

Table 2.2:   Organizational IT Issues in Argentina

Organizational IT Issues	Rank	Mean Rating*	Std. Deviation
Alignment between IT and business	1	1.53	0.60
IT reliability and efficiency	2	1.56	0.70
IT strategic planning	3	1.71	0.82
Security and privacy	4	1.71	0.81
Continuity planning and disaster recovery	5	1.76	0.88
Attracting and retaining IT professionals	6	1.77	0.97
Project management	7	1.77	0.70
Business agility & speed to market	8	1.82	0.73
Knowledge management	9	1.86	0.74
Revenue-generating IT innovations	10	2.05	0.82
Business process reengineering	11	2.08	0.87
Business productivity & cost reduction	12	2.14	0.79
IT service management (e.g., ITIL)	13	2.21	0.84
Globalization	14	2.31	0.80
Enterprise architecture	15	2.39	0.75
IT cost reduction	16	2.65	1.07
BYOD (Bring Your Own Computing Device)	17	2.97	1.12
Outsourcing	18	2.99	0.88

*Rating scale ranges from 1 to 5: 1 as most important and 5 as no importance.

epicenter (i.e., the US). Similarly, it is appropriate to mention that security and privacy also received a high score in our survey, but ranked in the fourth place, while in the SIM 2017 study, security and privacy ranked as the main concern.

With security and privacy taking fourth place and alignment of IT and business taking first place, the top five list of IT organizational issues is completed by IT reliability and efficiency (second place), IT strategic planning (third) and continuity planning and disaster recovery (fifth), which do not appear as notable topics in the SIM 2017 study. A possible interpretation of this situation is that even with cultural and educational maturity on these issues, political complexity, economic instability, and possible delays in matters of infrastructure are decisive and negative in their impact. Matters that seem basic (some even operational) take on more significance when they are not adequately available to most organizations.

The necessity of attracting and retaining IT professionals appears in sixth place. This is an extremely difficult task for many different reasons. There are global reasons, such as demand growth that enables people to have access to many different job opportunities. There are cultural reasons when dealing with a generation that gives less value to stability than did their predecessors. Finally, endogenous reasons (not necessarily exclusive to Argentina) should also be considered, such as difficulties in budgeting for training and other forms of employee retention.

In summary, it is possible to say that, in Argentina, the perception of which IT organizational issues are relevant shows that the evolution of organizational, cultural and educational thought has not been negatively impacted by economic restrictions and fluctuations of a highly volatile environment as might otherwise be assumed.

2.6 Technology and Infrastructure Issues

Our survey gauged the perceived importance of 16 technology topics in IT. Respondents valued topics between 1 and 5 (1 very important, 5 not important). Table 2.3 shows these topics, ranked according to their average score.

Again, it is reasonable to pay attention to similarities and differences between these results and that of the SIM 2017 study where possible. Enterprise application integration was ranked as the top-most important technology issue in this section of the survey. One interpretation of this result is that enterprise application integration ranks as the main concern because

Table 2.3: Technology and Infrastructure Issues in Argentina

Technology and Infrastructure Issues	Rank	Mean Rating*	Std. Deviation
Enterprise application integration	1	1.88	0.75
Business intelligence/analytics	2	1.93	0.63
Networks/telecommunications	3	1.96	0.89
Big data systems	4	2.06	0.91
Software as a service	5	2.09	0.82
Collaborative and workflow tools	6	2.11	0.73
Data mining	7	2.13	0.89
Mobile and wireless applications	8	2.14	0.95
Service-oriented architecture (SOA)	9	2.17	0.94
Enterprise resource planning (ERP) systems	10	2.19	0.79
Business process management systems	11	2.20	0.72
Mobile apps development	12	2.29	1.06
Customer relationship management (CRM) systems	13	2.34	0.78
Social networking/media	14	2.38	1.01
Cloud computing	15	2.39	0.93
Virtualization (desktop or server)	16	2.42	0.84

*Rating scale ranges from 1 to 5: 1 as most important and 5 as no importance.

it is an unresolved structural issue for companies, and hence tends to monopolize the conversation.

In second place, we find the topic of business intelligence/analytics. In this area, companies visualize a potential world of business opportunities at an affordable cost. For this reason, it ranks highly, ahead of other topics whose return-on-investment (ROI) is less clear. Related to business intelligence/analytics, we find big data systems fourth in the ranking. In Argentina, there are whole industries (such as media) which greatly value the processing of large amounts of information and the transformation of unstructured data into measurable data. Related to this, we highlight data mining, which ranks slightly below big data systems. Thus, there is an important cluster of interrelated and overlapping technology concerns.

In third place is networks/telecommunications, an always evolving and important topic. It can also be found in the SIM study, although it is lower in the ranking. Its positioning in the ranking may be explained similarly to the issue of enterprise application integration above.

Collaborative and work flow tools rank sixth in our survey. Topics surrounding productivity and the satisfaction levels of human resources are always areas of interest and conflict. In this case, the ranking reveals a real interest in improving general working conditions (for both human and economic reasons). The next positions are taken by issues such as SaaS, Mobile and Wireless Applications, and SOA which do not appear in SIM 2017s top 10. This may show that Argentina follows the lead of more developed markets but with a little delay.

To conclude, we highlight that Argentina presents two sides. One is evolving, vigorous and has a high cultural and educational level. The other has productivity constraints and limitations as a result of a weak economic context.

2.7 Individual IT Employee Issues

Another focus of attention was centered on the individual concerns of IT employees. This was surveyed by soliciting answers on a scale from 1 to 5, where 1 represents a high level of agreement with each statement (Table 2.4).

The IT segment seems to exist as an island that guarantees a certain wellness compared to other sectors in terms of employee wellbeing. The country has gone through repeated economics crises and the steady demand for IT personnel has bucked the trend of overall employment demand. This results in generally good working conditions in IT, which is why it is not surprising that most of the respondents report feeling comfortable with their working environment.

The survey reveals a medium level of work pressure. This is due to the fact that IT departments often act as mere service centers to provide support to other departments, receiving demands for quick problem resolution under strong pressure. The survey supports this view, showing responses that denote a moderate level of overexertion. However, the portion of the survey that inquires about the balance between work and personal life shows a more balanced scenario. Respondents show high levels of self-realization, as shown by the scores in that subset of statements. Responses also show limited concern about job security.

Table 2.4:　Individual IT Employee Issues in Argentina

Individual Issues	Mean Rating*	Std. Deviation
Job satisfaction		
In general, I like working here.	1.79	0.90
All in all, I am satisfied with my current job.	2.11	1.01
In general, I don't like my current job.	1.93	1.11
Work pressure		
I feel that the number of requests, problems or complaints that I deal with at work is more than expected.	3.10	1.11
I feel that the amount of work I do interferes with how well it is done.	2.86	1.15
I feel busy or rushed at work.	2.59	1.05
I feel pressured at work.	3.03	1.16
Work–life balance		
There is a blurring of boundaries between my job and my home life.	3.48	1.13
My work-related responsibilities create conflicts with my home responsibilities.	3.67	1.26
I do not get everything done at home because I find myself completing job-related work.	3.57	1.18
Workload and burnout		
I feel drained from activities at work.	3.51	1.18
I feel tired from my work activities.	3.32	1.25
Working all day is a strain for me.	3.49	1.28
I feel burned out from my work activities.	3.50	1.22
Sense of accomplishment		
I feel I'm making an effective contribution to what this organization does.	2.38	1.07
In my opinion, I do a good job.	1.70	0.64
I have accomplished many worthwhile things in this job.	1.94	0.95
At my work, I feel confident that I am effective at getting things done.	1.87	0.80
Threats to one's job		
I am worried that future technology advancements may pose a threat to my job.	3.68	1.41
I believe that other people may be able to perform my work activities.	2.17	1.02
I am concerned that my job may be eliminated son.	4.05	1.06
I am concerned that my job may be outsourced soon.	4.06	1.06
Career plans		
I will be with this organization 1 year from now.	3.45	1.40
I will take steps during the next year to secure a job at a different organization.	3.18	1.26
I will be with this organization 5 years from now.	2.63	1.35
I will be working in the IT field 1 year from now.	4.46	0.85
I will take steps during the next year to secure a job outside the IT field.	4.21	1.29
I will be working in the IT field 5 years from now.	1.66	1.04

*Rating scale ranges from 1 to 5: 1 as strongly agree, to 5 as strongly disagree.

The final section provides a summary by showing that employees are interested in staying in the IT profession over the next 5 years. The exception is the fourth statement in the section, possibly because respondents misunderstood the statement. The sector has been increasing its demand for personnel by nearly 10% every year on average, from 2006 to 2017, mostly decoupling itself from the already mentioned economic crises. Compensation and opportunities for growth seem to override the difficulties set by the challenging nature of typical IT tasks. As a result, the IT employees tend to exhibit moderately high levels of job satisfaction.

As a final thought, it can be considered that the IT sector has a privileged spot compared to others. The current situation, the consistently growing demand, the good working conditions, and an evolving role guarantee that satisfaction is here to stay despite the threats of larger economic conditions in Argentina.

2.8 Conclusion

We see in Argentina a software and service sector that is competitive on the world stage. However, the IT sector's development is constrained by its turbulent economic context. In the past 10 years, the number of employees in the IT sector has increased by nearly 10% every year on average, mostly decoupling itself from the economic crises that have troubled the country. The respondents' answers show they feel comfortable with their working environment and show little signs of being under excessive pressure or fear for their job security. This attitude matches the consistent growth presented in the IT sector of Argentina.

References

Adams, J., Khan, H., Raeside, R., & White, D. (2007). Research methods for graduate business and social science students. SAGE Publications India.

Argentina, G. D. (2018). https://www.argentina.gob.ar/pais. Retrieved from https://www.argentina.gob.ar/pais: https://www.argentina.gob.ar/pais.

CESSI (2018). Reporte anual sobre el Sector de Software y Servicios Informáticos de la República Argentina — Año 2017. Buenos Aires: CESSI - Cámara de Empresas de Software y Servicios Informáticos de la República Argentina.

CICOMRA (2018). CICOMRA — Cámara de Informática y Comunicaciones de la República Argentina. Retrieved from http://www.cicomra.org.ar/cicomra2/asp/estadistica_2017.asp.

IGN (2018). Instituto Geografico Nacional. Retrieved from Instituto Geografico Nacional: http://www.ign.gob.ar/.

Kappelman, L., Nguyen, Q., McLean, E., Maurer, C., Johnson, V., Snyder, M., & Torres, R. (2017). The 2016 SIM IT Issues and Trends Study. *MIS Quarterly Executive*, 16(1), 47–80.

Nobel Prize Organisation (2018). Nobel Prize Organisation. Retrieved from https://www.nobelprize.org/prizes/lists/all-nobel-prizes/.

Palvia, P., Jacks, T., Ghosh, J., Licker, P., Romm-Livermore, C., Serenko, A., & Turan, A. H. (2017). The World IT Project: History, trials, tribulations, lessons, and recommendations. *Communications of the Association for Information Systems*, 41(18), 389–413.

Palvia, P., Ghosh, J., Jacks, T., Serenko, A., & Turan, A. (2018). Trekking the globe with the World IT Project. *Journal of Information Technology Case and Application Research*, 20(1), 3–8.

The World Bank (GDP) (2018). The World Bank. Retrieved from https://data.worldbank.org/indicator/NY.GDP.MKTP.CD?view=chart&year_high_desc=true.

The World Bank (Pop.) (2018). The World Bank. Retrieved from https://data.worldbank.org/indicator/SP.POP.TOTL.

UNESCO (2018). UNESCO — WORLD HERITAGE CENTER. Retrieved from https://whc.unesco.org/en/list/.

Chapter 3

Information Technology Issues in Bangladesh

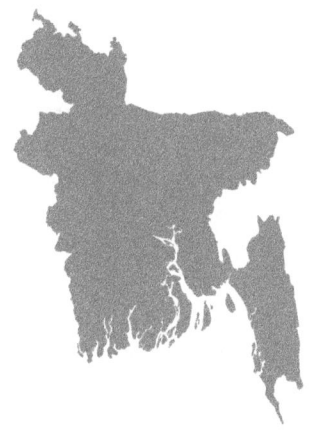

Md. Rakibul Hoque[*,‡] and Prashant Palvia[†,§]

*University of Dhaka, Dhaka, Bangladesh
Emporia State University, Emporia, KS, USA
†University of North Carolina at Greensboro,
Greensboro, NC, USA
‡rakibul@du.ac.bd
§pcpalvia@uncg.edu

Summary

This chapter shows that the information technology (IT) organizations in Bangladesh are mainly concerned about business productivity and cost reduction. Some medium and most large organizations of the country are rapidly looking for adopting business process management systems, such as enterprise resource planning (ERP), and using business intelligence and analytics for business growth and sustainability. Other major concerns of the IT organizations and the government are the digitalization of public service, e-governance and adaptability of the IT professionals. Although the IT sector and outsourcing sector are thriving in the country, the widespread coverage of broadband internet and cost of connectivity are still barriers for the IT industry growth. However, unlike before when connectivity coverage and cost were very high and there was little scope for IT jobs in the country, the youth of the current generation are very enthusiastic to choose the IT profession as their career. The government has taken multiple mega Information and Communication Technologies (ICT) projects with long-term vision to facilitate the IT industry in Bangladesh.

3.1   Introduction

Bangladesh was one of the least developed countries when information technology (IT) sector started its beginning in the 1960s, before its independence. Later, in the following decades, the large companies in the country started to adopt the use of computers. In the 1990s, the initiation of the mass adoption of computers for both official and personal use led to the creation of the IT industry. In 1997, the Bangladesh Association of Software and Information Services (BASIS) was started by the government. It acts as a national association for governing the software and IT service industry. Now, the IT industry has been showing a strong and consecutive upward trend for advancement in the last decade. The IT industry is mature enough to make revenues by providing services overseas as well as contributing to the national economy. One of the major accomplishments of the IT sector in Bangladesh lies in bringing forward new enthusiastic young entrepreneurs through different types of IT contests and funding schemes. IT companies are helping the country's IT industry through capturing a considerable share of international IT market. Currently, the evaluated size of the country's IT industry is around US$120 million (BIDA, 2017). Software development accounts for about 44% of the total industry revenue, whereas information technology enabled Services (ITES) accounts for about 56% (BIDA, 2017). Today, the country has invested about 50,000 crore (approximately US$6 billion) in IT industry, which is expected to be about 1% of the total GDP of the country in the next 5 years. According to a joint report with US-based Boston Consulting Group (BCG), by 2025, the revenue from the IT industry is projected to grow to US$4.6–4.8 billion from US$0.9–1.1 billion in 2017 (Sun, 2017).

3.2 Country Background and History

Bangladesh, officially known as the people's republic of Bangladesh, is a small $(1,47,570\,km^2)$ country, but one of the world's most densely populated, located in South Asia. The border of Bangladesh, formerly known as East Pakistan, was defined soon after the partition of India and Pakistan in 1947. The country achieved its independence from Pakistan in 1971 through a bitter war of nine months. Most of the parts of the country are low-lying and formed from the deposition of sediments carried by three big rivers named Padma, Jamuna, and Meghna. The country is very vulnerable to flooding, cyclones, and the rise in sea level.

The average life expectancy of the people of Bangladesh is 69 years for men and 70 years for women. Poverty is still widespread, however, the country has been maintaining a fast GDP growth (average 7% growth) and better index in life style, health service and education system. Although corruption has been always rampant, the country has shown improvement and growth in industrial, business and entrepreneurial activities. The garment industry and emigrant remittance are the main sources of economic sustainability and growth. Garment export accounts for about 14% of the GDP and 80% of the total exports, whereas the total value of export and import together is 38% of the GDP (Heritage.org, 2018). Of late in 2018, the country has graduated from the least developed country status to the low middle-income country status. Currently, the country is relentlessly working for achieving sustainable development goals (SDGs) to attain fast growth in resource mobilization, foreign direct investment (FDI), infrastructure and human resources development.

3.3 Information Technology in Bangladesh

According to BASIS, currently there are more than 1,500 IT companies in Bangladesh, of which about 400 companies export their services to more than 60 countries. The government is committed to the development of rural communities as per the level of their urban counterparts. In this regard, government has taken multiple visionary projects such as "Digital Bangladesh" and "Info-Sharker" in January 2009. Under the "Digital Bangladesh" project, the initial plan is to invest in Information and Communication Technologies (ICT) infrastructure to increase the IT exports and also to create a large number of IT jobs by 2021. One of the benefits of this project is building Business Process Outsourcing Centers (BPO) across the country to create jobs. The government has also set up the Bangladesh Hi-tech Park Authority (BHTPA) and extensively invested to build more than 30 Hi-tech parks and software technology parks to create 300,000 new jobs (Halim, 2018). These software technology parks will provide facilities like export processing zones (EPZs) where the investors from home and abroad would avail tax holiday for 10 years, along with exemption from income tax and import duties. According to a recent report by A.T. Kearney, a US-based global consulting firm, Bangladesh is among the top-50 IT destinations in the world (Bryon, 2016).

At government level, there are two main objectives behind leveraging the ICT industry in Bangladesh. These are: (1) to accelerate the IT growth to create new jobs and increase country's export through enabling IT and ITES services, and (2) to establish e-governance to support and modernize the public service sector. In this regard, Bangladesh Computer Council (BCC) is helping the government to adopt and build National ICT Infra-Network infrastructure to improve public services. This National ICT Infra-Network infrastructure is to be built in three phases. Phase 1 is known as BanglaGovNet Project that aims to improve the efficiency, accountability, and transparency of government services and decision making processes, as well as enhance interconnectivity among 18,130 government offices located at all ministries, 64 districts and 488 upazilas (administrative subunits of a district). Phase 2 aims to establish the internet backbone network with high capacity throughout the country, and phase 3 aims to spread the internet network from the internet backbone network to 2,600 union or rural administrative units throughout the country. The completion of phase 3 by 2018 would ensure fixed broadband connectivity to 15% of homes in the country.

The government has undertaken a flagship program called Access to Information (a2i) in 2007 that is financed by UNDP, USAID and the government itself (Khan and Jahan, 2018). Beside promoting and providing easy, affordable and reliable access to digitized government activities and public service, a2i is thought to bring changes through growing a culture of innovation among civil servants, citizens and entrepreneurs. Targeting socio-economic development, there are many initiatives of the a2i program, such as establishing digital centers in every union across the country to provide facilities regarding land records, birth records, telemedicine service, passport application, mobile financial services, rural e-commerce and e-learning to all the citizens. At government level, a2i aims to redesign the workflow within and among all ministries to increase efficiency. Inspired by the a2i program, Maldives and Bhutan signed MoUs with Bangladesh in

2015 and 2016, respectively. The success of the a2i programs brought the 'World Summit on Information Society (WSIS) Prize 2016' to Bangladesh in 2017.

As entrepreneurs are thought to be assets for any country, the a2i program specially works to harness entrepreneurship throughout the country. As an initiative of creating IT incubators, the a2i program has started the iDEA project to incubate 1,000 innovative initiatives and train IT entrepreneurs. Besides, a2i has also sponsored many ICT awareness and competition programs such as ICT Expo, National Hackathon, Startups Bangladesh and so on (Khan and Taher, 2017). Currently, more than 25% of the budget of a2i is allocated for Service Innovation Fund (SIF). The SIF is mainly available for innovative initiatives that would reduce the time, cost, and visits required in obtaining services related to the public service, healthcare, agriculture, socio-economic development and disaster management.

3.4   Methodology

Primary data collection for the World IT Project (Palvia *et al.*, 2017; Palvia *et al.*, 2018) in Bangladesh was challenging. One of the most common challenges is low response rate and making people understand about the research objective to obtain consent for participation. People and organizations in general are reluctant to share information with outsiders. However, to combat this challenge, the researchers had to directly approach the top-level managers and executives in different IT organizations. They had to seek consent through explaining the research purpose and the data collection plan and signing official letters of consent. In total, 284 participants were administered the survey over the period of 2015–2017. The respondents were properly explained the research objectives and the right to withdraw at any time, and given an assurance of anonymity.

Although the primary and national language of Bangladesh is Bengali, English was used for the interviews and the surveys as English is well understood by top-level managers and IT employees in major metropolitan cities. Questionnaires were given to the top-level managers along with envelopes to distribute to IT employees who worked under them. Upon completion, the managers collected the questionnaires from the respondents, signed and returned them to the local researchers. In some cases, timely reminders were sent to the managers. Once the data were collected, the

Table 3.1:   Descriptive Statistics

Characteristics	N	%	Characteristics	N	%
Education:			Years of work experience:		
High school or less	2	1	0–4 years	121	42.61
Associate degree	14	5	5–9 years	118	41.55
Bachelor's degree	129	45	10–19 years	34	11.97
Master's degree	136	48	20–29 years	5	1.76
Ph.D.	3	1	30+ years	6	2.11
Years of IT experience:			Organizational location:		
0–4 years	142	50.00	IT department employee	199	70.1
5–9 years	111	39.08	IT worker in non-IT	43	15.1
10–19 years	29	10.22	department		
20–29 years	1	.35	Contract employee	23	8.1
30+ years	1	.35	consultant	11	3.9
Work as:			Vendor employee	8	2.8
Mostly full time	245	86.3	Position:		
Mostly part time	20	7.0	Not part of management	87	31
Mostly over time	19	6.7	In lower management	54	19
Been laid off from IT job:			In middle management	100	35
Yes	49	17	In senior management	43	15
No	235	83			

local researchers entered the data in Excel Spreadsheet in a pre-defined format. The researchers also checked for data accuracy and validity. Table 3.1 shows the profile of the respondents.

3.5   Organizational IT Issues

Table 3.2 shows the organizational IT issues in Bangladesh, ranked in order of importance. Only two items in the top five, namely security and privacy and cost reduction, are ranked first and fifth in the top five of the 2017 SIM IT Key Issues and Trends study (Kappelman *et al.*, 2018). Further, only two items in the top-10 match items in the top 10 of the 2017 SIM issues list.

The highest-ranked item is business productivity and cost reduction. As ICT greatly reduces the costs of coordination, communications, and information processing (Brynjolfsson, 2000), the Bangladesh IT industry can contribute to economic development through employment generation, gaining share of service trade and most importantly by increasing production efficiency. IT has also created a new path for increasing interactions

Table 3.2: Organizational IT Issues in Bangladesh

Organizational IT Issues	Rank	Mean Rating*	Std. Deviation
Business productivity and cost reduction	1	1.62	0.82
Security and privacy	2	1.67	0.79
IT reliability and efficiency	3	1.73	0.92
Knowledge management	4	1.76	0.84
IT strategic planning	5	1.78	0.86
Revenue-generating IT innovations	6	1.81	0.79
Project management	7	1.84	0.82
Business agility and speed to market	8	1.85	0.85
Continuity planning and disaster recovery	9	1.86	0.89
Alignment between IT and business	10	1.89	0.83
Enterprise architecture	11	1.89	0.88
IT cost reduction	12	1.96	0.95
Globalization	13	1.99	0.91
IT service management (e.g., ITIL)	14	2.01	0.84
Attracting and retaining IT professionals	15	2.04	0.81
Business process reengineering	16	2.10	0.94
Outsourcing	17	2.11	0.99
Bring your own computing device (BYOD)	18	2.64	1.28

*Rating scale ranges from 1 to 5: 1 as most important and 5 as no importance.

with customers that helps in business productivity and cost reduction. For example, e-governance activities have been providing great facilities for improving total productivity of the firms in Bangladesh for many years. Furthermore, Bangladesh has a vast source of skilled and talented pool of human resources. Although, Bangladesh started to build its ICT infrastructure much later than India, in Bangladesh the average wage for IT professionals is half of its counterpart in India (Hossain, 2011). With the proper investment and utilization of the abilities of the young generation, such as language skills, analytical capability and IT adaptability, the use of IT can bring higher productivity and organizational transformation in Bangladesh.

The second highest-ranked item is security and privacy. The government passed the ICT Act, 2006, and proposed the Digital Security Act, 2016, which aims to ensure cybersecurity and to prevent cybercrime. IT professionals in Bangladesh focus on access control, anonymity, authentication and identification, cloud computing security, digital rights management, mobile computing security, privacy preserving system, secure banking and financial system, and web services from the perspective of security and privacy. As the technologies such as RFID, IoT, biometrics security, distributed systems security, intrusion detection and prevention systems are pervasive and ubiquitous today, the issues of security and privacy are also taken seriously in Bangladesh.

IT reliability and efficiency was ranked the third highest issue. In order for companies that are reliant on IT to run their operations, it is imperative that the underlying technology be reliable and efficient. As of today, more than 1,500 Bangladeshi IT companies are operating competitively and even taking a major share of the international market. In order for them to compete successfully in this global market, they are rightfully concerned about their operations being dependable and cost-effective.

Knowledge management and IT strategic planning were ranked fourth and fifth, respectively. Knowledge management is a systematic methodology to integrate and manage the identification, capture, evaluation, retrieval, and sharing of all business information as assets. From the perspective of Bangladesh, knowledge management is an important factor in sustainability and organizational performance. IT strategic plan is a roadmap for IT's projects and activities and must be integrated with business strategy. For example, to facilitate the implementation of IT strategic planning, the government has launched "Vision 2021" and "Digital Bangladesh" action plans to cope with the rapid change and to aid Bangladesh achieve the status of a Middle Income Country by 2021.

3.6   Technology and Infrastructure Issues

Table 3.3 shows the technology and infrastructure issues which are ranked in order of importance. Two of the top-five responses from Bangladesh match those in the top five of the 2017 SIM IT Key Issues and Trends Study (Kappelman *et al.*, 2018). Five of the top 10 in the Bangladesh list match the top 10 of the 2017 SIM issues list.

Software as a service (SaaS) is the top-ranked IT issue in Bangladesh. It is an Internet-dependent service distribution model where the customers' access and use centrally hosted software on subscription or on-demand basis. In Bangladesh, the use of SaaS is growing gradually, especially in the corporate sectors, where the users can use the most advanced and innovative technologies available in the industry for proactive as well as historical data analysis and generate insightful information from customers, financials, and operational activities. The use of SaaS in Bangladesh is mostly seen in conjunction with cloud computing and big data analysis.

Mobile and wireless applications ranked the second highest. Mobile telephony networks have demonstrated tremendous growth over the previous

Table 3.3: Technology and Infrastructure Issues in Bangladesh

Information Technology Related Issues	Rank	Mean Rating*	Std. Deviation
Software as a service	1	1.86	0.95
Mobile and wireless applications	2	1.88	0.90
Business intelligence/analytics	3	1.89	0.96
Enterprise application integration	4	1.91	0.83
Business process management systems	5	1.95	0.82
Collaborative and workflow tools	6	1.95	0.90
Big data systems	7	1.96	0.93
Enterprise resource planning (ERP) systems	8	1.98	0.91
Customer relationship management (CRM) systems	9	2.01	0.99
Networks/telecommunications	10	2.01	0.98
Data mining	11	2.02	1.00
Virtualization (desktop or server)	12	2.03	0.93
Social networking/media	13	2.05	0.98
Mobile apps development	14	2.10	0.95
Service-oriented architecture (SOA)	15	2.10	1.05
Cloud computing	16	2.15	1.14

*Rating scale ranges from 1 to 5: 1 as most important and 5 as no importance.

5 years. At the end of June 2018, total number of Mobile Phone subscribers had reached about 151 million (BTRC, 2018). The use of mobile applications has increased rapidly in the country. Wireless technologies such as Wi-Fi and 3G are used widely. Bangladesh's Mobile App Development Freelancers are highly skilled and talented. Even the government has launched a number of useful mobile applications to enhance the provision of better public services related to finance, business, game, security, e-commerce, travel and map, and healthcare for the people.

Business intelligence (BI)/analytics ranked third in the list. In Bangladesh, the demand for BI is increasing rapidly and BI is used in all businesses regardless of size and type. Job opportunities are also increasing in this area both in Bangladesh and globally. In order to capitalize on this trend, educational institutions, government agencies and non-government organizations may want to administer proper training for developing and improving the skills of IT professionals in this arena.

Enterprise application integration, related to middleware technologies, ranked fourth. IT professionals emphasize the use of EAI to ensure that information is used consistently by the business and that changes to core business data made by one application are correctly reflected in others. This is particularly important when applications have proliferated over the years. Business process management systems and collaborative and work-flow tools ranked as fifth and sixth in the list of IT issues. These applications are closely related to system optimization and project management and important for IT companies to complete projects successfully.

Ranked in the middle are technologies such as big data, ERP systems, CRM systems and networks/telecommunication. The potential of big data seems to be great in Bangladesh. The quick adoption of online platforms such as bikroy.com, ekhanei.com, as well as the proliferation of smartphones and 4G network point to the upcoming widespread application of big data in Bangladesh. As for ERP systems, Erp.com.bd, ERP2all, Extreme Solution and CMSN networks are well-known ERP service providers in the country. Networks/telecommunications now ranks lower at 10th place, as over the years Bangladesh has made huge progress in this sector.

3.7   Individual IT Employee Issues

Table 3.4 shows the summary of responses for the individual issues. Individual issues were categorized into seven categories: job satisfaction, work pressure, work–life balance, workload and burnout, sense of accomplishment, threats to job, and career plans. Once again, a lower average score means higher agreement with each statement.

The responses indicate that Bangladeshi IT professionals working in the IT sector have job satisfaction. However, they seem to indicate moderate levels of work pressure, although they seem to handle the workload well

Table 3.4:   Individual IT Employee Issues in Bangladesh

Measuring Job Related Issues	Mean Rating*	Std. Deviation
Job satisfaction		
In general, I like working here.	1.87	1.02
All in all, I am satisfied with my current job.	2.26	1.12
In general, I don't like my current job.	3.60	1.30
Work pressure		
I feel that the number of requests, problems or complaints that I deal with at work is more than expected.	2.60	1.11
I feel that the amount of work I do interferes with how well it is done.	2.53	1.12
I feel busy or rushed at work.	2.61	1.18
I feel pressured at work.	2.74	1.13

(*Continued*)

Table 3.4: (*Continued*)

Measuring Job Related Issues	Mean Rating*	Std. Deviation
Work–life balance		
There is a blurring of boundaries between my job and my home life.	2.94	1.23
My work-related responsibilities create conflicts with my home responsibilities.	3.02	1.37
I do not get everything done at home because I find myself completing job-related work.	2.95	1.21
Workload and burnout		
I feel drained from activities at work.	3.08	1.22
I feel tired from my work activities.	3.04	1.14
Working all day is a strain for me.	3.00	1.22
I feel burned out from my work activities.	3.16	1.18
Sense of accomplishment		
I feel I'm making an effective contribution to what this organization does.	2.06	0.96
In my opinion, I do a good job.	2.09	1.03
I have accomplished many worthwhile things in this job.	1.93	0.86
At my work, I feel confident that I am effective at getting things done.	1.86	0.88
Threats to one's job		
I am worried that future technology advancements may pose a threat to my job.	3.07	1.24
I believe that other people may be able to perform my work activities.	3.10	1.30
I am concerned that my job may be eliminated soon.	3.41	1.20
I am concerned that my job may be outsourced soon.	3.19	1.22
Career plans		
I will be with this organization 1 year from now.	2.82	1.37
I will take steps during the next year to secure a job at a different organization.	3.10	1.30
I will be with this organization 5 years from now.	2.73	1.16
I will be working in the IT field 1 year from now.	2.57	1.17
I will take steps during the next year to secure a job outside the IT field.	3.27	1.16
I will be working in the IT field 5 years from now.	2.40	1.17

*Rating scale ranges from 1 to 5: 1 as strongly agree and 5 as strongly disagree.

without much stress and maintain a balance between work and home life. Consistent with job satisfaction, the IT employees feel a strong sense of accomplishment at work and do not see imminent threats to their jobs. Finally, when asked about their career plans, the majority of them expect to continue in their jobs and in the IT industry, even in the long term. Some IT workers foresee being with the same employer in 5 years.

Enjoyment at the work place is considered significant by the IT employees in Bangladesh. Having a good salary is important as well, as is offered in the IT sector. While most IT employees are happy and satisfied with their jobs, on the flip side, IT workers often are dissatisfied with the fact that they have to provide service to international clients in western countries late at night. Late night shifts often put pressure on them and can sometimes lead to health issues. While the IT professionals in Bangladesh have higher levels of interest in IT jobs, because of growing funding and incubation support provided by the government and private investment options, the youth is showing a growing trend of IT entrepreneurship instead of opting for regular jobs.

3.8   Conclusion

The top-organizational concerns of IT in Bangladesh are business productivity and cost reduction, security and privacy, reliability and efficiency of IT systems, knowledge management and IT strategic planning respectively. The top-technology concerns include the application of Software as a service, mobile and wireless applications, business intelligence and analytics, Enterprise Application Integration, and business process management systems. Regarding individual concerns, the employees indicate that the IT industry continues to be a very attractive career option. They are, in general, satisfied with their jobs and expect to continue to work in the industry.

Today, the fast growth of local IT service industry and access to global markets because of the right policies taken by the government have paved the way for a long but prosperous journey for the growing IT industry in Bangladesh. In Bangladesh, the right use of the abundant amount of skilled and untapped manpower, along with the ICT initiatives and development schemes by the government, are strongly linked with the implementation plans for achieving the status of a middle-income country by 2021.

References

BIDA (2017). ICT sector in Bangladesh. Retrieved from bida.gov.bd: http://bida.gov.bd/wp-content/uploads/2018/01/ICT-sector-in-Bangladesh-1.pdf.

Bryon, A. (2016). Digital Bangladesh: An ICT revolution. Retrieved from the-worldfolio.com: http://www.theworldfolio.com/news/digital-bangladesh-a n-ict-revolution/3603/.

BTRC (2018). Bangladesh Telecommunication Regulatory Commission. Dhaka: BTRC.

Brynjolfsson, E., & Hitt, L. M. (2000). Beyond computation: Information technology, organizational transformation and business performance. *Journal of Economic Perspectives*, 14(4), 23–48.

Halim, H. (2018). What does 2018 hold for ICT in Bangladesh? Retrieved from dhakatribune.com: https://www.dhakatribune.com/feature/tech/201 8/01/04/2018-hold-ict-bangladesh.

Heritage.org (2018). Bangladesh Economy. Retrieved from heritage.org: https:// www.heritage.org/index/country/bangladesh.

Hossain, M. N. (2011). Integration of ICT Industries and Its Impact on Market Access and Trade: The Case of Ban. SANEI Working Paper, Nos. 11–16.

Hossain, N. S. (2011). Productivity and Performance of IT Sector in Bangladesh: Evidence from the Firm Level. XXXIV.

Khan, J., & Jahan, N. (2018). Scaling up the Service Innovation Fund Projects of Access to Information (A2i) Program of Bangladesh: The Way Forward for Private Innovators. *Global Journal of Management and Business Research: Interdisciplinary*, 10.

Khan, M., & Taher, F. (2017). ICT opens up new prospects for Bangladesh. Retrieved from thedailystar.net: https://www.thedailystar.net/drivers-eco nomy/ict-opens-new-prospects-bangladesh-1364893.

Kappelman, L., Johnson, V., McLean, E., & Maurer, C. (2018). The 2017 SIM IT Issues and Trends Study. *MISQ Exec*, 17(1), 53–88.

Palvia, P., Jacks, T., Ghosh, J., Licker, P., Romm-Livermore, C., Serenko, A., & Turan, A. H. (2017). The World IT Project: History, trials, tribulations, lessons, and recommendations. *Communications of the Association for Information Systems*, 41(18), 389–413.

Palvia, P., Ghosh, J., Jacks, T., Serenko, A., & Turan, A. (2018). Trekking the globe with the World IT Project. *Journal of Information Technology Case and Application Research*, 20(1), 3–8.

Sun, D. (2017). Bangladesh IT sector to grow five-fold by 2025. Retrieved from daily-sun.com: http://www.daily-sun.com/post/272250/2017/11/30/Bang ladesh-IT-sector-to-grow-fivefold-by-2025.

Chapter 4

Information Technology Issues in Brazil

Carlo Gabriel Porto Bellini*,§, Valter Moreno Jr.†,¶,
Alexandre Reis Graeml*,‖, and Tim Jacks§,**

*Federal University of Paraíba, Curitiba, Brazil
†State University of Rio de Janeiro (UERJ),
Rio de Janeiro, RJ, Brazil
‡Southern Illinois University Edwardsville, Edwardsville, USA
§cgpbellini@ccsa.ufpb.br
¶valter.moreno@gmail.com
‖alexandre.graeml@gmail.com
**tjacks@siue.edu

Summary

This chapter presents the importance that Brazilian information technology (IT) professionals assign to current IT-related organizational and technological issues, and a myriad of individual perceptions about the IT job. The organizational and technological issues come from a number of scales available in the literature, including reports sponsored by the Society for Information Management (SIM) and published annually regarding how US IT executives see the industry trends. As for the job issues, Brazilian professionals manifested their views on classical constructs. The Brazilian survey was done in 2015–2016 and gathered the answers of 348 professionals. Broadly, they see the proposed IT trends as important to themselves in the organizational setting, and they seem to be happy with their job appointments. A particular finding is that a very large array of different technologies that populate the modern work environment are considered important to the individual worker, probably due to technology integration, ubiquity, and pervasiveness being a reality in organizations.

4.1 Introduction

Brazil has become an iconic country in times of globalization, economic emergence, and the role of information technology (IT) to support both business activities and social interactions. In the last two decades, the country was acknowledged as an emergent political and economic power, but its rapid rise in the global landscape (*The Economist*, 2009) was followed by a sudden fall (*The Economist*, 2013). Especially after 2013, the country's reality was characterized by a multi-order crisis that included an unprecedented economic recession, social conflicts, urban violence, and institutional instability. The Brazilian IT sector, though, continues to be an important driver of the economy, while the country's IT market is among the largest in the world. Brazil has a solid IT knowledge base from the period when the country was isolated by internal legislation from the global technology suppliers — which thus made the country develop its own IT solutions.

In this chapter, we discuss how Brazilian IT professionals rate the importance of IT-related organizational and technological issues, and how they see their job positions according to classical theoretical constructs. The ratings of importance by Brazilians could be discussed *vis-à-vis* ratings expressed by US IT executives in the widely known annual surveys sponsored by the Society for Information Management (SIM). However, as the two surveys are not fully comparable, here we focus on key numbers from the Brazilian survey. As for the views of IT professionals about their job appointments, the overall understanding is that they are in jobs that satisfy their needs and align with their professional abilities, while also not involving excessive workload, work–home conflict, or exhaustion. They also do not feel threatened about losing their jobs, do not plan to leave the IT profession, but they are hesitant about staying in their current organization.

An important note is that our data were collected during one of the most dramatic crises in Brazilian history, and crises are expected to significantly influence one's job perceptions and intentions (Meneghel *et al.*, 2016; Boon and Biron, 2016; De Moura *et al.*, 2015; Murphy *et al.*, 2013). Thus, the findings on how Brazilian IT professionals see their jobs may reflect a particular moment of the country's economic cycle. A specific study on job-related perceptions and intentions of 291 of those Brazilian professionals across two moments of the crisis is available elsewhere (Bellini *et al.*, forthcoming).

4.2 Country Background and History

Brazil was discovered in 1500 during the Portuguese maritime expeditions. It became an independent kingdom in 1822, a republic in 1889, and today is one of the largest democracies in the world. The country's political boundaries did not change significantly in the last three centuries, especially considering its huge size. It is the fifth largest country in total area $(8,511,965 \, \text{km}^2)$ and fifth in population (210 million people, of which 86% are living in urban areas). Brazil is also the largest Portuguese-speaking nation, with Portuguese being the only language spoken by the vast majority of the population. Brazil is not as diverse a country in terms of ethnic groups as countries with similar sizes, like the USA, Canada, and Australia, but important differences in dominant ethnicities occur across the country's geographical regions. Depending on the region, ethnic groups are mostly represented by certain European and Asian people (descendants of immigrants), African people (descendants of slaves), and a variety of native Brazilians — some of them still living in their original conditions in reserves that are larger than many European countries.

In terms of the economy, as of 2016–2017, depending on calculations by the International Monetary Fund, the World Bank, or the United Nations, Brazil is the eighth or ninth GDP (gross domestic product) in the world, third among the BRICS (Brazil, Russia, India, China, South Africa) countries, and first in Latin America. Brazil is a highly industrialized country and among the world leaders in a number of economic sectors, such as in agribusiness. However, it is also home to precarious subsistence economies in vast regions of the country. Moreover, significant areas are dominated by untouched biomes, especially in the Amazon, Cerrado, and Semi-arid regions, where significant economic activity is almost absent.

Besides the mixed quality of sociodemographic numbers that character-ize the current Brazilian reality, in recent years the country also witnessed the emergence of scandals involving politicians and business leaders that led it into its most dramatic crisis in modern history. After the first half of 2013 and until today, the daily routine of Brazilians has changed dramatically, with frequent street riots and heated discussions in Congress, mass media, and social networks. There is no precedent for the current social reality in Brazil, once considered a peaceful country both in its international relations and internally. Physical and informational violence has spread among peo-ple, dividing the Brazilian society in two dominant groups: the far-right and the far-left supporters. The apex of the conflict occurred on September 6th, 2018 (the day before Independence Day in Brazil), when Jair Bolsonaro, the far-right presidential candidate who eventually won the elections in October, was stabbed during a campaign rally.[1]

4.3 Information Technology in Brazil

According to recent statistics of the Association for the Promotion of Excel-lence of the Brazilian Software (Softex, 2018), the Brazilian IT market is the ninth-largest in the world, accounting for 36% in Latin America and 2.1% of the country's GDP. The market is 52% represented by hardware, 22% by software, and 26% by IT services. The IT sector employs over 620,000 direct workers, and, depending on the method employed (MPS.BR, CMMI, MoProSoft, or IT Mark), it is ranked fourth, sixth, seventh or ninth in soft-ware process quality in the world. Part of the sector's standing is arguably due to decades of in-house development of software and hardware, as the Brazilian industry was protected against global competition until the 1990s (Bellini *et al.*, 2013). A case of note is the IT banking infrastructure, which is probably the most notorious illustration of Brazilian IT capabilities (Bellini and Pereira, 2009). However, while over 36,000 students per year complete their undergraduate studies in Computer Science and related fields, and another 180,000 students are enrolled in similar technical programs, it is

---

[1] https://www.bbc.com/news/world-latin-america-45441447.

expected that until 2022 the country will have a deficit of over 408,000 job positions in IT, with huge economic losses in missed opportunities (Softex, 2018).

4.4 Methodology

We surveyed Brazilian IT professionals as part of the World IT Project (Palvia *et al.*, 2017; Palvia *et al.*, 2018).[2] The project gathered data about the IT profession in 37 countries. According to Palvia *et al.* (2017), the project was motivated by the fact that most IT research is dominated by the perspectives of US and Western researchers, while researchers in other countries, particularly in the less developed ones, adopt theoretical models and empirical findings that are not applicable to their own context. As a consequence, the international community does not have access to potentially useful regional perspectives and cases. The problematic development of global knowledge in the IT field is due to a variety of reasons, including language barriers, unreliable access to mainstream literature in many parts of the world, lack of incentives, and lack of initiatives by both local researchers and the global community towards a richer and shared frame of reference.

The project leaders recruited researchers throughout the world to ensure understanding of local culture and to build bridges within the global IT community. A standard validated data collection instrument prepared in English was shared with all research teams that joined the project. Brazilian researchers joined the project in 2014. If necessary, a country's team translated the instrument to its dominant language (such as to Portuguese for Brazil), distributed the instrument to a large number of IT professionals in the country, and made use of local knowledge to interpret the results. The instrument included a variety of constructs related to organizational, technological, and individual issues of IT employees. It also included several contextual factors such as organizational variables, IT occupational culture, organizational culture, and national culture.

In Brazil, professionals were initially contacted through companies and universities via electronic mail, virtual social networks and academic meetings. They were asked to answer an online questionnaire about the World IT Project's constructs. After some difficulty to attract participants, the survey

---

[2]http://worlditproject.com/.

benefited from the help of Administradores, a popular web portal for the Brazilian business community that promoted the data collection instrument among IT professionals. We were then able to collect 385 questionnaires, resulting in 348 valid responses. Respondents answered the questionnaire in GoogleForms. We ensured theoretical consistency and professional colloquialism of the questions by having three Brazilian IT scholars independently translate the items and subsequently merging the translations after discrepancies were resolved. We also compared the translation with other instruments in the literature and submitted it to face validation by a Brazilian IT research team that included scholars and professionals.

After removing incomplete cases, outliers, identical answers and individuals who reported spending less than 50% of their time in IT-related activities at work, we obtained a final sample of 348 cases. The distribution of respondents is in Table 4.1. In terms of education, 84.2% of respondents reported having at least a bachelor's degree. This is higher than Joia and Mangia's (2017) study, where 67.5% of 323 Brazilian IT professionals had graduate degrees. The reason may be the increasing qualification of professionals in recent times, or idiosyncrasies of the two samples. In terms of age, 89.4% of the sample were in the 21–49 age range, 90% had from 5 to over 30

Table 4.1:   Descriptive Statistics

Characteristics	N	%	Characteristics	N	%
Education:			Years of work experience:		
High school or less	9	2.6	0–4 years	35	10.0
Associate degree	46	13.2	5–9 years	73	21.0
Bachelor's degree	107	30.8	10–19 years	109	31.3
Master's degree	166	47.7	20–29 years	96	27.6
Ph.D.	20	5.7	30+ years	35	10.1
Years of IT experience:			Organizational location:		
0–4 years	61	17.5	IT department employee	194	55.0
5–9 years	84	24.1	IT worker in non-IT dep.	8	2.3
10–19 years	106	30.5	Contract employee	5	1.4
20–29 years	78	22.4	Consultant	15	4.3
30+ years	19	5.5	Vendor employee	11	3.1
Work as:			Other	115	32.6
Mostly full time	213	60.3	Work position:		
Mostly part time	20	5.7	Not part of management	115	32.6
Mostly over time	115	32.6	In Lower management	95	26.9
Been laid off from IT job:			In middle management	82	23.2
Yes	92	26.4	In senior management	56	15.9
No	256	73.6			

years of total work experience, 82.5% had from 5 to over 30 years of work experience in IT, and a vast majority of 73.6% had no layoff experience. This means that our respondents were seasoned workers with seemingly good experience in the profession.

As for the respondents' primary professional roles, 43.4% were involved with programming, systems analysis or system administration, and 20.4% were involved with the more managerial or executive functions of project management and strategic planning. A large number of other primary roles are also represented in the sample, meaning that it is representative of most IT roles. A vast majority (83.9%) of respondents are not part of senior management, meaning that the IT job positions mostly involve operational and tactical functions.

As for the organizations where the Brazilian IT professionals work, again we have variety in the sample. The IT industry itself accounts for 35.4% of the professionals, while government and education are respectively represented by 14.4% and 10.3%. The significant participation of public and educational institutions is typical of certain regions of Brazil where industry activity is not properly developed. Important to note, our sample covers all geographical regions of the country. The organizations represented in our sample cover to a good degree all relevant sizes, and the IT department is also well represented. Additionally, according to 72.6% of the respondents, their organizations had medium to very high maturity in IT, which is additional evidence of the quality of the Brazilian IT sector.

4.5 Organizational IT Issues

The Brazilian survey asked respondents to "indicate the importance to you of the following IT-related organizational issues". The results are shown in Table 4.2. The items in the Brazilian questionnaire were developed specifically for the World IT Project with inputs from several instruments available in the literature, including the Society for Information Management (SIM) annual surveys of US IT executives.

The top-five issues for Brazilian IT professionals reveal that they are mostly concerned with broad issues on the role of IT in business similar to the SIM studies. On the other hand, bring your own device (BYOD) and outsourcing are consistently the least important issues in both studies. BYOD (#17) may have produced ambivalent feelings among respondents, as security is a highly praised issue (#6), and professionals expect that their companies will provide the needed resources for work. As for outsourcing

Table 4.2:   Organizational IT Issues in Brazil

Organizational IT Issues	Rank	Mean Rating*	Std. Deviation
IT reliability and efficiency	1	1.46	0.62
IT-business alignment	2	1.48	0.64
IT strategic planning	3	1.54	0.65
Knowledge management	4	1.58	0.69
Project management	5	1.61	0.69
Security and privacy	6	1.63	0.72
Attracting and retaining IT professionals	7	1.78	0.80
Continuity planning and disaster recovery	8	1.80	0.78
Revenue-generating IT innovations	9	1.81	0.80
Business productivity and cost reduction	10	1.84	0.68
IT service management	11	1.85	0.81
Business agility and speed to market	12	1.91	0.81
Business process reengineering	13	2.08	0.83
Enterprise architecture	14	2.16	0.89
IT cost reduction	15	2.16	0.89
Globalization	16	2.33	0.98
BYOD	17	2.89	1.14
Outsourcing	18	3.05	1.02

*Rating scale ranges from 1 to 5: 1 as most important and 5 as no importance.

(#18), ambivalence may have occurred due to the questionnaire not being clear about what was expected to be outsourced. At the same time, since the survey was done during the economic crisis in Brazil, individuals may have thought of their jobs as the target of outsourcing, thus strategically rating this issue as less important. The same line of thought may have influenced perceptions about IT cost reduction (#15), as respondents may have rated such costs as less important when thinking of the costs related to their job positions. Finally, the focus on globalization (#16) may have been of diminished importance due to professionals and companies being in need of first surviving the internal crisis.

4.6   Technology and Infrastructure Issues

The Brazilian survey asked respondents to "indicate the importance to you of the following IT-related technological issues". The items in the Brazilian questionnaire were developed specifically for the World IT Project with inputs from several instruments available in the literature, including the SIM surveys. Table 4.3 shows the technology concerns in our sample of IT professionals in Brazil.

Table 4.3:   Technology and Infrastructure Issues in Brazil

IT Related Issues	Rank	Mean Rating*	Std. Deviation
Enterprise app integration	1	1.77	0.73
Business intelligence/analytics	2	1.79	0.73
Networks and telecommunications	3	1.85	0.91
Business process management systems	4	1.89	0.79
Mobile and wireless applications	5	1.97	0.90
Software as a service	6	1.99	0.91
ERP systems	7	2.06	0.94
Cloud computing	8	2.07	0.92
Mobile app development	9	2.10	0.99
Collaboration and workflow tools	10	2.14	0.86
Virtualization	11	2.14	0.96
Service-oriented architecture	12	2.16	0.97
CRM systems	13	2.16	0.97
Big data systems	14	2.17	0.98
Data mining	15	2.19	0.97
Social networking	16	2.54	1.05

*Rating scale ranges from 1 to 5: 1 as most important and 5 as no importance.

For the technological issues, the difference in ratings of importance is smaller compared to the organizational issues. The overall perception is that most technologies are important. The impression is that, today, the personal routines in the work environment involve all types of IT tools, as technology integration, ubiquity, and pervasiveness are a reality throughout the organization. This is even the case for social networking technologies, as they serve to promote the organization, interact with clients, collect information about the competition, and ultimately develop business insights.

4.7   Individual IT Employee Issues

Table 4.4 shows descriptive statistics for Brazilian IT professionals' job-related perceptions and intentions. Overall, the professionals are satisfied with their jobs and believe that there is alignment between their abilities and the demands of the position (self-efficacy beliefs). As such, they do not feel insecure about their jobs. They also do not think that the workload is excessive or that there are work–home conflict and exhaustion due to work. Finally, they do not plan to leave the IT profession, but they are hesitant about staying in the current organization.

It is important to note that the survey was carried out in 2015–2016, that is, during one of the most dramatic crises in Brazilian history. As crises are expected to influence one's job perceptions and intentions about the self

Table 4.4:   Individual IT Employee Issues in Brazil

Individual Issues	Mean Rating*	Std. Deviation
Job satisfaction		
In general, I like working here.	1.79	0.79
All in all, I am satisfied with my current job.	2.11	0.94
In general, I don't like my current job.	4.00	1.07
Perceived work overload		
I feel that the number of requests, problems or complaints that I deal with at work is more than expected.	3.16	1.20
I feel that the amount of work I do interferes with how well it is done.	2.78	1.19
I feel busy or rushed at work.	2.91	1.18
I feel pressured at work.	3.10	1.21
Work–home conflict		
There is a blurring of boundaries between my job and my home life.	3.53	1.18
My work-related responsibilities create conflicts with my home responsibilities.	3.64	1.20
I do not get everything done at home because I find myself completing job-related work.	3.67	1.20
Work exhaustion/strain		
I feel drained from activities at work.	3.38	1.26
I feel tired from my work activities.	3.35	1.29
Working all day is a strain for me.	3.43	1.24
I feel burned out from my work activities.	3.51	1.24
Professional self-efficacy		
I feel I'm making an effective contribution to what this organization does.	1.96	0.86
In my opinion, I do a good job.	1.74	0.61
I have accomplished many worthwhile things in this job.	1.76	0.71
At my work, I feel confident that I am effective at getting things done.	1.75	0.62
Job insecurity		
I am worried that future technology advancements may pose a threat to my job.	2.95	1.36
I believe that other people may be able to perform my work activities.	2.07	0.83
I am concerned that my job may be eliminated soon.	3.78	1.17
I am concerned that my job may be outsourced soon.	3.89	1.18
Turnover intention		
I will be with this organization 1 year from now.	2.08	1.17
I will take steps during the next year to secure a job at a different organization.	3.19	1.34
I will be with this organization 5 years from now.	2.77	1.39

(Continued)

Table 4.4: (*Continued*)

Individual Issues	Mean Rating*	Std. Deviation
Turnover intention — IT profession		
I will be working in the IT field 1 year from now.	1.64	0.87
I will take steps during the next year to secure a job outside the IT field.	3.97	1.16
I will be working in the IT field 5 years from now.	1.94	1.08

*Rating scale ranges from 1 to 5: 1 as strongly agree and 5 as strongly disagree.

and the job (Meneghel *et al.*, 2016; Boon and Biron, 2016; De Moura *et al.*, 2015; Murphy *et al.*, 2013), Table 4.4 may reflect perceptions and intentions that are meaningful for that particular moment of the country. Interestingly, if taken as a whole, even under an economic crisis the respondents do not reveal a sense of job insecurity or an intention to leave the profession in search of more stability or better returns in other fields. However, they are not as decided about staying in the same organization. This may have to do with the nomadic culture of the IT profession, that is, workers may desire to cross organizational boundaries from time to time to be seen as successful (Moore and Burke, 2002).

4.8   Conclusion

This chapter addressed how Brazilian IT professionals rate the importance of certain IT-related organizational and technological issues, and how they see a number of aspects related to their current jobs. While any analysis of importance ratings is limited in many ways, it provides initial evidence of differences between the Brazilian context and that of other countries. We found that most IT-related organizational and technological issues are generally assumed as important, with only a few exceptions. As for job-related perceptions and intentions, our sample of Brazilian IT professionals revealed a significant fit between the professionals and their current job positions, notwithstanding the critical moment of the country's economy during which the survey was done.

Acknowledgments

We thank Administradores (www.administradores.com.br) and its CEO, Leandro Vieira, for promoting our online questionnaire, thus helping us to collect most of the data. We are also thankful to all respondents who kindly agreed to provide the empirical data.

References

Bellini, C. G. P. & Pereira, R. C. F. (2009). Editorial preface: IT management research in Brazil. *Journal of Global Information Technology Management*, 12(2), 1–4.

Bellini, C. G. P., Dantas, G. F. M., & Pereira, R. C. F. (2013). Are we still talking to ourselves? An analysis of the introspective information technology field by Brazilian experts. *International Journal of Human Capital & Information Technology Professionals*, 4(3), 11–25.

Bellini, C. G. P., Palvia, P., Moreno, Jr., V. A., Jacks, T., & Graeml, A. R. (forthcoming) Should I stay or should I go? A study of IT professionals during a national crisis. *Information Technology & People*. http://dx.doi.o rg/10.1108/itp-07-2017-0235.

Boon, C. & Biron, M. (2016). Temporal issues in person-organization fit, person-job fit and turnover: The role of leader–member exchange. *Human Relations*, 69(12), 2177–2200.

De Moura, Jr., P. J., Bellini, C. G. P., & Pereira, R. C. F. (2015). Cognition, behavior and structure of customer teams in enterprise systems implementation: A comparative study. *Journal of Global Information Technology Management*, 18(4), 271–291.

Joia, L. A. & Mangia, U. (2017). Career transition antecedents in the information technology area. *Information Systems Journal*, 27(1), 31–57.

Meneghel, I., Borgogni, L., Miraglia, M., Salanova, M., & Martínez, I. M. (2016). From social context and resilience to performance through job satisfaction: A multilevel study over time. *Human Relations*, 69(11), 2047–2067.

Moore, J. E. & Burke, L. A. (2002). How to turn around 'turnover culture' in IT. *Communications of the ACM*, 45(2), 73–78.

Murphy, W. M., Burton, J. P., Henagan, S. C., & Briscoe, J. P. (2013). Employee reactions to job insecurity in a declining economy: A longitudinal study of the mediating role of job embeddedness. *Group & Organization Management*, 38(4) 512–537.

Palvia, P., Jacks, T., Ghosh, J., Licker, P., Livermore, C., Serenko, A., & Turan, A. (2017). The World IT Project: History, trials, tribulations, lessons, and recommendations. *Communications of the AIS*, 41, 389–413.

Palvia, P., Ghosh, J., Jacks, T., Serenko, A., & Turan, A. (2018). Trekking the globe with the World IT Project. *Journal of Information Technology Case and Application Research*, 20(1), 3–8.

SOFTEX (2018). *Relatório anual*. Brasília, Brazil, Associação para Promoção da Excelência do Software Brasileiro. Available at: https://www.softex.br/bo oksoftex/.

The Economist (2009). Brazil takes off. *The Economist*, Nov 12, 2009. Available at: https://www.economist.com/leaders/2009/11/12/brazil-takes-off.

The Economist (2013). Has Brazil blown it? *The Economist*, Sep 27, 2013. Available at: https://www.economist.com/leaders/2013/09/27/has-brazil-blow n-it.

Chapter 5

Information Technology Issues in Canada

Alexander Serenko[*,†,¶], Nick Bontis[‡,**], Prashant Palvia[††,‡‡],
and Aykut Hamit Turan[‡‡]

[*]University of Toronto, 27 King's College Cir,
Toronto, Canada
[†]University of Ontario Institute of Technology, 2000 Simcoe St N,
Oshawa, Canada
[‡]McMaster University, Hamilton, Canada
[§]University of North Carolina at Greensboro,
Greensboro, North Carolina, USA
[¶]Sakarya University, Sakarya, Turkey
[∥]a.serenko@utoronto.ca
[**]nbontis@mcmaster.ca
[††]pcpalvia@uncg.edu
[‡‡]ahturan@sakarya.edu.tr

Summary

The information technology (IT) industry is an irrevocable part of the Canadian economy. It has adapted well to the needs of the global market. Security and privacy are considered top-organizational issues. This is not surprising since ensuring security is vital for the widespread adoption of IT products and services. The Canadian public has always been concerned with their privacy, which was reflected in the present study. Networks and telecommunications are considered top-technology issues, followed by virtualization and enterprise application integration. Overall, a majority of Canadian IT professionals are satisfied with their jobs and report an acceptable level of work pressure, work–life balance, workload, and burnout. Nevertheless, some are likely to leave their current organizations. Despite a high demand on IT jobs, some IT workers may also leave the IT profession.

5.1 Introduction

Canada is a large multi-cultural country which boasts a well-developed information technology (IT) sector. It is home to many leading multi-national IT corporations and domestic firms. The overall contribution of the IT sector to the Canadian economy has been continuously increasing at an average growth rate of 4% per year. Most importantly, around 80% of all Canadian IT products, particularly, computer and communications equipment, electronic components, and audio–video goods, are exported. In contrast, IT services, such as communications services, are domestically-oriented (ISEDC, 2017). The IT industry is the largest contributor to research and development (R&D) generating almost one-third of the private R&D expenditure in the country. The working environment in Canadian IT firms is generally positive, and IT employees report an acceptable level of job satisfaction (Statista, 2017). Almost all Canadian colleges and universities offer some form of IT education. Nevertheless, there is a shortage of skilled IT professionals in the country.

5.2 Country Background and History

Canada is the second largest country in the world by total area, covering almost 10 million km^2. It has 10 provinces and three territories. Canada's population is highly urbanized, with over 80% of the residents living in large and medium-sized cities located within 100 miles of the southern

US border. Canada has a population of over 37 million people, and its largest metropolitan areas include Toronto, Vancouver, Montreal, Calgary, and Ottawa (Statistics Canada, 2016). The southern part of Canada has four distinct seasons with hot summers and cold winters. Its official languages are English and French. French is spoken predominantly in Quebec and some parts of eastern Canada. Canada is a highly diverse multicultural nation due to its relatively open immigration policies attracting newcomers from all over the world. Canada is a federal parliamentary democracy, with several major political parties, and a constitutional monarchy, with Queen Elizabeth II being the formal head of state.

Prior to the French and English colonists, who came to North America in the 16th century, Canada had been inhabited by various indigenous peoples who currently represent around 5% of the country's population. During multiple armed conflicts, British North America gained territory, and in 1867, the colonies formed the federal dominion referred to as Canada. Newfoundland was the last province to join the union in 1949, and in 1999, Nunavut separated from Northwest Territories and formally became the third Canadian territory (Black, 2014). Canada has no official religion and is committed to the freedom of religion and religious pluralism.

Canada is considered a developed country with the 2016 GDP of over US$1.5 trillion representing the tenth largest economy in the world. It enjoys one of the highest levels of economic freedom (The Heritage Foundation, 2018) and has various social programs, such as free Medicare, the Pension Plan, and Student Loans. The service sector, represented by retail, finance, real estate, education, healthcare, and high tech, accounts for 70% of the GDP and 75% of total employment. Other critically important sectors include manufacturing, energy, raw materials, and agriculture. Canada is a member of the Group of Seven (G7), Organization for Economic Co-operation and Development (OECD), and the Asia-Pacific Economic Cooperation Forum (APEC).

5.3   Information Technology in Canada

The Canadian IT sector is represented by almost 40,000 individual organizations, 86% of which employ fewer than 10 workers, and only 110 organizations have over 500 employees (ISEDC, 2017). Almost 90% of IT organizations focus on software and computer services, and the rest specialize in IT wholesaling, manufacturing, and communications services. Most IT manufacturers are large in size, and they produce computer, communications, audio, video, and peripheral equipment, electronic components, and storage devices. Almost 60% of the leading IT companies are located in Ontario, followed by British Columbia (15%), Quebec (13%), and Alberta (11%) (Statista, 2017). Canada is home to many multinational IT companies, such as Samsung, Apple, Amazon, Alphabet (Google), IBM, and Dell. It also boasts a number of home-grown success stories, for example, Bell Canada, Telus, BlackBerry, and Shopify.

The IT sector generates almost 5% of the Canadian GDP. Even though this represents a small percentage of the Canadian economy and a fraction of the global IT market, the Canadian IT sector is very vibrant, innovative, and diverse, and it acts as an accelerator of the overall country's economy. As such, the competitiveness of most other Canadian industries depends on the success of the IT sector (Statista, 2017).

There is a steadily growing shortage of IT professionals in Canada, and the competition for IT talent is intense. As a result, an average IT salary has risen by around 20% in the last 5 years (Workopolis, 2017). As of 2018, the highest-demand IT jobs were project manager, software engineer, web developer, programmer, system analyst, system administrator, quality insurance analyst, and business analyst (Randstad Canada, 2018). Almost all Canadian universities and colleges offer IT-related programs, which are generally hosted in computer science or software engineering departments, specialized IS/IT faculties, and business schools. By following a recent trend of an analytics economy, various business analytics programs have been launched as HBA and MBA specializations or stand-alone undergraduate and graduate degrees. For example, the Schulich School of Business, York University has recently introduced the Master of Business Analytics program, and the Faculty of Science, the University of British Columbia has launched the Master of Data Science degree. As a result, the Canadian IT workforce is highly educated with over a half of all IT employees holding a university degree, compared to the 29% industry average (ISEDC, 2017).

5.4   Methodology

In Canada, knowledge workers are generally responsive to surveys because research surveys are frequently administered in universities and colleges as part of faculty research programs, which are done for course bonus points or pro bono. Even though Canada is a bilingual country, the survey was administered in English only because all Canadian IT workers are expected to be proficient in English. The original instrument (see Palvia *et al.*, 2017; Palvia *et al.*, 2018) with no modifications was used. As per the recommendation of the Research Ethics Board (a Canadian equivalent to the Institutional Review Board), no IP addresses or names of organizations were collected to fully ensure the respondents' anonymity. To encourage participation, the cover letter emphasized the importance of the study, explained anonymity safeguards, and described how the respondents may later receive a summary of the findings.

The online survey was created and administered on SurveyMonkey. A survey URL was sent to a number of IT managers, upper-level executives, and senior IT workers personally known to the researchers. They were asked to forward the link with instructions to their subordinates and/or colleagues. They were also asked to send two weekly reminders. As a result, 333 responses were received. Out of them, 22 surveys were incomplete and subsequently discarded, resulting in a dataset of 311 usable responses.

Table 5.1 presents the descriptive statistics of the Canadian IT respondents. They are generally very well-educated, have extensive IT and overall work experience, are mostly employed full-time, tend to work in an organizational IT department, and frequently hold managerial positions. Almost one-third of them have experienced an involuntary loss of an IT job.

Table 5.1: Descriptive Statistics

Characteristics	N	%	Characteristics	N	%
Education:			Years of work experience:		
High school or less	15	4.8	0–4 years	9	2.9
Associate degree	102	32.8	5–9 years	33	10.6
Bachelor's degree	149	47.9	10–19 years	87	28.0
Master's degree	41	13.2	20–29 years	98	31.5
Ph.D.	4	1.3	30+ years	84	27.0
Years of IT experience:			Organizational location:		
0–4 years	18	5.8	IT department employee	234	75.2
5–9 years	49	15.8	IT worker in non-IT department	34	10.9
10–19 years	118	37.9	Contract employee	16	5.1
20–29 years	90	28.9	Consultant	23	7.4
30+ years	36	11.6	Vendor employee	4	1.3
Work as:			Work position:		
Mostly full time	291	93.6	Not part of management	143	46.0
Mostly part time	6	1.9	In lower management	46	14.8
Mostly over time	14	4.5	In middle management	76	24.4
Been laid off from IT job:			In senior management	46	14.8
Yes	92	29.6			
No	219	70.4			

5.5 Organizational IT Issues

Table 5.2 offers a ranking of organizational IT issues, listed from the most to the least important ones. Security and privacy was indicated as the most important issue, which also received the top position in the 2017 SIM IT Issues and Trends Report (Kappelman *et al.*, 2018). The other issues included in top five where IT reliability and efficiency, knowledge management, continuity planning and disaster recovery, and alignment between IT and business. Out of them, in the SIM Report, alignment between IT and business was also ranked in top five, and business continuity as well as productivity/efficiency were ranked at the bottom of the list. At the same time, knowledge management, which was ranked third in the present study, was not mentioned in the SIM Report. This is not surprising since knowledge management is generally considered of high importance by the Canadian private and public organizations. For example, Canada is usually ranked in top-five countries based on the volume of knowledge management research (Serenko, 2013), and two of the most productive and influential knowledge management scholars are based in Canada (Gaviria-Marin *et al.*, 2018).

Table 5.2:   Organizational IT Issues in Canada

Organizational IT Issues	Rank	Mean Rating*	Std. Deviation
Security and privacy	1	1.86	0.82
IT reliability and efficiency	2	1.89	0.83
Knowledge management	3	2.16	0.82
Continuity planning and disaster recovery	4	2.18	0.89
Alignment between IT and business	5	2.20	0.78
IT strategic planning	6	2.23	0.80
Project management	7	2.25	0.79
Attracting and retaining IT professionals	8	2.26	0.85
IT service management (e.g., ITIL)	9	2.37	0.88
Business productivity and cost reduction	10	2.38	0.74
Business agility and speed to market	11	2.41	0.81
Enterprise architecture	12	2.47	0.83
IT cost reduction	13	2.51	0.82
Business process reengineering	14	2.59	0.78
Revenue-generating IT innovations	15	2.60	0.92
Globalization	16	2.93	1.01
Bring your own computing device (BYOD)	17	3.32	1.11
Outsourcing	18	3.35	1.01

*Rating scale ranges from 1 to 5: 1 as most important and 5 as no importance.

Many issues that were identified in the present study were not mentioned in the SIM Report (Kappelman *et al.*, 2018). Examples include IT strategic planning and project management, which were ranked in the top 10 in the present list. This shows that Canadian IT organizations approach IT planning from a strategic perspective and take project management activities very seriously. Attracting and retaining IT professionals is also more important in the Canadian environment due to a shortage of skilled IT talent. At the same time, business productivity and cost reduction as well as IT cost reduction were ranked approximately in the middle of both lists.

Interestingly, Canadian IT organizations are not worried about globalization and outsourcing as either a business opportunity or a threat. For example, the amount of IT outsourcing spending by private and public Canadian organizations showed no increase from 2013 to 2016, and it even exhibited a small decrease in 2014 (Statista, 2018). BYOD also received a very low ranking. On the one hand, BYOD may bring some hardware and software savings and increase employee effectiveness and efficiency. On the other hand, it can be associated with privacy and security risks, which may outweigh the benefits. To safeguard the implementation of BYOD initiatives, the Canadian government recently issued a report listing various recommendations (Office of the Privacy Commissioner of Canada, 2015). Thus, due to the existence of clear guidelines, BYOD has become less of a concern to the Canadian organizations.

5.6 Technology and Infrastructure Issues

Table 5.3 offers a ranking of technology and infrastructure issues in Canada.

The top-five issues were networks/telecommunications, virtualization, enterprise application integration (EAI), business intelligence/analytics, and collaborative and workflow tools. Networks/telecommunications was ranked the top issue, dramatically outperforming the nearest competitor. In contrast, this issue was not even included in the top-10 list in the 2017 SIM IT Issues and Trends Report (Kappelman *et al.*, 2018) even though it received the eighth spot in the lists of the largest IT investments. It is possible that the networks/telecommunications infrastructure is of primary importance to the Canadian telecommunications firms which represent a significant proportion of the country's IT sector. Virtualization and enterprise application integration received the second and third places, respectively, but they were not included in the top-10 issues in the SIM Report.

Table 5.3: Technology and Infrastructure Issues in Canada

IT Related Issues	Rank	Mean Rating*	Std. Deviation
Networks/telecommunications	1	2.17	0.82
Virtualization (desktop or server)	2	2.45	0.90
Enterprise application integration	3	2.47	0.82
Business intelligence/analytics	4	2.50	0.78
Collaborative and workflow tools	5	2.56	0.84
Mobile and wireless applications	6	2.57	0.95
Customer relationship management (CRM)	7	2.59	1.01
Business process management systems	8	2.61	0.88
Enterprise resource planning (ERP) systems	9	2.63	0.89
Software as a service	10	2.64	0.93
Service-oriented architecture (SOA)	11	2.67	0.93
Data mining	12	2.70	0.97
Big data systems	13	2.71	0.95
Cloud computing	14	2.86	0.97
Mobile apps development	15	2.94	1.04
Social networking/media	16	3.09	1.03

*Rating scale ranges from 1 to 5: 1 as most important and 5 as no importance.

Virtualization, which has become a key enabler of cloud computing, is now routinely taught as part of IT curriculum. Surprisingly, EAI, which is considered a mundane topic and is rarely given due attention in university and college IT curriculum, is in fact very important for Canadian IT organizations. This suggests that curriculum developers should not fully shift their attention to the trendy topics, such as virtualization and business intelligence/analytics at the expense of other equally important technology issues.

As expected, business intelligence/analytics was also ranked very highly in both reports. In addition to the importance of technical issues, skilled business intelligence/analytics professionals are still hard to find, as indicated in the SIM Report. This suggests that Canadian educational institutions offering business intelligence/analytics programs are on the right path. However, more advanced business intelligence platforms for big data and data mining are of lesser importance. It is possible that Canadian IT firms focus on smaller-scale business intelligence/analytics solutions and are delaying the implementation of more complicated business intelligence/analytics technologies. The fact that collaborative and workflow tools are included in top-five reflects the importance of knowledge management organizational issues — to facilitate intra-organizational knowledge flow, it is important to have appropriate collaborative and workflow tools in place.

5.7 Individual IT Employee Issues

Table 5.4 outlines the relative importance of individual IT employee issues. Several interesting findings emerged that warrant further elaboration.

First, the Canadian IT professionals are somewhat satisfied with their jobs — overall, they like their organizations and their work environments.

Table 5.4: Individual IT Employee Issues in Canada

Individual Issues	Mean Rating*	Std. Deviation
Job satisfaction		
In general, I like working here.	2.09	0.91
All in all, I am satisfied with my current job.	2.23	0.91
In general, I don't like my current job.	3.63	1.13
Work pressure		
I feel that the number of requests, problems or complaints that I deal with at work is more than expected.	3.05	1.05
I feel that the amount of work I do interferes with how well it is done.	3.00	1.04
I feel busy or rushed at work.	2.93	1.05
I feel pressured at work.	2.95	1.04
Work–life balance		
There is a blurring of boundaries between my job and my home life.	3.28	1.13
My work-related responsibilities create conflicts with my home responsibilities.	3.38	1.10
I do not get everything done at home because I find myself completing job-related work.	3.37	1.13
Workload and burnout		
I feel drained from activities at work.	3.00	1.10
I feel tired from my work activities.	2.94	1.08
Working all day is a strain for me.	3.20	1.06
I feel burned out from my work activities.	3.20	1.07
Sense of accomplishment		
I feel I'm making an effective contribution to what this organization does.	2.36	0.90
In my opinion, I do a good job.	1.89	0.70
I have accomplished many worthwhile things in this job.	2.08	0.77
At my work, I feel confident that I am effective at getting things done.	2.07	0.75

(Continued)

Table 5.4: (*Continued*)

Individual Issues	Mean Rating*	Std. Deviation
Threats to one's job		
I am worried that future technology advancements may pose a threat to my job.	3.26	1.11
I believe that other people may be able to perform my work activities.	2.69	1.00
I am concerned that my job may be eliminated soon.	3.34	1.09
I am concerned that my job may be outsourced soon.	3.50	1.13
Career plans		
I will be with this organization 1 year from now.	2.30	0.93
I will take steps during the next year to secure a job at a different organization.	3.29	1.07
I will be with this organization 5 years from now.	2.81	1.06
I will be working in the IT field 1 year from now.	2.00	0.87
I will take steps during the next year to secure a job outside the IT field.	3.57	1.02
I will be working in the IT field 5 years from now.	2.41	0.95

*Rating scale ranges from 1 to 5: 1 as strongly agree, to 5 as strongly disagree.

This finding is consistent with the recent report showing that almost 90% of the Canadian IT employees are at least somewhat satisfied with their job (Statista, 2017). In fact, despite a high velocity of change within the IT industry which requires life-long learning and continuous updating of one's skills, the Canadian IT workers enjoy high salaries, lucrative career opportunities, and pleasant working environment. Second, they experience relatively low work pressure. As such, they are not overwhelmed with requests, problems, or complaints, the amount of work does not force employees to compromise work quality, and they are not rushed or needlessly pressured. Work pressure may be quantitative and qualitative in nature. Quantitative work pressure results from a number of tasks exceeding the worker's ability to complete them. Qualitative work pressure means that employees are expected to produce the output of such high quality that they cannot possibly do regardless of the amount of time available at their disposal. However, the issues above are unlikely to have a major negative impact on the Canadian IT workers.

Third, the Canadian IT employees reported a relatively acceptable level of work–life balance. There are some boundaries between their professional and personal lives, and work responsibilities do not generally interfere with home activities. Achieving a healthy level of work–life balance is important both for employees' physical as well as mental well-being, which further results in higher job satisfaction and lower turnover intentions. Fourth, the Canadian IT professionals enjoy a reasonable level of workload and an acceptable level of burnout. All contemporary knowledge workers experience strain, pressure, and fatigue from work activities. Nevertheless, it is critical to maintain their levels within an acceptable range, otherwise the consequences may be devastating for both organizations and their employees. Fifth, most employees report a sense of accomplishment because they believe they are making an effective contribution to their organizations, realized their full potential, and are confident in their professional abilities. This is, again, a positive sign of a healthy climate within the Canadian IT industry.

Sixth, most Canadian IT professionals do not perceive any major threats to their employment resulting from other workers, technologies, or outsourcing. This results from a mismatch between a demand for and supply of skilled IT professionals in the country. Nevertheless, despite an acceptable level of work pressure, work–life balance, workload, burnout, and a sense of accomplishment, resulting in job satisfaction, some Canadian IT employees exhibit a tendency to move to another organization (turnover) or to leave the IT industry (turn away) in the future. Whereas turnover may be logically explained by the hot IT job market, the reasons for turn away are less obvious. For example, it may be assumed that due to a presently low level of unemployment in Canada, some IT workers may look for further professional challenge outside of the IT sector.

5.8 Conclusion

This chapter summarized organizational, technology, and individual issues reported by 311 Canadian IT workers. Overall, Canada has a well-developed IT industry which represents a vital part of the country's economy. There is a high demand of skilled IT workforce, and the Canadian educational institutions are trying to keep up with the needs of the job market. Security and privacy is considered the top-organizational issue which is consistent with the findings reported in other countries. Out of the technology and infrastructure issues, networks/telecommunications is ranked the most important. Canada is a large country, and it is vital to create an effective and efficient infrastructure to enable intra- and inter-organizational communication. With respect to individual issues, the Canadian IT workers do not experience any major negative effects, such as work pressure, work–life balance, workload, and burnout. They feel some sense of professional accomplishment, and are generally satisfied with their jobs. Nevertheless, some of them may change their employers or even move to other industries.

References

Black, C. (2014). *Rise to Greatness: The History of Canada from the Vikings to the Present*. Toronto: McClelland & Stewart.

Gaviria-Marin, M., Merigo, J. M., & Popa, S. (2018). Twenty years of the Journal of Knowledge Management: A bibliometric analysis. *Journal of Knowledge Management*, 22(8), 1655–1687.

ISEDC. (2017). Innovation, Science and Economic Development Canada. 2016 Canadian ICT sector profile. Available online at https://www.ic.gc.ca/eic /site/ict-tic.nsf/eng/h_it07229.html.

Kappelman, L., Johnson, V., Maurer, C., McLean, E., Torres, R., David, A., & Nguyen, Q. (2018). The 2017 SIM IT issues and trends study. *MIS Quarterly Executive*, 17(1), 53–88.

Office of the Privacy Commissioner of Canada. (2015). Is a Bring Your Own Device (BYOD) program the right choice for your organization? Available online at https://www.priv.gc.ca/en/privacy-topics/technology-and-priva cy/mobile-devices-and-apps/gd_byod_201508/.

Palvia, P., Jacks, T., Ghosh, J., Licker, P., Romm-Livermore, C., Serenko, A., & Turan, A. H. (2017). The World IT Project: History, trials, tribulations, lessons, and recommendations. *Communications of the Association for Information Systems*, 41(18), 389–413.

Palvia, P., Ghosh, J., Jacks, T., Serenko, A., & Turan, A. (2018). Trekking the globe with the World IT Project. *Journal of Information Technology Case and Application Research*, 20(1), 3–8.

Randstad Canada. (2018). Best technology and IT jobs in 2018. Available online at https://www.randstad.ca/best-jobs/best-it-and-technology-jobs/.

Serenko, A. (2013). Meta-analysis of scientometric research of knowledge management: Discovering the identity of the discipline. *Journal of Knowledge Management*, 17(5), 773–812.

Statista (2017). Information and communication technology (ICT) in Canada. Available online at https://www.statista.com/study/.

Statista (2018). Technology outsourcing industry spending by business and government in Canada from 2013 to 2018. Available online at https://www.sta tista.com/statistics/821790/canada-spending-tech-outsourcing-industry/.

Statistics Canada (2016). Census — population and demography. Available online at https://www150.statcan.gc.ca/n1/en/subjects/population_and_d emography.

The Heritage Foundation (2018). 2018 The Index of Economic Freedom. Available online at https://www.heritage.org/index/.

Workopolis (2017). The fastest-growing tech jobs and industries in Canada. Available online at https://careers.workopolis.com/advice/the-fastest-growing-tech-jobs-and-industries-in-canada/.

Chapter 6

Information Technology Issues in China

Jie Yu[*,§], Yue Guo[†,¶], and Prashant Palvia[‡,||]

*University of Nottingham Ningbo China, Zhejiang Sheng, China
†King's College London, London, UK
‡University of North Carolina at Greensboro,
Greensboro, NC, USA
§joseph.yu@nottingham.edu.cn
¶yue.guo@kcl.ac.uk
||pcpalvia@uncg.edu

Summary

In this chapter, we provide important information technology (IT) issues in China, such as organizational IT issues, technology issues, and individual issues. In the World IT Project survey, we recruited 310 IT workers in China. Most of the respondents were in their early career and worked full time in China's IT organizations. The findings show that the most important IT-related organizational issues are: IT reliability and efficiency, security and privacy, and IT strategic planning. Among technology issues, IT professionals identified the following issues as the top concerns: networks/telecommunications, big data systems, data mining, software as a service, and business intelligence/analytics. Most IT employees seem to be satisfied with their current jobs and felt a sense of accomplishment at work. Results further show that more than half of the IT workers would not change their jobs in the short term and felt secure in their current jobs. In addition, there were no significant work–life conflicts among the surveyed IT employees.

6.1 Introduction

Information technology (IT) is advancing rapidly worldwide. The development of IT plays an important role in the economic and social development of nations. On the one hand, IT has positively affected people's lives and has brought significant advantages for business. On the other hand, there are new challenges for IT management and staff. Previous studies have provided IT issues in organizations in some developed countries but there is little literature investigating IT-related issues in China. Since different countries have different economic strategies and market characteristics, IT issues may vary in different countries and regions. Therefore, it seems worthwhile to evaluate the important IT issues in China, such as organizational IT issues, technology issues, and individual issues. The current chapter is designed to provide insights into IT-related issues of Chinese organizations and their IT employees.

6.2 Country Background and History

China, officially the People's Republic of China (PRC), is a country located in East Asia with a population of 1.4 billion. China (PRC) was established in 1949 and has reformed the economy and moved forward rapidly from a traditional agricultural society to a modern industrial society since 1978. In the past 40 years, the Chinese government has made every effort to develop its industry and technology. To some extent, China has become an industrialized country with a relatively low level of technology. China has also been investing in high-tech and advanced manufacturing industries in recent years. For example, China has actively accelerated industrialization with informatization and has attained great achievement in the IT industry in recent years, e.g., companies like Alibaba, JD, Tencent, and Xiaomi have emerged. Along with the development of technology, there is increasing growth in China's economy. According to nominal GDP in 2017, China was the world's second-largest economy (Fang *et al.*, 2018).

6.3 Information Technology in China

China has gone through an impressive growth in the IT industry over the last two decades. Due to reforms and opening-up policy, China has developed informatization rapidly by allocating considerable resources to the IT industry. In recent years, the IT industry has become one of the pillar industries to promote economic development and social progress (Chismar, 1996; Hanna and Qiang, 2011). Over the past 20 years, China has implemented several 5-year plans to promote the country to be a world-leader and high-tech society through the development of its IT industry.

According to a recent survey, the number of China's Internet users has grown steadily in the last decade and China had 772 million netizens by the end of 2017 (China Internet Network Information Center, 2018). Mobile netizens accounted for 97.5% of the total Internet users. As of December 2017, the number of Internet companies that had gone public in either domestic stock exchange markets or overseas stock exchange markets has reached 102, with a total market value of RMB 8.97 trillion. There have been 77 Chinese Internet and IT Unicorn Companies.[1] The number of artificial intelligence (AI) enterprises has reached 592, accounting for 23.3% of the AI companies in the world.

In terms of e-commerce, the total retail sales of consumer goods in China was RMB 7.18 trillion, accounting for 40% of the global online retail sales (China Internet Network Information Center, 2018). Digitization has brought both opportunities and challenges to Internet companies. For example, with advanced information technologies, online payment has quickly replaced traditional payment means of cash, while the issues of online privacy and security have surged.

To conclude, China's IT companies are continuously growing and the IT industry has become the pillar of China's economy. The IT industry is large and complex and important to the world; so it is worthwhile to investigate the IT-related issues in Chinese organizations.

6.4 Methodology

The standard English instrument from the World IT Project (Palvia *et al.*, 2017; Palvia *et al.*, 2018) was translated into Chinese by a research assistant and then the Chinese version was back translated into English by another research assistant for assuring linguistic validity. The back-translation was compared with the original instrument by two researchers. After several discussions, discrepancies were resolved and the Chinese instrument was finalized.

---

[1]Internet and IT Unicorn Companies refer to Internet startups or IT startups valued at more than $1 billion each.

Given the length of the survey, early attempts to collect data by hosting the survey on an online service and inviting IT employees from companies were not very successful. Therefore, one of the researchers contacted the MBA/EMBA center from the School of management of a renowned Chinese university. Alumni of this center include many senior executives of various industries in major metropolitan cities of China (e.g., Wuhan, Suzhou, Shenzhen, Jinan, Nanjing, and Guangzhou). The researcher obtained their e-mail addresses from the Center's alumni database. An invitation letter was drafted which explained the objectives and target respondents for the survey. The senior executives were requested to forward the invitation to their IT departments or IT teams. Another researcher contacted an IT professional headhunter company. With their help, every time they contacted a candidate, they invited the candidate to participate in the survey. The data collection started in October 2016 and lasted for about 2 months.

There was no specific incentive or bonus offered to participants. Ethical approval was obtained from the Hohai University Research Ethics Committee. It was presented during the face-to-face interview and was made available if requested by online participants. Communication between the investigators and the core World IT Project team were mainly via emails.

The Chinese sample consisted of 310 IT employees. Nearly all subjects were born in China and 65% were males. Most of the respondents were relatively young. More specifically, more than half of these subjects (50.3%) were between 21 and 29 years of age and 31.9% were between 30 and 39 years of age. The most common job roles were programming (31.9%), system analysis and design (13.2%), financial (7.4%) and system administrator (7.1%). The majority of the respondents (56.1%) were non-management employees. Detailed statistics are shown in Table 6.1.

6.5   Organizational IT Issues

Around 18 organizational IT issues were rated by the respondents using a 5-point Likert scale ranging from 1 (of most importance) to 5 (of no importance). Based on the average ratings, the eighteen organizational issues are ranked from 1 to 18 and are shown in Table 6.2.

The findings indicate that IT reliability and efficiency is the most significant IT-related organizational issue. Security and privacy is the second important organizational IT issue, followed by

Table 6.1: Descriptive Statistics

Characteristics	N	%	Characteristics	N	%
Education:			Years of work experience:		
High school or less	7	2.3	0–4 years	126	40.6
Associate degree	41	13.2	5–9 years	89	28.7
Bachelor's degree	169	54.5	10–19 years	80	25.8
Master's degree	83	26.8	20–29 years	10	3.2
Ph.D.	10	3.2	30+ years	5	1.6
Years of IT experience:			Organizational location:		
0–4 years	162	52.3	IT department employee	212	68.4
5–9 years	83	26.8	IT worker in non-IT department	43	13.9
10–19 years	62	20.0	Contract employee	15	4.8
20–29 years	2	0.6	Consultant	34	11.0
30+ years	1	0.3	Vendor employee	6	1.9
Work as:			Work position:		
Mostly full time	272	87.7	Not part of management	174	56.1
Mostly part time	29	9.4	In lower management	59	19.0
Mostly over time	9	2.9	In middle management	51	16.5
Been laid off from IT job:			In senior management	26	8.4
Yes	76	24.5			
No	234	75.5			

IT strategic planning, project management and continuity planning and disaster recovery. Globalization, bring your own computing device (BYOD) and outsourcing are the least important organizational issues in rank.

In the era of informatization, reliable and efficient information technologies are needed to improve employee's work performance and enhance organizations' competitiveness. In China, economic and market reforms have given rise to companies with distinct types of ownership and cultures (Tsui *et al.*, 2006). The complex and nuanced organizational structures may pose challenges to the development of reliable and efficient information systems. Privacy and security are also very important with the booming development of IT and digitization. Privacy and security concerns exist wherever personal sensitive information is collected, used, stored and shared. The era of IT and big data has arrived in China and is contributing in many ways to economic growth. As is the case, a large quantity of data regarding consumers' personal information is a key asset for business firms and other organizations. However, the unauthorized or unintended sharing of such data has led to privacy and security concerns. In recent years, China

Table 6.2: Organizational IT Issues in China

Organizational IT Issues	Rank	Mean Rating*	Std. Deviation
IT reliability and efficiency	1	1.98	0.84
Security and privacy	2	2.02	0.83
IT strategic planning	3	2.04	0.86
Project management	4	2.05	0.82
Continuity planning and disaster recovery	5	2.09	0.84
Alignment between IT and business	6	2.11	0.81
Knowledge management	7	2.12	0.84
Business productivity and cost reduction	8	2.13	0.81
Revenue-generating IT innovations	9	2.14	0.79
IT service management (e.g., ITIL)	10	2.15	0.85
Attracting and retaining IT professionals	11	2.15	0.84
Enterprise architecture	12	2.16	0.80
Business agility and speed to market	13	2.17	0.85
Business process reengineering	14	2.20	0.77
IT cost reduction	15	2.23	0.88
Globalization	16	2.25	0.89
Outsourcing	17	2.55	0.88
BYOD	18	2.61	0.93

*Rating scale ranges from 1 to 5: 1 as most important and 5 as no importance.

has enacted regulations and laws in data privacy and security due to the increasing awareness among people and organizations.

Driven by IT penetration and market-oriented development, IT strategic planning plays an important role in enhancing economic competitiveness in Chinese enterprises. Strategic planning is critical to organizations because it can provide the planners with a road map to get to where they want to be. Strategic planning is also an important tool for motivating a team to work together and make decisions to achieve specific goals via a variety of techniques. It is thus ranked high in our sample.

BYOD, which means employees can bring their own devices to their workplace, is the least important issue in the study. BYOD is an increasing trend in the business world. One survey shows that 75% of respondents in high growth markets such as Brazil, Russia, India, UAE and Malaysia use their personal device for work, while only 44% of employees in the more mature markets do so (Drury and Absalom, 2012). In our sample, the participants believe the issue of BYOD is much less important than other priorities. Perhaps in the Chinese culture, employees are happy to accept the flexibility provided by BYOD and do not care about the blurred boundaries between work and their own life.

6.6 Technology and Infrastructure Issues

Once again, the IT employees were asked to rate 16 technology related issues on a 5-point Likert scale ranging from 1 (of most importance) to 5 (of no importance). Results are shown in Table 6.3. The top-five most important issues are networks/telecommunications, big data systems, data mining, software as a service, and business intelligence/analytics.

China's networks and telecommunications have been evolving since the early 1990s and commercial access to the Internet has been available since then (Hanna and Qiang, 2011). With the increasing number of Internet access points and services, networks and telecommunications play a critical role in the development of the Chinese economy. In the context of informatization, networks and telecommunications are important tools for business. Telecommunication-based services can enable companies to communicate effectively with customers and suppliers, and allow employees to collaborate with each other regardless of their location.

Big data systems and data mining are ranked second and third highest. In the data-driven economy, the use of big data has been a vital way for leading corporations to outperform their competitors. Big data and data mining enable companies to analyze massive amounts of data and create

Table 6.3: Technology and Infrastructure Issues in China

IT Related Issues	Rank	Mean Rating*	Std. Deviation
Networks/telecommunications	1	2.14	0.91
Big data systems	2	2.16	0.89
Data mining	3	2.19	0.90
Software as a service	4	2.20	0.86
Business intelligence/analytics	5	2.22	0.84
Customer relationship management (CRM) systems	6	2.23	0.87
Cloud computing	7	2.23	0.87
Collaborative and workflow tools	8	2.23	0.82
Business process management systems	9	2.24	0.83
Enterprise resource planning (ERP) systems	10	2.24	0.85
Mobile and wireless applications	11	2.24	0.88
Service-oriented architecture (SOA)	12	2.25	0.86
Mobile apps development	13	2.26	0.90
Enterprise application integration	14	2.28	0.84
Virtualization (desktop or server)	15	2.28	0.86
Social networking/media	16	2.32	0.91

*Rating scale ranges from 1 to 5: 1 as most important and 5 as no importance.

valuable patterns to help make better decisions. At present, big data and data mining are enormously popular trends in business and research development in China. China's internet companies such as Baidu, Alibaba, and Tencent have collected massive amounts of data and employed the data for their business (Swanson, 2015).

Another important technology in China, ranked fourth, is software as a service (SaaS). SaaS is based on cloud computing and allows software delivery through an internet connection. Employees are able to utilize the software and access the data from anywhere and thus work more effortlessly and efficiently. Given a web browser and internet connection, users can perform their work immediately and do not need to install any software.

With the large amounts of data available to companies, traditional business models are not effective any more. Business intelligence and analytics (ranked fifth) can help decision-makers find revealing insights and make informed decisions. Appropriate analysis supported by business intelligence tools is able to support companies in making a variety of operational and strategic decisions. Chinese organizations have now become aware of the power of business intelligence and analytics, and have put forth great effort to promote the development and use of such tools.

Social networking and media are ranked last in our sample. This is a bit surprising as social networking and media can be considered effective ways to reach potential customers and clients at a very low cost. Weibo and Wechat, which are the two largest social networking platforms in China, seem to have exerted important influence on individual lives. The two applications are also useful tools for branding on social media in China. However, Weibo and Wechat marketing started only a few years ago and may not have directly impacted individual IT employees. In addition, for most Chinese people, social networking sites are thought to be tools to provide individuals with instant message services and entertainment.

6.7   Individual IT Employee Issues

Table 6.4 shows the summary of the responses about individual issues. Subjects were asked to rate how much they agree with general statements about individual job issues based on a 5-point Likert scale ranging from 1 (strongly agree) to 5 (strongly disagree). Thus a score below 3 shows agreement with the corresponding item and above 3 shows disagreement. This assessment was made on seven areas of concern and constituted 28 items. The seven areas are: job satisfaction, work pressure, work–life balance, workload and burnout, sense of accomplishment, threats to one's job, and career plans.

The percentages of employees who agree or disagree on individual issues are shown in Table 6.5. As can be seen from Tables 6.4 and 6.5, most respondents are satisfied with their current jobs and feel a sense of accomplishment at work. Generally, employees in the IT occupation receive higher pay and enjoy a better work environment than peers in other occupations. In addition, most IT companies provide their employees with better benefits, such

Table 6.4:   Individual IT Employee Issues in China

Items	Mean Rating*	Std. Deviation
Job satisfaction		
In general, I like working here.	2.23	0.83
All in all, I am satisfied with my current job.	2.33	0.79
In general, I don't like my current job (reversed).	1.65	1.10
Work pressure		
I feel that the number of requests, problems or complaints that I deal with at work is more than expected.	3.02	1.07
I feel that the amount of work I do interferes with how well it is done.	3.07	1.01
I feel busy or rushed at work.	2.71	0.91
I feel pressured at work.	2.69	0.89
Work–life balance		
There is a blurring of boundaries between my job and my home life.	2.90	1.00
My work-related responsibilities create conflicts with my home responsibilities.	3.04	1.04
I do not get everything done at home because I find myself completing job-related work.	3.07	1.04
Workload and burnout		
I feel drained from activities at work.	3.16	0.98
I feel tired from my work activities.	3.14	0.99
Working all day is a strain for me.	3.20	1.01
I feel burned out from my work activities.	3.27	0.97
Sense of accomplishment		
I feel I'm making an effective contribution to what this organization does.	2.32	0.85
In my opinion, I do a good job.	2.31	0.85
I have accomplished many worthwhile things in this job.	2.28	0.85
At my work, I feel confident that I am effective at getting things done.	2.29	0.86

(*Continued*)

Table 6.4: (*Continued*)

Items	Mean Rating*	Std. Deviation
Threats to one's Job		
I am worried that future technology advancements may pose a threat to my job.	2.91	0.97
I believe that other people may be able to perform my work activities.	2.76	0.94
I am concerned that my job may be eliminated soon.	3.09	0.99
I am concerned that my job may be outsourced soon.	3.20	1.01
Career plans		
I will be with this organization 1 year from now.	2.58	0.96
I will take steps during the next year to secure a job at a different organization.	3.11	0.99
I will be with this organization 5 years from now.	2.95	0.99
I will be working in the IT field 1 year from now.	2.10	0.98
I will take steps during the next year to secure a job outside the IT field.	3.50	1.09
I will be working in the IT field 5 years from now.	2.43	0.97

*Rating scale ranges from 1 to 5: 1 as strongly agree, to 5 as strongly disagree.

Table 6.5: Percentages of Employees in Agreement or Disagreement

Areas of Concern	Agree (mean ratings < 3), %	Neutral (mean ratings = 3), %	Disagree (mean ratings > 3), %
Job Satisfaction	88.4	7.7	3.9
Work Pressure	45.2	16.1	38.7
Work–Life Balance	39.7	21.6	38.7
Workload and Burnout	31.6	17.1	51.3
Sense of Accomplishment	77.7	11.3	11.0
Threats to One's Job	33.9	30.3	35.8
Career Plans	57.7	21.9	20.3

as an employee assistant program. They also make a strong effort to further develop employees' skills by providing professional learning opportunities and comprehensive training. Furthermore, IT organizations generally offer employees two pathways for their development; a management development pathway and a professional development pathway. With such an environment and development opportunities, most employees are pleased to work in the IT industry and have a high level of satisfaction. Not only satisfaction, these employees have a high sense of accomplishment and they feel they are doing a good job and making an effective contribution to the organization.

In terms of work pressure, a moderate number of the respondents (45.2%) reported feeling pressured and rushed at work. Chinese companies have to face increased competition and as such, companies may encourage or require employees to work overtime and take responsibilities for unexpected tasks. In terms of work–life balance, the participants did not report significant conflicts. Similar results are found in the area of workload and burnout. Basically, many companies organize a variety of employee recreational activities to help relax their staff and enrich their after-work lives. Some companies even allow employees to work flexibly from anywhere.

The average ratings for threats to job were in the middle and about one-third were concerned, one-third not concerned, and another one-third neutral. On the one hand, IT workers are worried to be replaced due to the increasing number of IT graduates and the rapid development of technologies every year. On the other hand, a large workforce in the IT industry is still needed to promote economic growth and keep IT systems running and growing. There are also a great number of employment opportunities for IT employees and it is easy for a skilled IT worker to find a high-paying job.

The career plans and turnover of IT employees are mirrored by their high levels of satisfaction and accomplishment. A majority (57.7%) indicated that they would not change their current jobs in the IT field and only 20.3% were likely to change their career in the near future. Individuals face both challenges and opportunities in the IT field and yet they can choose their own career path if they are talented and competent in information technologies. Generally, most employees in IT companies are satisfied with their jobs for the relatively high salaries and comfortable work environment and do not have plans to change their jobs any time soon. Only a minority may be concerned about unemployment due to fierce competition in the job market.

6.8 Conclusion

This chapter presented the organizational, technological and individual issues of IT employees in China. Among IT-related organizational issues, IT reliability and efficiency, security and privacy and IT strategic planning are among the pressing concerns. The top issues in technology included: networks/telecommunications, big data systems, data mining, software as a service, and business intelligence/analytics. The Chinese IT workers seem to be satisfied with their current jobs and had little desire to change jobs. Most of them had a high sense of self-efficacy and sense of accomplishment at work. The IT employees believed they had a moderate amount of pressure at work with a healthy work–life balance. In general, these results bode well for the Chinese IT worker.

References

China Internet Network Information Center (2018). Statistical report on internet development in China. Retrieved from http://www.cnnic.cn/hlwfzyj/hlwx zbg/hlwtjbg/201803/P020180305409870339136.pdf.

Chismar, W. G. (1996). Telecommunication and an information infrastructure in China. In Glasson, B. C., Vogel, D. R., Bots, P. W. G., & Nunamaker, J. F. (Eds.), *Information Systems and Technology in the International Office of the Future*. Boston, MA: Springer, pp. 91–98.

Drury, A. & Absalom, R. (2012). BYOD: An emerging market trend in more ways than one. Retrieved from http://www.us.logicalis.com/globalassets/united -states/whitepapers/logicalisbyodwhitepaperovum.pdf.

Fang, C., Garnaut, R., & Song, L. (2018). 40 years of China's reform and development: How reform captured China's demographic dividend. In Fang, C., Garnaut, R., & Song, L. (eds.), *China's 40 Years of Reform and Development: 1978–2018*. Acton ACT, Australia: ANU Press, pp. 5–26.

Hanna, N. K., & Qiang, C. Z. W. (2011). China's evolving informatization Strategy. In Hanna, N., & Knight, P. (eds.), *Seeking Transformation Through Information Technology*. New York: Springer, pp. 89–137.

Palvia, P., Jacks, T., Ghosh, J., Licker, P., Romm-Livermore, C., Serenko, A., & Turan, A. H. (2017). The World IT Project: History, trials, tribulations, lessons, and recommendations. *Communications of the Association for Information Systems*, 41(18), 389–413.

Palvia, P., Ghosh, J., Jacks, T., Serenko, A., & Turan, A. (2018). Trekking the globe with the World IT Project. *Journal of Information Technology Case and Application Research*, 20(1), 3–8.

Swanson, A. (2015). How Baidu, Tencent and Alibaba are leading the way in China's big data revolution [South China Morning Post]. Retrieved from https://www.scmp.com/tech/innovation/article/1852141/how-baidu -tencent-and-alibaba-are-leading-way-chinas-big-data.

Tsui, A. S., Wang, H., & Xin, K. (2006). Organizational culture in China: An analysis of culture dimensions and culture types. *Management and Organization Review*, 2(3), 345–376.

Chapter 7

Information Technology Issues in Egypt

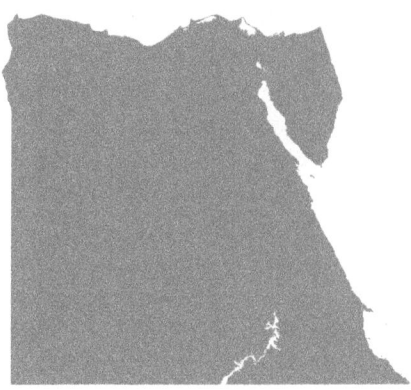

Myriam Raymond*,¶, ShiKui Wu†,‖, and Alexander Serenko‡,§,**

*Université d'Angers, Campus Belle-Beille,
Angers France
†Lakehead University, Thunder Bay, Ontario, Canada
‡University of Toronto, Toronto, Canada ON
§University of Ontario Institute of Technology,
Oshawa, Canada
¶myriam.raymond@univ-angers.fr
‖shikui.wu@lakeheadu.ca
**a.serenko@utoronto.ca

Summary

Egypt has attracted a large number of multinational organizations and become an information technology (IT) hub in the Middle East region. This chapter presents the key issues in Egypt from organizational, technological, and individual perspectives. The top-IT issues in organizations are security and privacy, IT reliability and efficiency, and alignment between IT and business, while globalization, outsourcing, IT cost reduction, and "bring your own device" (BYOD) are the least concerns. The IT infrastructure in Egypt has been developed with recent upgrade of national networks, while networks and telecommunications, virtualization, and enterprise application integration are still highly ranked issues. Social networking/media appears to be the least important issue. The Egyptian IT workers are highly satisfied with their jobs, and they are confident and willing to accomplish their tasks successfully. Egypt has a healthy and growing IT workforce with higher education, strong interests, and both short- and long-term career plans in the IT filed.

7.1 Introduction

Egypt has long been one of the regional leaders in the Middle East and North Africa. Since 1970s, its "open-door" policy has both encouraged domestic organizations to enter into global market and attracted multinational organizations to invest in Egypt. As an information technology (IT) hub in this region, the country has been attractive for IT outsourcing (Willcocks et al., 2009). The government has continuously invested in the national IT infrastructure and e-government initiatives (Gebba and Zakaria, 2012; MCIT Report, 2015). The domestic organizations work closely with their international business partners to offer a wide range of IT products and services, from call centers and technical services to software design and research and development centers (MCIT Report, 2015). IT has become one of the key sectors driving economic growth in Egypt and contributing to the country's GDP, revenue, export, and employment. According to the annual report (ITIDA, 2013) by the IT Industry Development Agency (ITIDA), the Egyptian IT sector represented 4.4% of the country's GDP and offered 268,000 jobs in 2012. Egypt's economy has been growing at an annual rate of 4–6%, and further expansion and structural reformation are expected by the year 2020 (World Bank, 2018).

7.2 Country Background and History

Egypt, officially the Arab Republic of Egypt, with a population of over 95 million, is the most populous country in North Africa, the Middle East, and the Arab world. It is a transcontinental country across those regions and has been considered as a regional leader. Due to the large areas of desert, the great majority of its people live near the banks of the Nile River (about 40,000 km^2) and about half reside in urban areas and densely populated centers. A large number of Egyptians live abroad (about 2.7 million); about 70% of emigrants live in other Arab countries, and most of the remaining 30% reside in Europe and North America.

In Egypt, Islam is the official religion, and Arabic is the official language. It has one of the longest histories over the world, tracing its heritage back to the 6000–4000 BC (Midant-Reynes, 2000). From the 16th to the early 20th century, Egypt was ruled by foreign imperial powers. The country gained independence from the British Empire and formed a republic in the late 1950s. In its recent history, Egypt has endured challenges with political instability and economic underdevelopment.

Economic conditions have improved considerably after the implementation of liberal economic policies and major reforms. The "open-door" policy has also attracted a large amount of foreign investment, and the country has been rebounding from the recent 2011 revolution and its aftermath. Its economy is one of the largest and most diversified in the Middle East and is projected with fast and continuous growth (World Bank, 2018). The major industries in Egypt are agriculture, energy, and tourism. Exports of goods and services have contributed positively to its GDP, while the gas, telecom, and construction sectors are fast growing. The main economic activity is still driven by large public investment in state-led projects.

7.3 Information Technology in Egypt

Similar to other industries, the IT sector in Egypt has long been led by government investments. In 1854, the county launched its first telegram line connecting Cairo and Alexandria, followed by the first telephone line between the two cities installed in 1881. The government has committed to develop its IT sector since the national project for a technological renaissance in 1999. E-government initiatives have also been implemented despite the challenges and critics in public adoption (Abbassy and Mesbah, 2016; Gebba and Zakaria, 2012).

The IT sector has expanded rapidly in recent years, with many startups providing outsourcing services to multinational companies from North America and Europe. Egypt has become an attractive region as an IT hub for outsourcing, especially since the creation of the Smart Villages (Willcocks *et al.*, 2009). Initially, the IT outsourcing practice was mainly limited to call centers and technical services. Egypt's most prominent multinational companies are Xceed BPO service with its 7,000+ work stations and the Raya Contact Center, which has more than 5,000 work stations serving global Fortune 1,000 companies in 25 different languages. Recently, the government has led several programs to diversify the outsourcing services and to support small- and medium-sized enterprises (ITIDA, 2013). More advanced products and services in IT outsourcing have been initiated and developed, including: business process, software design and development, and research and development.

The Egyptian IT sector is one of the most dynamic and vibrant employment areas. Skilled IT professionals are well-paid following the Egyptian job market standards. There are about 480,000 university graduates annually in Egypt; among them, 110,000 are foreign language graduates and 50,000 hold an IT-related degree. The government facilitates the Education Development Program for Egyptian Universities (EDUEgypt) that collaborates with universities and leading companies involved in business process outsourcing and multinational clients. They aim to create job opportunities and prepare the graduates with a range of technical competencies.

Social media has been widely used by Egyptian IT professionals, and this albeit a freedom of press limitation. A recent law granting public authorities the right to monitor social media users in the country has been ratified by the government (Channel NewsAsia, 2018). Egypt remains the first country in the Middle East and North Africa (MENA) region with 17 million Facebook users. Since November 2017, its citizens have enjoyed Twitter Lite, which is designed to make Twitter more accessible in developing regions where data plans are expensive. Snapchat expanded its Egyptian operations exponentially with the user base of approximately 3 million (Radcliff and Lam, 2018).

7.4   Methodology

The country investigator (the first author) formed a local research team for data collection with Dr. Hisham Salah (a Research Scientist at Wadsworth Center, NYSDOH), Dr. Mahmoud Allam (the former Computer Engineering Program Director at Nile University), and a junior assistant researcher. The standard World IT survey instrument (Palvia *et al.*, 2017; Palvia *et al.*, 2018) was used, and it was translated into the Arabic language by a professional translator. The team then compared notes with the Saudi Arabia Arabic instrument, and minor adjustments to the wording of some questions were made. The team found that it was difficult to collect primary data on IT professionals in Egypt. Some of the common challenges included distrust of research objectives and reluctance to share information. Reaching out through key people or senior management did not prove to be an efficient approach.

At the initial stages of data collection, paper surveys were printed, distributed, and collected. The team found that people often refused to meet with the junior assistant for paper survey and appeared to prefer answering questions online anonymously. The team then created the survey on the SurveyGizmo online platform in dual languages (English and Arabic). An email campaign was launched on LinkedIn giving participants the option to respond to the survey in dual languages. The campaign started in December 2015 and ended in October 2017. Most of the data were collected during the summer of 2016. Responses reached a peak during the Ramadan period (June 6–July 5, 2016).

The research team used the SurveyGizmo platform for data entry consolidation, data aggregation, and data export. The researchers also created and used a Google sheet to manage the contacts and to track the data collection progress. In total, 284 responses were received, out of which 175 were completed. This reached a 61.6% completion rate, a high rate when taking into account the length of the survey. The average time for completing a paper survey was about half an hour, while the average time for finishing an online survey was approximately 23 minutes.

Table 7.1 shows the descriptive statistics of the IT professionals who took part in the study. Note that the total number with regard to the level

Table 7.1:   Descriptive Statistics

Characteristics	N	%	Characteristics	N	%
Education:			Years of work experience:		
High school or less	0	0	0–4 years	24	13.7
Associate degree	4	2.3	5–9 years	44	25.1
Bachelor's degree	138	78.9	10–19 years	76	43.4
Master's degree	30	17.1	20–29 years	27	15.4
Ph.D.	3	1.7	30+ years	4	2.3
Years of IT experience:			Organizational location:		
0–4 years	33	18.9	IT department employee	132	75.4
5–9 years	49	28	IT worker in non-IT department	16	9.1
10–19 years	68	38.9	Contract employee	7	4
20–29 years	21	12.0	Consultant	20	11.4
30+ years	4	2.3	Vendor employee	0	0
Work as:			Work position:		
Mostly full time	155	88.6	Not part of management	52	29.7
Mostly part time	1	0.6	In lower management	27	15.4
Mostly over time	19	10.9	In middle management	54	30.9
Been laid off from IT job:			In senior management	41	23.4
Yes	7	4.0			
No	168	96.0			

of management is less than 175, which is due to one missing response to this question in the survey.

Most of the survey respondents held bachelor's degrees (78.9%), and more than 80% had at least 5 years of IT experience. The respondents were mostly working full-time (88.6%), and only one worked part time. Only a few (4%) experienced an involuntary loss of an IT job. In terms of their job positions in the organization, more than 75% worked in IT departments, and about 10% either worked in non-IT departments or were external consultants. They frequently worked at different levels of managerial positions.

7.5   Organizational IT Issues

The participants were asked to rank 18 organizational IT issues in order of importance. Table 7.2 presents the ranking result from the respondents from Egypt. Security and privacy was ranked as the most important issue, which also received the top position in the 2017 SIM IT Issues and Trends Report (Kappelman *et al.*, 2018). The other issues included in the top five were IT reliability and efficiency, alignment between IT and business, project management, and IT strategic planning. Among them, alignment between IT and business was also ranked among the top five in the SIM Report.

Security and privacy is ranked globally among the top issues as IT companies face more and more challenges with emerging technologies, such as cloud computing, mobile apps, and social networks. The respondents also indicated the importance of aligning IT and business models and strategies, and pointed out misalignment situations due to the discord between business needs and IT implementations.

A number of issues ranked in the top five in the present study were not listed in the SIM Report (Kappelman *et al.*, 2018). Many respondents reported a major concern regarding IT reliability and efficiency, particularly the IT infrastructure that failed to deliver accurate, timely, and reliable data and information for business decision-making. They also felt that IT was often not strategically planned and that project management was not undertaken professionally. As a result, IT investments and projects often failed to deliver their promises.

The top-five issues identified in the present study pinpoint to the fact that IT still performs a "separate" business function in the Egyptian organizations. The respondents noticed the important role of IT in cross-functional management. IT professionals lack a comprehensive understanding of the business mandates and requirements, while business professionals

Table 7.2:   Organizational IT Issues in Egypt

Organizational IT Issues	Rank	Mean Rating*	Std. Deviation
Security and privacy	1	1.41	0.64
IT reliability and efficiency	2	1.43	0.58
Alignment between IT and business	3	1.49	0.58
Project management	4	1.50	0.60
IT strategic planning	5	1.54	0.59
Continuity planning and disaster recovery	5	1.54	0.61
IT service management (e.g., ITIL)	7	1.55	0.70
Knowledge management	8	1.58	0.63
Attracting and retaining IT professionals	9	1.61	0.69
Business agility and speed to market	10	1.63	0.68
Revenue-generating IT innovations	11	1.72	0.71
Business process reengineering	12	1.76	0.63
Business productivity and cost reduction	13	1.79	0.76
Enterprise architecture	14	1.80	0.74
Globalization	15	2.11	0.86
Outsourcing	16	2.29	0.80
IT cost reduction	17	2.31	0.97
BYOD	18	3.24	1.31

*Rating scale ranges from 1 to 5: 1 as most important and 5 as no importance.

also need to know the various technical solutions and their implications. This may be partially due to the fact of the separation between pure IT degrees and pure business degrees in Egypt education. Currently, very few educational programs are offered to prepare future business professionals to manage IT in organizations.

It is interesting to note that globalization and outsourcing were not reported as major organizational IT issues in this study. Egypt has long been immersed in globalization. In recent years, the "open-door" policy adopted by President Sadat and his successors has encouraged local organizations to expose them to the global market. Egypt, as an IT hub, has been attractive for IT outsourcing (Willcocks *et al.*, 2009), and the local companies have often become suppliers and partners of international organizations. Hence, outsourcing has been managed without serious concerns.

Finally, IT cost reduction and BYOD appeared to be the least concerns in organizational IT issues in the present study. This indicates that the everlasting cost reduction efforts do not impact the IT function and its spending in Egyptian organizations. Although it has been observed that organizations tend to cut down their IT budgets with economic downturns, the respondents indicated that this did not constitute a major concern for their organizations. This may be possibly due to the budget increase for

IT infrastructure upgrades and investments, and thus there is no need for
reducing costs and using own devices.

7.6 Technology and Infrastructure Issues

Table 7.3 shows a ranking of technology and infrastructure issues in Egypt,
listed from the most to the least important ones. The top-five issues were
networks/telecommunications, virtualization, enterprise application integration, business intelligence/analytics, and collaborative and workflow
tools.

Networks/telecommunications was the highest-ranked technology and
infrastructural issue in Egypt. In the SIM report (Kappelman *et al.*, 2018),
it was among the top-10 IT investments but was not concerned as a top
issue. In Egypt, the network structure remains shaky and unreliable despite
a countrywide use of the Internet. At the national level, more effort has
recently been announced to upgrade the existing Egyptian network infrastructure. At the organizational level, IT professionals seek more stable
technologies and solutions to address the networks and telecommunications
issue.

Virtualization and enterprise application integration (EAI) were highly
ranked in the present study, while they were not reported as top issues by

Table 7.3: Technology and Infrastructure Issues in Egypt

IT Related Issues	Rank	Mean Rating*	Std. Deviation
Networks/telecommunications	1	1.37	0.51
Virtualization (Desktop or Server)	2	1.57	0.61
Enterprise application integration	3	1.64	0.71
Business intelligence/analytics	4	1.69	0.76
Collaborative and workflow tools	5	1.70	0.72
Mobile and wireless applications	6	1.72	0.69
Customer relationship management (CRM)	7	1.74	0.75
Business process management systems	7	1.74	0.76
Enterprise resource planning (ERP) systems	9	1.76	0.77
Software as a service	10	1.78	0.71
Service-oriented architecture (SOA)	10	1.78	0.75
Data mining	12	1.86	0.83
Big data systems	13	1.87	0.74
Cloud computing	14	1.92	0.90
Mobile apps development	15	2.12	0.93
Social networking/media	16	3.07	1.30

*Rating scale ranges from 1 to 5: 1 as most important and 5 as no importance.

Kappelman *et al.* (2018). Many organizations in Egypt are migrating to cloud computing, which often requires virtualization implementations. The respondents also reported the importance and challenges in cross-functional integration of business applications. It indicates great effort is still required to bring together different functions of the organization with connected modules or applications. This may also require more intra-organizational collaborations, and thus appropriate collaborative and workflow tools would be essential.

Consistent with the SIM report, business intelligence/analytics was also highly ranked in this study. This shows that the Egyptian IT professionals acknowledge the importance and trend in the advanced usage of data and information. More advanced business intelligence platforms for data mining and big data, however, were rated relatively less important. This may indicate another technological area due for development in Egypt as it lacks infrastructure and support for larger-scale and complicated business intelligence practices.

It is worth noting that the least problematic area reported by the respondents was social networking and media, after other least important issues such as mobile apps development, cloud computing, big data systems, data mining, and service-oriented architecture. Social media has been widely used by the Egyptian IT professionals, mirroring a nationwide trend (Channel NewsAsia, 2018). Egypt remains the leading MENA country with 17 million Facebook users.

7.7 Individual IT Employee Issues

The respondents were also asked to rate several individual IT issues related to their job and career. Table 7.4 presents the relative importance of these issues to the Egyptian IT professionals. Note that a lower average score means higher agreement with each statement in the survey.

Meanwhile, the respondents showed very high accomplishments in their work. They were confident and happy to make valuable contributions to their organizations. Also, the Egyptian IT workers did not find any threats to their jobs, neither due to technological advancements nor because of outsourcing practices.

In terms of their career plans, the respondents showed strong interests in working in the IT field. Whereas some indicated that they might move to other organizations, most of them would keep working in the IT field in both short and long term.

Table 7.4:   Individual IT Employee Issues in Egypt

Individual Issues	Mean Rating*	Std. Deviation
Job satisfaction		
In general, I like working here.	1.89	0.77
All in all, I am satisfied with my current job.	2.20	0.88
In general, I don't like my current job.	2.10	0.94
Work pressure		
I feel that the number of requests, problems or complaints that I deal with at work is more than expected.	3.01	1.05
I feel that the amount of work I do interferes with how well it is done.	3.41	0.98
I feel busy or rushed at work.	3.05	1.03
I feel pressured at work.	2.88	1.04
Work–life balance		
There is a blurring of boundaries between my job and my home life.	3.10	1.12
My work-related responsibilities create conflicts with my home responsibilities.	3.38	1.06
I do not get everything done at home because I find myself completing job-related work.	3.16	1.10
Workload and burnout		
I feel drained from activities at work.	3.00	1.10
I feel tired from my work activities.	2.88	1.05
Working all day is a strain for me.	2.72	1.08
I feel burned out from my work activities.	2.93	1.12
Sense of accomplishment		
I feel I'm making an effective contribution to what this organization does.	1.73	0.62
In my opinion, I do a good job.	1.80	0.68
I have accomplished many worthwhile things in this job.	1.67	0.62
At my work, I feel confident that I am effective at getting things done.	1.70	0.64
Threats to one's job		
I am worried that future technology advancements may pose a threat to my job.	3.66	0.98
I am concerned that my job may be eliminated soon.	3.86	0.97
I am concerned that my job may be outsourced soon.	3.74	0.96
Career plans		
I will be with this organization 1 year from now.	2.41	0.86
I will take steps during the next year to secure a job at a different organization.	2.94	1.04
I will be with this organization 5 years from now.	2.82	1.00
I will be working in the IT field 1 year from now.	1.76	0.82
I will take steps during the next year to secure a job outside the IT field.	3.67	1.03
I will be working in the IT field 5 years from now.	1.99	0.87

*Rating scale ranges from 1 to 5: 1 as strongly agree and 5 as strongly disagree.

In summary, the result shows a very healthy workforce in the Egyptian IT sector. This may be attributed to the continuously growing IT industry in Egypt. On the demand side, the government invested in the IT infrastructure nationwide, and a large number of multinational companies were attracted to Egypt to offer a range of IT products and services. On the supply side, there is a large base of new IT graduates, and universities and colleges offer various IT-related programs. The Egyptian IT sector has thus become one of the most dynamic and vibrant job markets.

7.8 Conclusion

Unlike many other sectors of the Egyptian economy, the IT industry has remained remarkably resilient in the face of political uncertainty and macroeconomic instability wrought by the 2011 revolution. Due to its established national Internet infrastructure, Egypt has a number of advantages to attract potential IT investors and outsourcing partners, such as strong support from government agencies and a large base of new graduates with multiple linguistic and technical skills.

In recent years, the IT sector has been expanding its existing IT infrastructure and outsourcing services. The present study reveals several important issues that exist in the IT sector, despite a healthy and growing workforce supporting its development. Currently, the Ministry of Communications and Information Technology (MCIT) is working on the IT 2020 Strategic plan. This plan focuses on IT infrastructure, digital content, electronics design and manufacturing, IT industrial programmes and initiatives, and legislative and policy frameworks. A number of technology parks and subsidy programmes will be coupled with initiatives such as the large-scale infrastructure upgrade and enhanced EDUEgypt programs. These efforts will help sustain Egypt's regional leadership in IT.

References

Abbassy, M., & Mesbah, S. (2016). Effective e-government and citizens adoption in Egypt. *International Journal of Computer Applications*, 133(7), 7–13.

Channel NewsAsia (2018). Egypt President Approves Law Clamping down on Social Media.

Gebba, T. R., & Zakaria, M. R. (2012). E-government in Egypt: An analysis of practices and challenges. *International Journal of Business Research and Development*, 4(2), 11–25.

ITIDA (2013). Impact Review 2013 Report. Information Technology Industry Development Agency, Cairo, Egypt. Available online at http://www.itida. gov.eg/En/Brochures/Impact%20Review.pdf.

Kappelman, L., Johnson, V., Maurer, C., McLean, E., Torres, R., David, A., & Nguyen, Q. (2018). The 2017 SIM IT issues and trends study. *MIS Quarterly Executive*, 17(1), 53–88.

MCIT Report (2015). Measuring the Digital Society in Egypt: Internet at a Glance — Report. Arab Republic of Egypt, Ministry of Communication and Information Technology. Available online at http://www.mcit.gov.eg/Upcont/Documents/Publications_1272015000_Measuring_the_Digital_Society_in_Egypt_12_.pdf.

Midant-Reynes, B. (2000). *The Prehistory of Egypt: From the First Egyptians to the First Kings*. Oxford: Blackwell Publishers.

Palvia, P., Jacks, T., Ghosh, J., Licker, P., Romm-Livermore, C., Serenko, A., & Turan, A. H. (2017). The World IT Project: History, trials, tribulations, lessons, and recommendations. *Communications of the Association for Information Systems*, 41(18), 389–413.

Palvia, P., Ghosh, J., Jacks, T., Serenko, A., & Turan, A. (2018). Trekking the globe with the World IT Project. *Journal of Information Technology Case and Application Research*, 20(1), 3–8.

Radcliff, D., & Lam, A. (2018). *The Middle East — The Story of 2017: Key Developments, Stories and Research Findings*.

Willcocks, P. L., Griffiths, C., & Kotlarsky, J. (2009). *Beyond BRIC: Offshoring in Non-BRIC Countries: Egypt — A New Growth Market*. London, UK: London School of Economics and Political Science.

World Bank (2018). *Egypt's Economic Outlook — October 2018*: The World Bank.

Chapter 8

Information Technology Issues in Finland

Mikko Ruohonen*,**, Nicholas Mavengere[†,††], A.K.M. Najmul Islam[‡,‡‡],
Alexander Serenko[§,§§], Ulla-Riitta Ahlfors[¶,¶¶],
and Aykut Hamit Turan[∥,∥∥]

Tampere University, Tampere, Finland
[†]*Bournemouth University, Bournemouth, the UK*
[‡]*University of Turku, Turku, Finland*
[§]*University of Toronto, Toronto, Canada*
University of Ontario Institute of Technology,
Oshawa, Canada
[¶]*University of Jyväskylä, Jyväskylä, Finland*
[∥]*Sakarya University, Sakarya, Turkey*
***Mikko.Ruohonen@tuni.fi*
[††]*nmavengere@bournemouth.ac.uk*
[‡‡]*najmul.islam@utu.fi*
[§§]*a.serenko@utoronto.ca*
[¶¶]*ullahlf@hotmail.com*
[∥∥]*ahturan@sakarya.edu*

Summary

Information technology (IT) industry is essential in Finland because of its significant export contribution, extensive workforce, and research and innovation contributions. This chapter highlights key issues in this important industry. For instance, IT reliability and efficiency are the top issues necessary for the Finnish IT industry's competitiveness in a global context. Furthermore, business intelligence and analytics tools, techniques and skills are central to the Finnish IT industry. The industry has a very experienced workforce that is however aging and thus there is a need for training of young personnel to join the industry. Generally, the Finnish IT workers are satisfied with their jobs and exhibit low turnover intentions.

8.1   Introduction

Information technology (IT) has changed the way we work, communicate and socialize. Basically, IT has influenced the way we do things. The work environment in different industries has also been changing in line with the technological advances. IT offers both opportunities and challenges. Companies that manage to leverage technology both at strategic and operational levels are the current leaders in different industries. The organizational IT issues are of interest in defining success in the current competitive global environment. It is important that these IT issues be elaborated at firm, industry, national, regional and global levels. This chapter specifically notes IT issues in Finland and also relates to global factors, as technology knows no boundaries.

The advances and impact of IT from technical and social dimensions are evident in Finland. The global aspects related to IT, such as outsourcing, also have a bearing on the Finland IT landscape. In addition, the IT work space in Finland has received attention because of past and present successful Finnish companies. For example, Doz and Kosonen (2008) coined fast strategy in relation to Nokia's dominance in the telecommunication industry. Furthermore, proposals to promote job satisfaction in the IT industry in Finland have been suggested. For example, Islam *et al.* (2018) with respect to the IT industry in Finland noted how professional self-efficacy mitigates the effect of stress and strain on job satisfaction.

8.2   Country Background and History

Finland (in Finnish, Suomi) is located in northern Europe, with a population of approximately 5 million. Finland's neighboring countries are Sweden to the west, Norway to the north, Russia to the east and Estonia to the South, across the Gulf of Finland. It is one of the Nordic countries together with Sweden, Norway, Denmark, and Iceland. Finland joined the European Union in 1995 and is part of the Economic and Monetary Union (EMU). Finland marked 100 years of independence in 2017. It gained its independence from Russia on the December 6, 1917. Finland was part of Sweden until 1809. When Sweden lost its position as a great power in the early 18th century, Russia conquered Finland in the 1808–1809 war with Sweden. Then Finland was an autonomous grand duchy within the Russian empire from 1809 to 1917. Some important dates from Finnish history:

- **1809:** Sweden surrenders Finland to Russia. Finland becomes a semi-autonomous Grand Duchy within the Russian Empire.
- **1917:** Finland declares independence from Russia on December 6.
- **1918:** Finnish civil war for control and leadership during the transition from Grand Duchy of the Russian Empire.
- **1919:** Finland becomes a republic with a president as head of state as the present constitution is adopted.
- **1939–1940:** The Winter War against Russia.
- **1941–1944:** Fighting between the Finnish and Russian forces resumes in the campaign known as the Continuation War.
- **1955:** Finland joins the United Nations.
- **1995:** Finland becomes a member of the European Union.

The capital city is Helsinki and official languages are Finnish and Swedish. Various agencies consider Finland a safe, free and stable country. The Finnish political system is a parliamentary republic with many parties and usually with coalition governments. Ultimate political power is vested in the 200-member unicameral parliament. About three quarters of Finland is covered by forests and thus it is known as the land of forests. Finland has spectacular lakes and islands. Lakes and other bodies of water cover 10% of the $338,424\,\text{km}^2$ of its national territory.

8.3 Information Technology in Finland

The IT industry is important in Finland in terms of job creation, export, research and innovation. The Finnish IT industry is one of the main contributors to the economy. IT companies contribute over 50% of Finnish exports and the industry is the main export contributor in Finland. About 700,000 people are directly or indirectly involved in the IT sector, which is nearly

15% of the population. 70% of investment in research and development in Finland is through the IT industry. In summary, the IT industry plays a huge role in Finland in terms of historical, present and future success of the country (Oksanen, 2018).

At the turn of the 1970s and 1980s, there were rampant challenges and opportunities that led Finland to rethink its national strategies. In 1979, a technology committee to assess technical development was appointed. The committee's tasks included the assessment of IT impact and proposing actions to strengthen the role of IT. Based on the committee's recommendations, the government introduced numerous initiatives, which would secure the competitiveness of Finland's industry and service sectors. A step into this direction was taken in 1980 when an IT action programme for Finland was launched. The development of information and communication technologies was boosted further by the establishment of a new funding agency Tekes, the Technology Development Center in 1983 (now Business Finland, https://www.businessfinland.fi/en/do-business-with-finland/home/). Overall, during the 1980s, a number of decisions were made, which later proved to be particularly successful and favorable from the perspective of Finnish Information Society development (Oksanen, 2018). The Technology industry currently comprises five subsectors (Technology Industries, 2018):

1. Electronics and the Electrotechnical Industry,
2. Mechanical Engineering,
3. Metals Industry,
4. Consulting Engineering,
5. Information Technology.

8.4 Methodology

An online survey of IT professionals in Finland was conducted from September to December 2016 with the help of the Finnish Information Processing Association, TIVIA (www.tivia.fi). It is a nationwide independent society of ICT professionals and has about 10,000 individual members, 400 organization members and 28 member associations. Its areas of interest include (TIVIA, 2019):

- The Governmental service processes and the structure of the Finnish Information Society.
- The future of the Finnish Software Industry.

- Improving cooperation and understanding between Business and ICT Management.
- Overall improvement of ICT skills.

TIVIA offered an excellent channel to contact Finnish IT professionals from different industries, companies and trades. The survey was sent to a selected group of members, around 3,000 professionals. The questionnaire was administered in English. Altogether, 149 responses were collected, out of which 144 were completed and retained for further analysis. Table 8.1 illustrates the descriptive statistics of the IT professionals who took part in the World IT Project (Palvia *et al.*, 2017; Palvia *et al.*, 2018) in Finland.

Table 8.1 shows that the IT personnel in Finland are quite educated with nearly half with a master's degree. In addition, above 70% have over 20 years of working experience, reflecting a well-experienced group. About 65% have over 20 years of IT experience. However, this could also be an indication of aging personnel, which in turn shows a need for rampant training of new and young IT workers. About 60% of Finnish IT professionals are located in IT departments, and about 30% are IT workers in non-IT departments, such as consultants and vendor employees. Almost 90% of the Finnish IT

Table 8.1:   Descriptive Statistics

Characteristics	N	%	Characteristics	N	%
Education:			Years of work experience:		
High school or less	8	5.6	0–4 years	3	2.2
Associate degree	31	21.5	5–9 years	5	3.6
Bachelor's degree	37	25.7	10–19 years	24	17.4
Master's degree	64	44.4	20–29 years	53	38.4
Ph.D.	4	2.8	30+ years	53	38.4
Years of IT experience:			Organizational location:		
0–4 years	3	2.1	IT department employee	81	56.3
5–9 years	5	3.5	IT worker in non-IT department	17	11.8
10–19 years	43	29.9	Contract employee	3	2.1
20–29 years	57	39.6	Consultant	19	13.2
30+ years	36	25.0	Vendor employee	9	10.4
Work as:			Work position:		
Mostly full time	125	87.4	Not part of management	79	54.9
Mostly part time	5	3.5	In lower management	18	12.5
Mostly over time	13	9.1	In middle management	28	19.4
Been laid off from IT job:			In senior management	19	13.2
Yes	30	20.8			
No	114	79.2			

professionals are full-time employees and almost 80% have never been laid off from an IT job. About a half of the IT professionals in Finland are part of management.

8.5   Organizational IT Issues

The organizational IT issues in Finland, illustrated in Table 8.2, are ranked in order of importance. The Society for Information Management (SIM)'s 38th Anniversary IT Trends Study (Kappelman *et al.*, 2018) also ranks IT issues in order of importance based on the responses from IT executives in 769 organizations in the US. These rankings are important for the awareness of critical IT issues that organizations should be conscious about, and they also help stakeholders keep track of the changes over the years. In addition, the rankings could be used for comparative purposes and suggest ways to improve at different levels, such as organizational, industry and national levels. We will briefly review the Finland IT issues, shown in Table 8.2, in relation to the SIM issues list. There are several issues in both lists but with different ranks, such as security and privacy, alignment between IT and business, agility, innovations and cost reductions.

Table 8.2:   Organizational IT Issues in Finland

Organizational IT Issues	Rank	Mean Rating*	Std. Deviation
IT reliability and efficiency	1	1.67	0.741
Security and privacy	2	1.78	0.834
Alignment between IT and business	3	1.86	0.784
IT strategic planning	4	2.03	0.769
Project management	5	2.12	0.792
Knowledge management	6	2.12	0.835
Continuity planning and disaster recovery	7	2.13	0.855
Business agility and speed to market	8	2.13	0.874
Revenue-generating IT innovations	9	2.25	0.862
Business productivity and cost reduction	10	2.36	0.804
Enterprise architecture	11	2.38	0.926
Business process reengineering	12	2.42	0.907
Attracting and retaining IT professionals	13	2.43	0.902
IT service management (e.g., ITIL)	14	2.48	0.903
IT cost reduction	15	2.70	0.872
Globalization	16	2.95	1.070
Outsourcing	17	3.46	1.008
Bring your own computing device (BYOD)	18	3.57	0.968

*Rating scale ranges from 1 to 5: 1 as most important and 5 as no importance.

Globalization and advances in technology could make IT reliability and efficiency be perceived as the most important issue in Finland. This indicates that there is an increased risk of attack from all angles, at the global scale and thus security and privacy issues are second in Table 8.2. Security is also ranked first in the 2017 SIM list, but IT reliability and efficiency is not in the list.

Business and IT alignment includes the application of IT appropriately and timely in harmony with business strategies, goals and needs (Luftman and McLean 2004). It received the third rank as an organizational IT issue. Business and IT alignment is important because technology alone does not provide businesses with a competitive advantage (Carr, 2003). However, business competitive advantage could be promoted by how well the organization aligns its business with technology. Alignment of IT with business is also ranked third in the SIM list, making it the only issue that is highly regarded in both US and Finland. This is because business and IT alignment is constantly important and challenging to organizations because of continuous IT improvements and the dynamic business environment. Therefore, the alignment efforts both at strategic and operational levels need constant checks and updates.

IT is increasingly important in different industries. Therefore, there is a need to value IT strategic planning, which provides a clear direction of strategic input in business. IT strategic planning is ranked fourth in Table 8.2. Project management and knowledge management are ranked fifth and sixth, respectively. All these three IT issues are not in the top-10 2017 SIM issues list.

8.6 Technology and Infrastructure Issues

Technology and infrastructure issues in Finland are illustrated in Table 8.3. These are ranked in order of importance. The Society for Information Management (SIM)'s 38th Anniversary IT Trends Study (Kappelman *et al.*, 2018) also ranked technology and infrastructure issues from three perspectives, namely investment size, perceived as requiring more investment and worrisome. Four of the top-10 issues are common in both the Finland and US investment size lists. The SIM is perceived as requiring more investment and the worrisome list also relates to the Finland list as explained as follows.

Gartner (2018) defined business intelligence as an "umbrella term that includes the applications, infrastructure and tools, and best practices that

M. Ruohonen et al.

Table 8.3:   Technology and Infrastructure Issues in Finland

IT Related Issues	Rank	Mean Rating*	Std. Deviation
Business intelligence/analytics	1	2.27	0.914
Enterprise application integration	2	2.31	1.012
Networks/telecommunications	3	2.35	1.045
Collaborative and workflow tools	4	2.37	0.879
Mobile and wireless applications	5	2.38	1.050
Virtualization (Desktop or Server)	6	2.42	1.047
Software as service	7	2.54	0.932
Customer relationship management (CRM) systems	8	2.57	1.091
Cloud computing	9	2.61	1.064
Business process management systems	10	2.75	0.979
Service oriented architecture	11	2.77	0.962
Enterprise resource planning (ERP) systems	12	2.79	1.041
Data mining	13	2.82	0.976
Big data systems	14	2.84	1.058
Mobile apps development	15	2.84	1.167
Social networking	16	3.01	0.964

*Rating scale ranges from 1 to 5: 1 as most important and 5 as no importance.

enable access to and analysis of information to improve and optimize decisions and performance." Business intelligence and analytics is the top issue in both the Finland and US investment size lists. This is supported in literature, for instance, Chen *et al.* (2012) highlighted that business intelligence and analytics are significant in solving problems in contemporary business organizations. This is an important area because of factors such as globalization, information overload, technological advances, increased business competition and changing regulatory environment. Globalization which includes the interaction, collaboration and relationships at an international scale rapidly increased due to factors, such as the ease of IT use and decreased cost of doing IT. This means that companies need global understanding derived from analytics in interacting and relating. Moreover, globalization means that there is huge data from all over the world, meaning information overload. Business intelligence seeks to draw meaning from this complex information matrix. In addition, technological advances bring opportunities and challenges to companies. For example, there is a need to keep track of tools to enhance analytics and make informed decisions. Companies need to offer both quality services and products due to increased business competition. Business intelligence and analytics could aid in these efforts. On the other hand, the changing regulatory environment to protect the citizens means that companies should also be aware of the changing laws, for instance, that relate to analytics. The US SIM perceived as requiring more investment list also ranked business intelligence and analytics as first and in the worrisome list it is ranked third. This shows that business intelligence is important in Finland, US and the rest of the world.

Enterprise application integration (EAI) is the process of incorporating systems in an organization so that processes are linked and information resources are shared. EAI seeks to promote business integration by offering different solutions, such as data integration and business processes integration. EAI is ranked number two in Table 8.2. This could be a pointer of advanced isolated systems in Finland and thus valuing the integration offered by EAI. In the US, legacy applications maintenance, updating and consolidation is ranked tenth in the SIM list. Moreover, it is ranked 37th in the list of the applications that should get more funding and 16th most personally worrisome. This is an indication of the nature of the environment in terms of integration requirements, which already exists in the US.

Networks and telecommunication is ranked third in Finland and eighth in the US. Nokia, a Finnish company, is one of the main players in business networks provision. However, in the US list of those applications that should get more funding, networks are the 30th and 19th most personally worrisome. After networks, there are collaborative and workflow tools and mobile and wireless applications, which are fourth and fifth, respectively, in Table 8.3, but these are not in the US lists. Customer relationship management systems, cloud computing and ERP systems which are ranked eighth, ninth and 12th in Finland are the only other technologies in the US lists. This points to the ever-increasing power of customers and reveals a need to offer improved services and quality products.

8.7 Individual IT Employee Issues

Table 8.4 shows a summary of the responses about individual issues. Note that, once again a lower average score means higher agreement with each statement. The responses indicate that the Finnish IT workers are generally happy with their work environment and they face moderate work pressure. It seems that workload is relatively light and the workers experience little burnout. Furthermore, the IT employees feel a strong sense of accomplishment and they believe to be doing a good job. The data illustrate that IT professionals are generally pleased with the working environment and are confident about the future.

Table 8.4:   Individual IT Employee Issues in Finland

Individual Issues	Mean Rating*	Std. Deviation
Job satisfaction		
In general, I like working here.	2.06	0.939
All in all, I am satisfied with my current job.	2.22	1.025
In general, I don't like my current job.	3.99	1.035
Work pressure		
I feel that the number of requests, problems or complaints that I deal with at work is more than expected.	3.27	0.961
I feel that the amount of work I do interferes with how well it is done.	2.72	0.944
I feel busy or rushed at work.	2.75	1.055
I feel pressured at work.	2.77	1.061
Work–life balance		
There is a blurring of boundaries between my job and my home life.	3.35	1.118
My work-related responsibilities create conflicts with my home responsibilities.	3.70	1.141
I do not get everything done at home because I find myself completing job-related work.	3.52	1.205
Workload and burnout		
I feel drained from activities at work.	3.35	1.085
I feel tired from my work activities.	3.19	1.146
Working all day is a strain for me.	3.63	0.977
I feel burned out from my work activities.	3.82	1.027
Sense of accomplishment		
I feel I'm making an effective contribution to what this organization does.	2.17	0.807
In my opinion, I do a good job.	1.80	0.622
I have accomplished many worthwhile things in this job.	1.86	0.669
At my work, I feel confident that I am effective at getting things done.	2.03	0.739
Threats to one's job		
I am worried that future technology advancements may pose a threat to my job.	3.58	1.168
I believe that other people may be able to perform my work activities.	2.57	1.002
I am concerned that my job may be eliminated soon.	3.89	1.076
I am concerned that my job may be outsourced soon.	4.07	0.983
Career plans		
I will be with this organization 1 year from now.	2.33	1.141
I will take steps during the next year to secure a job at a different organization.	3.56	1.240

(*Continued*)

Table 8.4: (*Continued*)

Individual Issues	Mean Rating*	Std. Deviation
I will be with this organization 5 years from now.	3.07	1.270
I will be working in the IT field 1 year from now.	1.71	0.822
I will take steps during the next year to secure a job outside the IT field.	4.20	0.957
I will be working in the IT field 5 years from now.	2.25	1.168

*Rating scale ranges from 1 to 5: 1 as strongly agree to 5 as strongly disagree.

A high sense of accomplishment among the Finnish IT professionals could be explained by an experienced workforce with many years in the industry. However, the workforce feels some work pressure. For instance, the workers feel moderately rushed at work and thus the amount of work done interferes with how well it is done. Nevertheless, the workers perceive that there is little blurring of boundaries between job and home life, and that work-related responsibilities do not generally create conflict with home duties.

Table 8.4 also illustrates that the Finnish IT workers do not feel that much pressure in terms of threat to their jobs. For example, there is little worry that future technology advancement may pose a threat to one's job, and there is little concern that the job may be eliminated or outsourced. As a result, the IT workers believe that they will continue with the same organization and will not be looking for a job in other organizations in 1 or 5 years' time. There is strong belief among IT workers that they will continue working in the IT industry.

8.8 Conclusion

The top-IT organizational concerns in Finland are in security and privacy, IT reliability and efficiency, and alignment between IT and business. Top technology concerns are business intelligence and analytics, enterprise application integration, and networks. IT professionals in Finland indicated that they do not feel any threat to their jobs, are satisfied with work–life balance, and have a very strong sense of accomplishment. They are generally happy with the work environment due to low work pressure. The IT industry is very important in Finland because of its significant export contribution, high workforce, and research and innovation contributions.

Acknowledgments

We are grateful for the financial support of TSR, The Finnish Work Environment Fund (https://www.tsr.fi/en/frontpage) and the fruitful collaboration with TIVIA, the Finnish Information Processing Association (www.tivia.fi). We also thank the international consortium of the World IT Project and look forward to further research and development activities.

References

Carr, N. G. (2003). IT doesn't matter. *Harvard Business Review*, 81, 41–49.

Chen, H., Chiang, R. H. L., & Storey, V. C. (2012). Business intelligence and analytics: From big data to big impact. *MIS Quarterly*, 36(4), 1165–1188.

Doz, Y., & Kosonen, M. (2008). Fast Strategy: How Strategic Agility Will Help You Stay Ahead of the Game, Wharton School Publishing, PA.

Gartner (2018). Business intelligence, https://www.gartner.com/it-glossary/business-intelligence-bi/ retrieved 31.7.2018.

Islam, A. K. M, Mavengere, N. B., Ahlfors, U., Ruohonen, M., Serenko, A., & Palvia, P. (2018). A stress-strain-outcome model of job satisfaction: The moderating role of professional self-efficacy. In *Twenty-fourth Americas Conference on Information Systems*, New Orleans.

Kappelman, L., Johnson, V., McLean, E., & Maurer, C. (2018). The 2017 SIM it issues and trends study. *MISQ Exec*, 17(1), 53–88.

Luftman, J., & McLean, E. (2004). Key issues for IT executives. *MIS Quarterly Executive*, 3(2), 89–104.

Oksanen, J. (2018). Information society governance and its links to innovation policy in Finland, VTT Studies.

Palvia, P., Jacks, T., Ghosh, J., Licker, P., Romm-Livermore, C., Serenko, A., & Turan, A. H. (2017). The World IT Project: History, trials, tribulations, lessons, and recommendations. *Communications of the Association for Information Systems*, 41(18), 389–413.

Palvia, P., Ghosh, J., Jacks, T., Serenko, A., & Turan, A. (2018). Trekking the globe with the World IT Project. *Journal of Information Technology Case and Application Research*, 20(1), 3–8.

Technology Industries (2018). Technology industries of Finland, https://teknologiateollisuus.fi retrieved 1.8.2018

TIVIA (2019). Finnish Information Processing Association — TIVIA, Network for Finnish ICT Professionals. https://tivia.fi/in-english retrieved 28.2.2019.

Chapter 9

Information Technology Issues in France

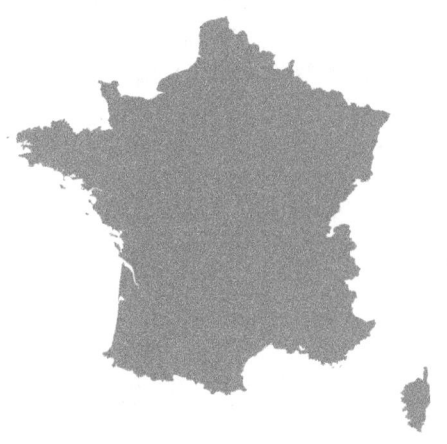

Hajer Kefi[*,§], Michel Kalika[†,¶], and Prashant Palvia[‡,||]

*PSB, Paris School of Business, France
†Business Science Institute & IAE Lyon, France
‡The University of North Carolina at Greensboro, USA
§h.kefi@psbedu.paris
¶kalika.michel@gmail.com
||pcpalvia@uncg.edu

Summary

The information technology (IT) industry in France realizes the importance of developing highly innovative IT-based business solutions in order to meet the challenges of fierce global competition. The new FrenchTech eco-system, strongly supported by the French government, is enhancing this trend. Against this backdrop, we examined the important organizational, technological and individual concerns of IT employees in France. The three topmost organizational IT issues are: Revenue generating IT innovations, security and privacy, and project management. The top-three technology issues are: Business intelligence and analytics, customer relationship management systems, and mobile and wireless systems. In general, the IT employees are satisfied with their jobs and the IT profession, and plan to stay with the current employer in the near term.

9.1 Introduction

In France, information technology (IT) is a high-performance sector but has to cope with many issues and difficulties. Given the highly competitive environment, price pressure, especially from aggressive new comers from emergent economies, and the on-going consolidation in the global market, the French IT sector has persistently suffered from structural low margins. Therefore, it had to move towards more flexible and diversified product and service offerings combined with extensive use of outsourced resources. Providing highly innovative products and services was found to be a solution for the difficulties that were faced, especially during the 2009 economic collapse (Atradius, 2016). The results of the World IT Project survey conducted in France show how these challenges have been transcended by the French IT professionals. In particular, organizational IT issues, technology issues, and individual issues were investigated. In this chapter, these results are put into perspective and interpreted with regard to major trends in the world.

9.2 Country Background and History

France is considered as a "great power" in the world, playing a premier political, economic and cultural role. It is one of the five permanent members of the United Nations Security Council and a leading member state of the European Union and the Eurozone. It is also a member of the Group of 7, North Atlantic Treaty Organization (NATO), Organization for Economic Co-operation and Development (OECD) and the World Trade Organization (WTO).

Through history, France has been a precursor in instilling the culture of human rights, since the Declaration of the Rights of Man and of the Citizen in 1789 during the French revolution, and a leading and inspiring democracy that contributed during centuries to defend the values of freedom, justice and equality between human beings.

The French lifestyle is also very rich, consisting of high gastronomy, fashion, arts and literature. It disseminates through the world, especially via La Francophonie (the international organization representing countries and regions where French is a lingua franca or customary language, i.e., 57 member states and governments).

From an economic perspective, France is ranked as the sixth largest economy in the World and the third in the European Union, with a GDP estimated at $2.925 trillion (nominal; 2018) (World Economic Prospects Monthly, 2017). The most flourishing French industries include machinery, chemicals, aircraft, electronics and tourism (as France is the most visited destination in the world) (Turner and Freiermuth, 2017). According to the Global Competitiveness Report 2017–2018, three factors contribute predominantly to strengthen the French economy: infrastructure, a large and globally integrated market, and a world top-ranked innovation ecosystem (Schwab, 2017). On the other hand, a fairly rigid labor market remains an obstacle to attracting foreign investments and international talent and is likely to foster domestic companies' relocation to lower wage areas of the world.

9.3   Information Technology in France

France has the third largest IT sector in Europe (European Information Technology Observatory, 2018). According to the Outlook Digital Economy report (OECD, 2017), the French economy ranks eighth in the world in terms of IT services exportations, which represents 3% of its foreign trade. The digital economy accounts for 3.7% of French jobs, 5.2% of GDP and 7.9% of total private-sector value added and more than a quarter of private-sector R&D activity (OECD, 2017).

In 2015, an overwhelming majority (83.4%) of IT specialists employed in France were men (83.9% in the EU). 76.9% hold a tertiary level education (60.5% in EU), and more than 1 IT specialist out of 3 (37.7%) was aged less than 35 (36.4% in EU) (Eurostat, 2016).

Since digital transformation has become an economic and business priority in all sectors, the country's economic growth attributed to IT and digital development is expected to grow exponentially, despite many competitiveness issues that the sector is facing at the international level. The fiercest competition comes from prime movers in the emerging digital world: North

America, China and Eastern Asia. Digital progress in France is essentially in a middle gear, with digital GDP likely to grow steadily to 7% by 2020 (€180 billion).

The country is moving forward with clear strengths. It has a strong user base, solid infrastructure, and quality mathematics and science education. It is also one of the countries where innovation is considered the strongest provider of sustainable competitiveness. In McKinsey's top-250 Internet-related firms database, there are many highly innovative French firms: Alcatel-Lucent, Cap Gemini S.A., Atos Origin S.A., Dassault Systèmes SA, Orange/France Telecom, Gemato, Group Bull SA, Groupe Steria SCA, Illad SA, and Vivendi (McKinsey, 2011).

More recently, the "French Tech", an innovative public policy initiated by the French Government in 2013, has created a booming start-up ecosystem driven by a new generation of entrepreneurs, investors, engineers, and other talented people. The number of IT-driven startups is exploding, many of them are now international market leaders, such as BlaBlaCar, DBV Technologies, and Criteo to name a few.

9.4 Methodology

Data were collected for the World IT Project (Palvia *et al.*, 2017; Palvia *et al.*, 2018) in France from February 2015 to June 2016. First of all, the World IT Project questionnaire originally written in English was translated to French by the first author of this chapter, then back-translated to English by a bilingual specialist, who reformulated the items presenting divergence in meaning.

The questionnaire was administered exclusively in an online mode using Google forms. The alumni databases of several schools and universities in France served as the primary mechanism to reach IT Workers. Participants were asked to fill the questionnaire only if they worked in France. Many respondents expressed privacy concerns and required clarification about the purpose of the study, particularly with regard to questions that investigate individual IT issues and how they relate to professional and social domains. So, the two country investigators had to send several emails to the targeted population with more details about the World IT Project, saying that it is a nonprofit international project and that the data will be aggregated, anonymized and used only for research purposes.

The country investigators received 293 valid responses, after eliminating 8 that were inadvertently collected from non-IT workers. Table 9.1 shows

Table 9.1: Descriptive Statistics (France)

Characteristics	N	%	Characteristics	N	%
Level of education:			Years of work experience:		
High school or less	20	6.8	0–4 years	50	17.1
Associate degree	21	7.2	5–9 years	70	23.9
Bachelor's degree	34	11.6	10–19 years	80	27.3
Master's degree	169	57.7	20–29 years	76	25.9
Ph.D.	49	16.7	30+ years	17	5.8
Years of IT experience:			Organizational location:		
0–4 years	73	24.9	IT department employee	84	28.7
5–9 years	59	20.1	IT worker in non-IT department	39	13.3
10–19 years	100	34.1	Contract employee	39	13.3
20–29 years	49	16.7	Consultant	94	32.1
30+ years	12	4.1	Vendor employee	37	12.6
Employment status:			Work position:		
Mostly full time	224	76.5	Not part of management	71	24.2
Mostly part time	51	17.4	In lower management	82	28.0
Mostly over time	18	6.1	In middle management	59	20.1
Been laid off:			In senior management	81	27.6
Yes	4	1.4			
No	289	98.6			

descriptive statistics of the sample. Note that, because of missing data, some totals can be less than 293.

9.5 Organizational IT Issues

Table 9.2 reports results concerning the organizational IT issues experienced by the respondents from France ranked in order of importance. Two of them match the top five of the US based 2017 SIM IT Key Issues and Trends study (Kappelman *et al.*, 2018), namely Security/privacy and IT/Business Alignment. Innovation which is ranked first in the French sample belongs to the top 10 of 2017 SIM list, along with three other items: IT talent retention, business agility and continuity planning.

Among the highest-ranked issues, security/privacy appears first in the 2017 SIM issues list and second for the respondents in France. This issue which is faced by IT companies all over the world is receiving an increasing interest by IT decision-makers in France due to the severe and recurrent security attacks experienced by many big companies and governmental agencies in the country. Furthermore, intra- and inter-organizational hyperconnectivity, due notably to IoT and new work practices allowed by services

Table 9.2:   Organizational IT Issues in France

Organizational IT Issues	Rank	Mean Rating*	Std. Deviation
Revenue generating IT Innovations	1	1.71	0.63
Security and privacy	2	1.72	0.72
Project management	3	1.73	0.67
IT reliability and efficiency	4	1.77	0.84
Alignment between IT and business	5	1.88	0.81
Knowledge management	6	1.90	0.82
Business agility and speed to market	7	1.91	0.81
Business productivity and cost reduction	8	2.07	0.71
IT cost reduction	9	2.21	0.83
IT strategic planning	10	2.32	0.86
Attracting and retaining IT professionals	11	2.35	1.06
Business process reengineering	12	2.40	0.77
Continuity planning and disaster recovery	13	2.45	1.10
Enterprise architecture	14	2.53	0.83
IT service management	15	2.69	1.03
Globalization	16	3.07	1.15
Outsourcing	17	3.38	1.21
Bring your own computer device (BYOD)	18	3.64	1.19

*Rating scale ranges from 1 to 5: 1 as most important and 5 as no importance.

such as BYOD or corporate portals, are also exacerbating security concerns (Debar, 2017). Investments devoted to cybersecurity are therefore expected to grow exponentially according to the National Agency of Information Systems Security (ANSSI, 2019). Since May 2018, the General Data Protection Regulation (GDPR) is providing high standards of privacy protection in all the countries belonging to the EU and has raised the legal constraints that companies have to comply with.

Not surprisingly, IT innovation is ranked first by the IT professionals in France. The CIGREF (a premier French Information Systems Professionals Association) reports regularly a similar result and echoes a major concern expressed by executives in big, medium and small-sized organizations from public and private sectors, i.e., using IT as an enabler to innovation through product and process differentiation, business agility (which was ranked seventh in this study and 9th in the 2017 SIM key issues list) in a highly competitive environment (CIGREF, 2017).

Strategic alignment which has remained in the top three of the SIM key issues list for more than a decade now is also a recurrent major concern for IT French professionals, as regularly reported by many studies (Kefi, 2011; Mourrain and Deltour, 2016). The French model of corporate governance is therefore increasingly putting IT issues, and especially IT-enabled innovation issues, within its strategic priorities.

Project Management (ranked fourth) and IT reliability and efficiency (ranked fifth) complete the list of top five most important IT organizational issues. Both issues relate to the ability of the firm to keep the IT processes under control (with respect to time, budget and quality). The two issues together ensure leadership and clear focus on the objectives and deliverables of the projects that extensively use outsourced resources and involve different stakeholders, including project sponsors, customers, consultants, subcontractors — located either inside or outside of the country. According to CIGREF (2017), technical and organizational skills related to project management are the most difficult to find and also the most important for organizations in France.

9.6 Technology and Infrastructure Issues

Table 9.3 presents the ranks in order of importance of the technology and infrastructure issues for IT professionals in France. Many similarities can be noticed between the five highest-ranked topics in this list and the US based 2017 SIM IT Key Issues and Trends study (Kappelman *et al.*, 2018).

Table 9.3: Technology and Infrastructure Issues in France

IT Related Issues	Rank	Mean Rating*	Std. Deviation
Business intelligence/analytics	1	1.87	0.74
Customer relationship management (CRM) systems	2	1.94	0.70
Mobile and wireless applications	3	2.03	0.92
Collaborative and workflow tools	4	2.06	0.72
Enterprise resource planning (ERP) systems	5	2.10	0.71
Networks/telecommunications	6	2.18	1.03
Enterprise application integration	7	2.19	0.76
Business process management systems	8	2.31	0.81
Mobile apps development	9	2.32	1.15
Big data systems	10	2.37	1.19
Data mining	11	2.37	1.17
Software as a service	12	2.38	0.93
Social networking/media	13	2.38	1.20
Cloud computing	14	2.58	1.06
Virtualization (desktop or server)	15	2.67	1.04
Service-oriented architecture	16	2.79	0.98

*Rating scale ranges from 1 to 5: 1 as most important and 5 as no importance.

For example, business intelligence/analytics and ERP appear in the top five of both lists. If we consider the top 10 highest-ranked issues in both lists, then additional common issues are networks/telecommunications and CRM. There are notable differences as well. For example, cloud computing is ranked third in the SIM list but low in our list.

Analytics and business intelligence is ranked first in both the lists and is also a major concern for companies all over the world since competing in the age of data-driven business and digital transformation has become a matter of survival. In France, the most innovative companies constituting the FrenchTech ecosystem are fostering disruptive business models built on extracting data, automating processes and creating value using AI and machine learning systems, especially in the marketing and financial sectors. Moreover, the incumbent big companies (including public companies) in the banking, transportation and energy sectors (to cite a few) are striving to initiate their digital transformation journey.

According to the eCAC maturity barometer 2017 (a survey conducted within the 32 most important companies in France to assess the efficiency of their digital strategies), data concerns are not equally important in all business areas. Retail is ahead of the other sectors, while manufacturing is lagging. Small- and mid-sized companies are focusing more on digitalizing their business processes based on relatively less innovative IT-enabled integrative solutions, namely CRM and ERP systems, which explains their high ranking in this study. Data issues are addressed by these companies mostly from a legal lens due to the new GDPR imperatives on data protection.

The other technology and infrastructure issues are related to collaborative and workflow tools and mobile and wireless applications. This reflects the need of these companies to manage virtual teams, and to develop and implement remote processes to reach their clients and business partners at any time and place.

In summary, the technology and infrastructure issues in France suggest a pragmatic, integrative and collaborative digital strategy followed by a majority of the companies. They seem to be clearly aware of the imperative to engage with data-driven digital transformation processes.

9.7 Individual IT Employee Issues

Table 9.4 shows a summary of the responses about individual issues. Note that a lower rating below the mid-point of 3 in our scale corresponds to a stronger agreement with the listed statement. Thus, on average, IT

Table 9.4: Individual IT Employee Issues in France

Individual Issues	Mean Rating*	Std. Deviation
Job satisfaction		
In general, I like working here.	2.09	0.71
All in all, I am satisfied with my current job.	2.40	0.98
In general, I don't like my current job.	3.54	0.93
Work pressure		
I feel that the number of requests, problems or complaints that I deal with at work is more than expected.	2.67	0.96
I feel that the amount of work I do interferes with how well it is done.	2.93	1.16
I feel busy or rushed at work.	2.77	1.11
I feel pressured at work.	2.77	0.97
Work–life balance		
There is a blurring of boundaries between my job and my home life.	2.93	1.09
My work-related responsibilities create conflicts with my home responsibilities.	3.15	1.17
I do not get everything done at home because I find myself completing job-related work.	3.10	1.12
Workload and burnout		
I feel drained from activities at work.	3.42	1.03
I feel tired from my work activities.	2.83	1.12
Working all day is a strain for me.	3.44	1.18
I feel burned out from my work activities.	2.89	1.06
Sense of accomplishment		
I feel I'm making an effective contribution to what this organization does.	2.15	0.82
In my opinion, I do a good job.	1.95	0.51
I have accomplished many worthwhile things in this job.	2.05	0.62
At my work, I feel confident that I am effective at getting things done.	1.98	0.55
Threats to one's job		
I am worried that future technology advancements may pose a threat to my job.	3.45	1.17
I believe that other people may be able to perform my work activities.	2.26	0.82
I am concerned that my job may be eliminated soon.	3.57	1.16
I am concerned that my job may be outsourced soon.	3.28	1.12
Career plans		
I will be with this organization 1 year from now.	2.65	1.37
I will take steps during the next year to secure a job at a different organization.	3.36	1.34

(Continued)

Table 9.4:   (*Continued*)

Individual Issues	Mean Rating*	Std. Deviation
I will be with this organization 5 years from now.	3.03	1.34
I will be working in the IT field 1 year from now.	2.29	1.07
I will take steps during the next year to secure a job outside the IT field.	3.59	1.13
I will be working in the IT field 5 years from now.	2.25	1.01

*Ratings scale ranges from 1 to 5: 1 as strongly agree and 5 as strongly disagree.

employees in France are relatively satisfied with their work, even though they express experiencing moderate levels of work pressure. On work–life balance, they do not report any significant issues, yet they have mixed feelings about workload and job burnout. They do not perceive any major threats from future technology advancements and other worrisome issues like outsourcing, yet they acknowledge that other people can do their jobs. The lowest scores obtained concern the sense of accomplishment which appears to be quite strong among the French IT employees. As for their career aspirations, given the high levels of satisfaction and accomplishment, these employees plan to stay in their current organizations at least in the short term. Long term, they intend to stay in the IT sector although they may move from their current organizations.

When drilling the data deeper, it was observed that there are significant differences due to age and gender within the respondents. Mid-career and older IT workers have higher levels of job satisfaction than their younger colleagues, while they do not necessarily experience more work pressure and workload. Age does not seem to have any significant impact on sense of accomplishment, which is relatively high for all generations. But older workers seem to worry more about the future and seem to doubt their ability to fit within the new work environment and work practices due to advancements in technology. Older employees are clearly much more reluctant to leave their actual organizations and the IT sector than their younger colleagues.

Concerning the gender effect, it appears relatively insignificant with regard to the sense of accomplishment, perceived threats to job security and career plans, whereas there are some differences concerning job satisfaction, perceived pressure and work–life balance. Female IT workers seem less satisfied with their jobs and more sensitive to work–life balance, work pressure and workload. This could be explained by the fact that the French society still expects women to handle a larger part of the housework than men. While important progress has been made in this area, more work has to be done toward gender equality. This is a paradox, since France has fostered audacious feminist theories since the fifties, e.g., with Simone de Beauvoir and Virginie Despentes (Mossuz-Lavau, 2009), while the mainstream culture still appears to be sexist. The under representation of women in politics, the high sphere of corporate governance and the scientific community in the education sector reflect this reality. In the IT sector, many initiatives have been taken to enhance the presence of women in this field, such as Femme@numérique (women@digital) developed by the CIGREF (https://femmes-numerique.fr/l-initiative/), and more young women now seem more willing to embrace different IT careers.

9.8   Conclusion

In France, the top-IT organizational issues are about IT-driven innovation, security & privacy, and project management. The top technology and infrastructure concerns are about data analytics, integrative enterprise systems, namely CRM and ERP systems, and technology tools to support collaborative work and mobile applications. IT employees seem happy at work, although they may struggle to maintain a certain balance between the private and professional spheres of life; this is especially true for women. Overall, the IT sector in France is flourishing but has to face many challenges due to the intensive international competition and a certain delay in fully entering the digital transformation age.

References

ANSSI (2019). The French National Digital Security Strategy. https://www.ssi. gouv.fr/en/cybersecurity-in-france/ Accessed February 2019.

Atradius (2016). Market Monitor — ICT industry — France. https://atradi us.fr/rapports/market-monitor---ict-industry---france.html#, accessed in 09/25/2018.

CIGREF (2017). *Une Réponse Aux Challenges de l'entreprise.*

Kefi, H. (2011). Processus organisationnels et systemes d'information et de communication: Alignement et performance. (How to Align ICT and Organizational Processes to Achieve Performance. With English summary.). *La Revue Des Sciences de Gestion*, 46(251), 189–200. https://doi.org/10.3917 /rsg.251.0189

Mossuz-Lavau, J. (2009). De Simone de Beauvoir à Virginie Despentes: les intellectuelles et la question du genre. *Modern & Contemporary France*, 17(2), 177–188.

Mourrain, A., & Deltour, F. (2016). La PME face au choix d'un Système de Gestion Intégré?: les risques du processus de pré-implémentation. *Revue Internationale P.M.E.: Économie et Gestion de La Petite et Moyenne Entreprise*, 29(1), 27. https://doi.org/10.7202/1036769ar

OECD (2017). OECD Digital Economy Outlook 2015. Paris: OCDE Editions. h ttps://doi.org/https://doi.org/10.1787/9789264276284-en.

Palvia, P., Jacks, T., Ghosh, J., Licker, P., Romm-Livermore, C., Serenko, A., & Turan, A. H. (2017). The World IT Project: History, trials, tribulations, lessons, and recommendations. *Communications of the Association for Information Systems*, 41(18), 389–413.

Palvia, P., Ghosh, J., Jacks, T., Serenko, A., & Turan, A. (2018). Trekking the globe with the World IT Project. *Journal of Information Technology Case and Application Research*, 20(1), 3–8.

Schwab, K. (2017). *The Global Competitiveness Report 2017-2018*. World Economic Forum. https://doi.org/92-95044-35-5

Turner, R., & Freiermuth, E. Travel & Tourism — ECONOMIC IMPACT 2017 (2017). Retrieved from https://www.wttc.org/-/media/files/reports/econo mic-impact-research/countries-2017/saudiarabia2017.pdf.

World Economic Prospects Monthly (2017). *Economic Outlook*, 41, 1–29. https://doi.org/10.1111/1468-0319.12316.

Chapter 10

Information Technology Issues in Germany

Elisabeth F. Mueller[*,§], Jaideep Ghosh[†,¶], and Prashant Palvia[‡,∥]

University of Passau, Germany
†*Shiv Nadar University, India*
‡*University of North Carolina at Greensboro, USA*
§*elisabeth.mueller@uni-passau.de*
¶*jghosh20770@gmail.com*
∥*pcpalvia@uncg.edu*

Summary

Germany has the largest information technology (IT) market in Europe and one of the most important markets in the world. While it is home to some globally leading IT companies, the German IT industry is more accurately characterized by highly specialized small- and medium-sized enterprises (SMEs) and many vibrant IT start-up companies. A major challenge for the German IT industry that runs as a common theme through the present study is the use of new technologies as companies need to continue investing in innovations and technologies to keep up with the global pace of innovation. In this context, the companies seek to recruit highly skilled IT professionals. However, the supply of skilled labor is short, which is another theme that arises in this study.

10.1 Introduction

High productivity and innovative talent have a long tradition in the German economy and shape one of its most important sectors, the information technology (IT) sector. The IT industry distinguishes itself through a high innovation rate. In addition, since the innovations in other industries are based to a great extent on information technologies, the IT sector also drives growth and inventions in other industries. Importantly, IT is a German government priority as well. The government's Digital Agenda sets out key fields of action for achieving transition to a digital world. Among other things, the agenda focuses on the expansion of the digital infrastructure, networked production, and digital integration in society, education and science, as well as in security and data protection. According to a study conducted by the German association of the IT industry BITKOM (2018), the number of employees working in the IT sector has almost doubled from 2007 to 2018. BITKOM expects the number of IT professionals to further increase within the next few years. In a recent survey, more than two thirds of the companies indicated their intention to create additional jobs. However, they do not always find qualified workers in Germany. Thus, the number of job vacancies for IT specialists has recently shown a marked increase. One option to encounter the shortage of skills in the IT industry is to activate the untapped potential of underrepresented vocational groups, such as female employees.

The present study gives an overview of current information systems management and technology issues in Germany. Based on a survey of 308

IT specialists, we show the top organizational and IT infrastructure issues and describe the IT professionals' individual issues. The characteristics of the German IT industry serve as a basis for interpreting the results.

10.2   Country Background and History

Germany was first unified in 1871 as a modern federal state. After World Wars I and II, two German states were formed in 1949: the western Federal Republic of Germany (FRG) and the eastern German Democratic Republic (GDR). The democratic FRG embedded itself in key western economic and security organizations, the EC and NATO, while the communist GDR was on the front line of the Soviet-led Warsaw Pact. With the reunification of the two German states in 1990, the country strengthened its role as a key member of the European Union and of the continent's economic, political, and security organizations (The World Factbook, 2018). Today, with a population close to 82 million, Germany is the most populous member state of the European Union and a member of the United Nations, NATO, the G7, the G20, and the OECD. As a country with a very high standard of living, Germany upholds social security and universal health care system as well as tuition-free university education.

Germany has a strong economy. In 2017, it had the world's fourth largest economy by nominal GDP and the fifth largest by PPP (World Bank, 2018). In the first quarter of 2018, Germany's GDP expanded 2.3% over the same quarter of the previous year. GDP growth rate averaged 2.02% from 1971 until 2018, reaching an all-time high of 7.20% in the first quarter of 1973 and a record low of –6.80% in the first quarter of 2009. The service sector contributes around 70% of the total GDP, industry 29.1%, and agriculture 0.9% (Statista, 2018). As a global leader in several industrial and technological sectors, Germany is both the world's third largest exporter and importer of goods. Among the top-10 exports of Germany are vehicles, machinery, chemical goods, electronic products, and pharmaceuticals. The vast majority of German companies belong to the German "Mittelstand," SMEs, which are mostly family-owned. The German employment rate is very high. In the last quarter of 2017, 66.6% of 15–74 year-olds were in the work force — the highest rate since the reunification in 1990. The unemployment rate of 3.6% is substantially below the OECD average of 5.5% (OECD, 2018).

10.3 Information Technology in Germany

Germany's IT market is the largest in Europe. In 2017, IT industry revenues amounted to 86 billion euros (US$98 billion), of which around 45% were generated within the IT services segment (Statista, 2018). The IT hardware and software segments generated 28% and 27% of total IT industry revenues, respectively (Statista, 2018). Recent industry reports estimate that the IT sector in Germany has expanded at a moderate compound annual growth rate of 2.95% over the period 2006–2017 (Statista, 2018). German companies employ almost 900,000 IT professionals, and an increasing number of companies base their business model on IT (BITKOM, 2018). Germany is home to 12 out of 34 so-called European ICT hubs which are geographical agglomerations of best performing ICT production, R&D, and innovation activities, located in the European Union, that play a central role in global international networks (De Prato and Nepelski, 2014). These ICT hubs are located in the largest metropolitan areas such as Berlin and Munich, but interestingly also include small-sized but vibrant regions and cities such as Darmstadt and Karlsruhe.

Big national and international players such as IBM, Microsoft, or SAP (one of the largest business software companies originally from Germany) are present on the German IT market. However, the market is best characterized by the large number of highly specialized SMEs, the German "Mittelstand," and a creative and vibrant start-up scene in hot-spots such as Berlin. Increased business demand for smart data products, services in the cloud, and security technology are driving domestic IT market growth, which is further stimulated by a far-reaching government program of digitization.

The German IT sector faces significant challenges for sustained success such as the shortage of skills. The rapid expansion of IT since the late 1990s has caused a growing demand for IT professionals. The supply of such skilled employees is limited. In a survey by the German Federal Statistical Office, 58% of the participating companies emphasize that they have severe problems with recruiting and retaining IT professionals (Destatis, 2018). Another challenge is the use of new technologies in German companies. A recent BITKOM (2018) study reveals that German companies need to intensify the use of new technologies to keep up with the global pace of innovation. While surveyed CEOs and managing directors believe that, overall, the German economy is relatively well positioned in terms of digitization, they feel that German companies are lagging behind their international competitors in the use of technologies associated with artificial intelligence, 3D printing, Blockchain, robotics, internet of things, big data, etc. As these technologies have the potential to disrupt entire industries, integrating them in the companies' business model is deemed crucial for sustained success.

10.4   Methodology

The goal of this study was to gather comprehensive survey data on information systems management and technology in Germany. We translated the World IT Project instrument (Palvia *et al.*, 2017; Palvia *et al.*, 2018) from English to German, following translation and back-translation procedures (Brislin, 1980) and adapted the instrument where necessary to take country-specific characteristics into account (for example, the educational qualification). We subsequently implemented a web version of the questionnaire. The questionnaire was pre-tested by German IT professionals and academics to ensure that the IT context in the translated instrument is understood properly. We invited IT professionals to participate in the web survey by placing the invitation, including the link leading to the web survey, in the newsletter of one of Germany's largest and most widely read IT news sites. The participants were guaranteed anonymity and were assured that the data would only be used at an aggregate level. No incentives were provided.

After four weeks and one reminder, 308 IT professionals had completed the survey satisfactorily. Data were analyzed using Stata 15. Table 10.1 shows the descriptive statistics of the IT professionals who took part in the study. Most of the participants are between 30 and 39 years (32.25%) or between 21 and 29 years old (29.64%). Moreover, most of them (93.77%) are male, emphasizing the underrepresentation of women in the IT profession. More than half of the respondents (54.55%) have earned a university degree (bachelor's degree or higher). The majority have 10 years of professional experience or more (61.26%) and 10 years of IT experience or more (55.85%). They most often work as programmers (27.60%) or system administrators (22.73%). Around 70% of the respondents are not part of management.

The organizations represented in the sample are mostly established firms, with around a quarter (26.38%) that are between 10 and 19 years old and another quarter (26.71%) that are 50 years or older. They show some variation in size, as 16.23% have more than 10,000 employees, 11.69% have between 101 and 200 employees, 11.69% have between 201 and 500

Table 10.1:   Descriptive Statistics

Characteristics	N	%	Characteristics	N	%
Education:			Years of work experience:		
High school or less	48	15.58	0–4 years	49	16.23
Vocational education	92	29.87	5–9 years	68	22.52
Bachelor's degree	40	12.99	10–19 years	97	32.12
Diploma/master's degree	115	37.34	20–29 years	69	22.85
Ph.D.	13	4.22	30+ years	19	6.29
Years of IT experience:			Organizational location:		
0–4 years	53	17.21	IT department employee	229	74.59
5–9 years	83	26.95	IT worker in non-IT department	28	9.12
10–19 years	102	33.12	Contract employee	19	6.19
20–29 years	61	19.81	Consultant	27	8.79
30+ years	9	2.92	Vendor employee	4	1.30
Work as:			Work position:		
Mostly full time	288	93.81	Not part of management	212	69.06
Mostly part time	14	4.56	In lower management	43	14.01
Mostly over time	5	1.63	In middle management	32	10.42
Been laid off from IT job:			In senior management	20	6.51
Yes	50	16.45			
No	254	83.55			

employees, and 11.04% have up to 10 employees. Accordingly, the organizations also vary in the size of their IT departments. Almost a quarter of the companies (22.80%) employ only up to 5 IT specialists, whereas 15.64% of the companies have an IT department with more than 1,000 employees. More than half of the organizations have a centralized IT department (51.62%), whereas 18.83% have a decentralized and 29.55% have a federated IT department.

10.5   Organizational IT Issues

Table 10.2 shows the organizational IT issues in Germany ranked in order of importance. Note that a lower average score means greater importance. Mean rating scores range from 1.70 to 4.15 with high density at moderate levels of importance (2.5–3.5). Only one item in the top five, namely security and privacy, matches an item in the top five of the 2017 SIM IT Key Issues and Trends study (Kappelman *et al.*, 2018). Beyond security and privacy, three other items in the top 10, namely alignment between IT and the business, IT innovations, and business agility, match items in the top 10 of the 2017 SIM list (Kappelman *et al.*, 2018).

Table 10.2: Organizational IT Issues in Germany

Organizational IT Issues	Rank	Mean Rating*	Std. Deviation
IT reliability and efficiency	1	1.70	0.65
Security and privacy	2	1.88	0.84
Knowledge management	3	2.01	0.88
Attracting and retaining IT professionals	4	2.18	1.01
IT strategic planning	5	2.18	0.85
Project management	6	2.27	0.90
Continuity planning and disaster recovery	7	2.27	0.89
Alignment between IT and business	8	2.57	0.86
Revenue-generating IT innovations	9	2.77	1.06
Business agility and speed to market	10	2.82	0.97
Enterprise architecture	11	2.82	0.99
Business productivity and cost reduction	12	2.89	0.93
IT service management (e.g., ITIL)	13	2.89	1.09
Business process reengineering	14	2.94	0.98
IT cost reduction	15	3.32	0.91
Globalization	16	3.37	1.14
Bring your own computing device (BYOD)	17	3.66	1.20
Outsourcing	18	4.15	0.92

*Rating scale ranges from 1 to 5: 1 as most important and 5 as no importance.

IT reliability and efficiency was ranked the highest. This issue is important to IT companies of all segments, be they IT service firms or software or hardware firms. Providing reliable and, in particular, high-quality IT solutions in an efficient way is key to German companies who are selling products or services "Made in Germany", a brand that stands for innovation, quality, and expertise with an excellent reputation in global markets. "Made in Germany" is a promise that shapes customer expectations and therefore drives German companies to develop reliable and efficient IT products or services in order to meet these expectations.

The second highest-ranked issue is security and privacy which is on top of the 2017 SIM issues list, showing very plainly that this issue is a key concern for companies across the world and not specific to a certain country. In Germany, security and privacy are important issues for the companies as they operate predominantly in knowledge-intensive industries. Protecting the knowledge with the help of an IT infrastructure that ensures security and privacy is crucial for the companies to maintain their competitive edge.

Knowledge management was ranked the third highest. As the rate of change in the IT industry is high and the industry itself is knowledge-centric and dynamic, IT-related businesses regularly generate new knowledge and implement new technologies. In keeping with the motto, "what is good today is outdated tomorrow," the companies need to manage knowledge in a way that supports their employees in constantly upgrading their skills and learning new technologies. For German companies, generating and using new knowledge and keeping up with the global pace of innovation are major challenges, which emphasize the need to develop appropriate knowledge management systems and processes. Besides, knowledge management issues are closely linked to IT reliability and efficiency, the highest-ranked issue, as they underline the companies' need to manage knowledge in a way that enables them to deliver up-to-date, well-functioning and high-quality products or services.

The fourth highest-ranked issue is that of attracting and retaining IT professionals, mirroring one of the main challenges of today's German IT industry. Despite their global importance, German companies have reported an imminent shortage of labor as well as difficulties in recruiting qualified employees, which endangers their growth opportunities. Around 55,000 positions for IT professionals were vacant in 2017, an increase of 8% compared to 2016 (BITKOM, 2018). Especially, employees with adequate technical and science-related skills are scarce. In addition, the share of female IT professionals has been growing only slowly, reaching 28% of the total IT workforce in 2017 (BITKOM, 2018). Furthermore, IT professionals show relatively high turnover rates. As they are in high demand, IT specialists easily find new jobs and are less loyal to the companies they work for, in particular when they are not satisfied with their jobs (Ertürk and Vurgun, 2015). As a reaction to the sustained shortage of skilled IT workers, German companies have underscored the necessity to motivate young people, especially women, to study in IT or IT-related degree programs.

IT strategic planning completes the top-five organizational IT issues in Germany. The IT strategic plan details the technology-enabled business management processes an organization employs to run its operations and guides IT-related decision-making. Currently, the digitization of business processes and operations is transforming the German economy and IT has found its way into all kinds of organizations. Therefore, IT strategic planning remains important to align the IT strategy with the organization's goals and mission.

In summary, most of the top organizational IT issues emphasize the IT-related knowledge and skills that today's businesses rely on in the knowledge-intensive, high-technology oriented industries for which Germany is renowned. Attracting, managing and retaining talent and further developing the appropriate human capital therefore can be seen as the overarching theme that the surveyed IT professionals have concerns about.

10.6   Technology and Infrastructure Issues

Table 10.3 shows the technology and infrastructure issues ranked in order of importance. Mean rating scores range from 2.07 to 3.66 with, again, a high density at moderate levels of importance (2.5–3.5). These values indicate that the IT professionals consider most issues as moderately important and, interestingly, no issues as issues of utmost importance or of no importance. Four items in the top 10, namely networks/telecommunication, business intelligence/analytics, software as a service, and customer relationship management (CRM) systems, match items in the top 10 of the 2017 SIM IT Key Issues and Trends study (Kappelman *et al.*, 2018). While this number of matches may suggest some level of similarity, one has to point out that remarkable differences remain, particularly on the top of the list. For example, networks and telecommunications was ranked highest in the German list of technology and infrastructure issues, while it only ranks eighth on the 2017 SIM list.

Table 10.3:   Technology and Infrastructure Issues in Germany

IT Related Issues	Rank	Mean Rating*	Std. Deviation
Networks/telecommunications	1	2.07	0.93
Virtualization (desktop or server)	2	2.21	1.02
Enterprise application integration	3	2.44	0.97
Collaborative and workflow tools	4	2.66	1.01
Mobile and wireless applications	5	2.77	1.15
Business intelligence/analytics	6	2.81	1.05
Software as a service	7	2.89	1.21
Service-oriented architecture (SOA)	8	2.96	1.11
Customer relationship management (CRM) systems	9	3.02	1.10
Business process management systems	10	3.05	1.08
Enterprise resource planning (ERP) systems	11	3.07	1.21
Mobile apps development	12	3.14	1.25
Data mining	13	3.23	1.20
Cloud computing	14	3.31	1.23
Big data systems	15	3.36	1.15
Social networking/media	16	3.66	1.09

*Rating scale ranges from 1 to 5: 1 as most important and 5 as no importance.

Networks and telecommunications is the top technology and infrastructure issue as identified by the German IT professionals. While this may seem surprising, given that Germany's connectivity with the rest of the world is stable and reliable, Germany is not a leading country in terms of the overall quality of its IT infrastructure, which furthermore is not equally rolled out across different parts of the country. For example, according to the OECD Broadband statistics (2017), only 2% of the total broadband subscriptions in Germany are fiber connections compared to nearly 80% in South Korea or Japan. Moreover, IT infrastructure in rural areas is weaker than in larger cities or metropolitan areas, which leads to spatial concentration of IT companies in agglomerations and regional wealth inequalities. For these reasons, expanding the digital infrastructure is one key goal of Germany's federal government.

The second highest-ranked issue is desktop or server virtualization. Desktop virtualization is a response to an increasing number of employees working remotely and from multiple devices, whereas server virtualization is an answer for companies that need to maximize server efficiency. German companies engage in virtualization of IT hardware to react to their employees' demands for flexible working conditions to take advantage of the opportunities offered by the digital era.

Three application-focused issues, namely, enterprise application integration, collaborative and workflow tools, and mobile and wireless applications, complete the top-five technology and infrastructure issues. These issues are examples of how IT affects internal business processes and underline the variety and importance of IT applications in today's business operations. A number of German IT companies have developed expertise in these services and build their business model on them by specializing in providing customized applications.

Business intelligence and analytics appears at the sixth position, while it is ranked first in the SIM 2017 list. However, in the SIM 2017 list, business intelligence and analytics is combined with data mining and big data, which are ranked separately in the German infrastructure and technology issues list, but only on positions 13 and 15, respectively. Despite not being among the top-five issues, business intelligence and analytics is considered a growing and highly knowledge-intensive segment demanding specialized know-how and expertise. Large companies as well as small- and medium-sized companies belonging to the "Mittelstand" provide services related to analytics and seek to capitalize on the increase in demand for these services. As the economy's demand for individuals skilled in business intelligence and analytics is growing, German IT companies have called for extending and continuously updating related curricula at educational institutions.

Software as a Service and service-oriented architecture, ranked at positions seven and eight, reflect areas of operation of small- and medium-sized IT businesses but also of major German IT service companies (e.g., SAP). The application areas of CRM systems and business process management systems round out the top 10 issues, followed by enterprise resource planning at the eleventh position. These areas again offer opportunities to German IT companies that are specialized in these applications and provide development and maintenance services.

In summary, the technology and infrastructure issues emphasize that the German economy is undergoing significant changes in the digital era which affects all kinds of processes and operations within companies. The ranking reflects major challenges of the German IT industry such as a somewhat reluctant use of new technologies. For example, in contrast to the US list, big data systems were ranked second last, which indicates that the German big data market still appears to be at an early stage. However, although big data technology use may have its origins in North America, Europe (including Germany) is quickly catching up in their use, especially in developing skills and expertise in pioneering technologies.

10.7 Individual IT Employee Issues

Table 10.4 shows the summary of the responses on individual IT employee issues. A lower average score means higher agreement with each statement. Overall, the responses indicate that the IT professionals are quite happy

Table 10.4: Individual IT Employee Issues in Germany

Individual Issues	Mean Rating*	Std. Deviation
Job satisfaction		
In general, I like working here.	1.97	0.92
All in all, I am satisfied with my current job.	2.22	1.07
In general, I don't like my current job.	3.98	1.09
Work pressure		
I feel that the number of requests, problems or complaints that I deal with at work is more than expected.	3.46	1.06
I feel that the amount of work I do interferes with how well it is done.	3.28	1.18

(*Continued*)

Table 10.4:   (*Continued*)

Individual Issues	Mean Rating*	Std. Deviation
I feel busy or rushed at work.	3.35	1.07
I feel pressured at work.	3.41	1.10
Work–life balance		
There is a blurring of boundaries between my job and my home life.	3.31	1.34
My work-related responsibilities create conflicts with my home responsibilities.	3.60	1.17
I do not get everything done at home because I find myself completing job-related work.	3.79	1.17
Workload and burnout		
I feel drained from activities at work.	2.99	1.16
I feel tired from my work activities.	2.93	1.18
Working all day is a strain for me.	3.57	1.12
I feel burned out from my work activities.	3.73	1.13
Sense of accomplishment		
I feel I'm making an effective contribution to what this organization does.	1.78	0.75
In my opinion, I do a good job.	1.76	0.61
I have accomplished many worthwhile things in this job.	2.04	0.89
At my work, I feel confident that I am effective at getting things done.	2.13	0.76
Threats to one's Job		
I am worried that future technology advancements may pose a threat to my job.	4.13	0.98
I believe that other people may be able to perform my work activities.	2.27	0.99
I am concerned that my job may be eliminated soon.	4.23	0.97
I am concerned that my job may be outsourced soon.	4.22	1.05
Career plans		
I will be with this organization 1 year from now.	2.29	1.21
I will take steps during the next year to secure a job at a different organization.	3.43	1.28
I will be with this organization 5 years from now.	3.15	1.16
I will be working in the IT field 1 year from now.	1.48	0.76
I will take steps during the next year to secure a job outside the IT field.	4.38	0.91
I will be working in the IT field 5 years from now.	1.77	0.98

*Rating scale ranges from 1 to 5: 1 as strongly agree and 5 as strongly disagree.

with their jobs although they feel some but not excessive pressure at work. Achieving a work–life balance does not seem to be much of a problem either. Beyond that, the respondents perceive the workload to be high and straining but do not seem to be experiencing burnout. Furthermore, the IT employees feel a sense of accomplishment which, however, is not as strong as in other countries, indicating a more pragmatic approach of German IT professionals towards their work. They do not feel threats to their jobs, reflecting the generally favorable career prospects in the IT field. Finally, while the IT professionals do not intend to search for a job outside the IT industry, they are less loyal to their current employer.

In the early 2000s, people working in the German IT sector were often stereotyped as nerds, who were singularly focused on computers, displaying a high aptitude towards IT-related matters and sometimes lacking in interpersonal skills. Typically, these IT professionals were intrinsically motivated by their interests and passion for IT matters. Given the knowledge intensity of the IT industry and the ephemerality of the technologies, IT companies focused on attracting this type of employees, as they were considered willing to keep abreast of new knowledge and skills. Contrary to other professions where basic knowledge remains the same over decades, the half-life of knowledge and skills in the IT profession is estimated at less than 2 years (Ang and Slaughter, 2000), which requires constant education in order to keep them and the company competitive.

Today, the German IT industry not only offers an opportunity for people to satisfy their intrinsic urge to gain knowledge and engage in cognitive challenging work, but it also offers the possibility to meet extrinsic expectations regarding one's career. For example, the high demand for IT professionals due to constant growth of the industry offers security to individuals, in that, they are very likely to find another job in the field. Moreover, wages at the entry-level compared to other industries are relatively high, which sets a financial incentive for people to join the industry. Statistics show that the wage at the entry-level in the IT industry within the first 2 years is among the highest of the top-10 wages (Staufenbiel, 2017).

Due to the shortage of highly skilled labor, German companies are increasingly accommodating extrinsically motivated IT professionals in terms of working conditions, salary, work–life balance, etc. If they do not and if the employees are not happy, they are likely to search for other IT jobs in other companies. Besides the general cost of employee turnover, the disruption of processes and loss in organizational productivity (Huselid, 1995; Thatcher *et al.*, 2002), IT employee turnover is further associated with a loss of highly specialized skills and tacit knowledge about business applications and operations (Kaplan and Lerouge, 2007; Moore and Burke, 2002). As these detriments can result in a loss of competitive advantage, retaining IT employees is an important goal for German IT companies.

Overall, the German IT sector continues to exhibit a positive image and is perceived to be a thriving, future-oriented and promising industry that offers white collar jobs for highly skilled workers. Nevertheless, the supply of skilled labor is short and endangers the further growth and development of the industry. In this context, the German government, the association of the German IT industry BITKOM, and even the companies themselves are running campaigns to attract more students to IT degree programs and more workers to the IT industry. Some of these campaigns (e.g., "Women in IT," "Girls go STEM") particularly aim at unlocking the potential of female IT workers and are designed to increase the share of the female IT workforce.

10.8   Conclusion

The results of the study of the German IT industry show that the top organizational IT issues are reliability and efficiency of IT systems, security and privacy, knowledge management, and attracting and retaining IT employees. The top technology and infrastructure issues include networks/telecommunications, virtualization, and enterprise application integration. Employees indicate that the IT industry continues to offer attractive career prospects. They are, in general, satisfied with their jobs and show relatively low intention to leave the IT industry but are open to moving to other employers. Since the digitization is fundamentally transforming not only the German IT industry but the entire economy, building, developing and managing the IT human capital is among the most important issues confronting the industry in order to keep up with the global pace of innovation.

References

Ang, S., & Slaughter, S.A. (2000). The missing context of information technology personnel: A review and future directions for research. In R. Zmud (Ed.), *Framing the Domains of IT Management: Projecting the Future Through the Past.* Cincinnati, OH: Pinnaflex Education Resources, Inc., pp. 305–327.

BITKOM (2018). http://www.bitkom.org. Accessed August 2018.

Brislin, R.W. (1980). Translation and content analysis of oral and written material. In H.C. Triandis & J.W. Berry (Eds.), *Handbook of Cross-cultural Psychology.* Boston, MA: Allyn and Bacon, pp. 389–444.

De Prato, G., & Nepelski, D. (2014). *Mapping the European ICT Poles of Excellence. The Atlas of ICT Activity in Europe. JRC Scientific and Policy Reports EUR 26579 EN.* Seville: JRC-IPTS.

Destatis (Federal Statistical Office) (2018). https://www.destatis.de/DE/Zahle
nFakten/GesamtwirtschaftUmwelt/UnternehmenHandwerk/IKTUnt
ernehmen/Tabellen/05_IT_Fachkraefte_IKT_Unternehmen.html. Accessed
August 2018.

Ertürk, A., & Vurgun, L. (2015). Retention of IT professionals: Examining the
influence of empowerment, social exchange, and trust. *Journal of Business
Research*, 68(1), 34–46.

Huselid, M.A. (1995). The impact of human resource management practices on
turnover, productivity, and corporate financial performance. *Academy of
Management Journal*, 38(3), 635–672.

Kaplan, D.M., & Lerouge, C. (2007). Managing on the edge of change.
Human resource management of information technology employees. *Human
Resource Management*, 46(3), 325–330.

Kappelman, L., Johnson, V., Maurer, C., McLean, E., Torres, R., David, A., &
Quynh, N. (2018). The 2017 SIM IT Issues and Trends Study. *MIS Quarterly Executive*, 17(1), 53–88.

Moore, J.E., & Burke, L.A. (2002). How to turn around turnover culture' in IT.
Communications of the ACM, 45(2), 73–78.

OECD (2017). http://www.oecd.org/sti/broadband/broadband-statistics. Accessed August 2018.

OECD (2018). https://www.oecd.org/germany/Employment-Outlook-Germany
-EN.pdf. Accessed August 2018.

Palvia, P., Jacks, T., Ghosh, J., Licker, P., Romm-Livermore, C., Serenko, A., &
Turan, A. H. (2017). The World IT Project: History, trials, tribulations,
lessons, and recommendations. *Communications of the Association for
Information Systems*, 41(18), 389–413.

Palvia, P., Ghosh, J., Jacks, T., Serenko, A., & Turan, A. (2018). Trekking the
globe with the World IT Project. *Journal of Information Technology Case
and Application Research*, 20(1), 3–8.

Statista (2018). https://de.statista.com/themen/1373/it-branche-deutschland.
Accessed August 2018.

Staufenbiel (2017). https://www.staufenbiel.de/magazin/gehalt/bestbezahlte-jo
bs.html. Accessed August 2018.

Thatcher, J.B., Stepina, L.P., & Boyle, R.J. (2002). Turnover of information
technology workers. Examining empirically the influence of attitudes, job
characteristics, and external markets. *Journal of Management Information
Systems*, 19(3), 231–261.

The World Factbook (2018). http://www.iiasa.ac.at/~marek/fbook/04/print/g
m.html. Accessed August 2018.

World Bank (2018). https://data.worldbank.org/indicator/. Accessed August
2018.

Chapter 11

Information Technology Issues in Ghana

Jerry Godwin Diabor[*,¶], Alexander Serenko[†,‡||], and Jaideep Ghosh[§,**]

*University of Greenwich, & NCC, UK;
IPMC College of Technology, Ghana
†Faculty of Information, University of Toronto, Canada
‡Faculty of Business and IT,
University of Ontario Institute of Technology, Canada
§Department of Decision Sciences, Operations Management &
Information Systems, School of Management and
Entrepreneurship (SME), Shiv Nadar University,
Gautam Buddha Nagar, Uttar Pradesh, India
¶cj_wondergod@yahoo.co.uk,
Jerry.Godwin@ipmcghana.com
||a.serenko@utoronto.ca
**jghosh20770@gmail.com

Summary

This project focuses on organizational, technological, and individual information technology (IT) issues in Ghana. IT has been widely adopted in Ghana, and it is growing at a relatively fast pace as more firms are trooping into the country. In this study, it was noticed that security and privacy became the top issues. This was not surprising because this has been a matter of concern to many organizations around the world as the adoption of IT must provide the assurance to the issues surrounding privacy and security. The government institutions in Ghana have started to rely heavily on IT, and security and privacy have been a major concern for employees.

Business intelligence/analytics (BI/BA) is considered one of the major technology issues as indicated by a high demand for BI/BA professionals. Both small and large IT firms are providing BI/BA services in Ghana, and the demand is very strong. BI/BA is followed by customer relationship management systems and enterprise application integration in that order. Most of the IT professionals in the country are somewhat satisfied with their jobs, and they enjoy lower work pressures, workload, and burnout. They also exhibit a strong sense of professional accomplishment. Nevertheless, they are only somewhat loyal to their organizations and are more loyal to the IT profession because currently there is a high level of unemployment in Ghana, and it is difficult for IT professionals to easily change jobs.

11.1   Introduction

Ghana is located on the West coast of Africa, and it is the first country within the Sub-Saharan region to launch the use of the Internet for both organizations and individual users. According to the Global Digital Agencies annual report, there are 10,110,000 individual Internet users, and this figure represents 35% of the entire population of 29,150,000 (Zurek, 2018). In addition to landline Internet services, the use of wireless technologies also shows a phenomenal increase in telecommunication accessibility (Fosu Ignatius, 2016). Currently, there are almost 60,000 Internet hosts in the country. In terms of Internet hosting, Ghana is rated 102nd in terms of the number of IP addresses assigned (countryipblocks.net, 2016).

In 2016, the number of authorized Internet providers stood at 54 according to Ghana National Communication Authority (NCA). The Internet Service Providers (ISP) in the country are on the increase for the past several years. Africa Online, Ghana Internet Service, and Network Computers, just to mention a few, are some of the first providers of Internet in Ghana.

Recently, others have joined and are bringing more competition into the ISP market in the country (Shemcy Shem, 2018). Internet usage in Ghana is mainly through mobile phones. Due to this, ISPs are growing rapidly to gain a larger market share in the Ghanaian telecommunications market (Shemcy Shem, 2018). The telecommunications sector in Ghana is becoming one of the most attractive destinations for investment in the country. The present government has provided the environment for firms to invest and benefit from various tax reliefs.

11.2 Background and History of Ghana

The name Ghana was changed by its first president, Dr. Kwame Nkrumah, on March 5, 1957 from Gold Coast, the same year that independence was attained. Ghana was the first African country in the Sub-Saharan region to attain independence. There were many colonies that came through to rule the country, and the last was Britain. After gaining independence, Ghana joined the Commonwealth and later earned its republic on July 1, 1960, and it currently constitutes a system of parliament as part of governance.

Ghana has experienced many insurgences through coup d'état which led to overthrowing of the first president in 1966. After the overthrow, there have been further coups, and the last one took place on December 22, 1979. As a result, Flt. Lt. Jerry John Rawlings ruled the country as a military leader for almost 11 years, he was later elected as a civilian president in 1992, and was later re-elected as the president of Ghana.

The democratic system in Ghana has been remarkable over the years. Many successful elections have been conducted, and the presidents were elected peacefully. Currently, Ghana stands tall among others in the sub-region, serving as a role model for the other African countries to emulate.

The Government of Ghana has instituted a system that provides control of operators and entrepreneurs in the IT sector. The government has developed an open market allowing private individuals and organizations to invest in the IT sector. The IT sector in Ghana is highly controlled and monitored by the National Communication Authority. IT operators and new entrants are obliged to have or obtain a license to operate in the country. The vision of the Government of Ghana is to use IT to industrialize the country through economic measures that will see more IT firms established in the country.

In Ghana, English is considered an official language, whereas many other languages, such as Twi, Ewe, Ga, and Hausa are frequently spoken and used in schools, businesses, and government offices. The gross domestic product (GDP) of the country in the first quarter of 2018 increased by 1.5% from the previous quarter. The Ghana's Growth Rate was cupped at an average of 1.83% from the year 2006 to 2017, with the highest rate of 7.4% in the initial quarter of 2011. The growth rate is largely due to the sustainable level of demand of the domestic service segment and the manufacturing industry.

11.3   Information Technology in Ghana

Ghana is a developing country that is doing its best to use IT as a necessary foundation for its medium- and long-term socio-economic growth and progress. Subsequently, Ghana had setup policies and strategies to encourage the use of IT in the country. Currently, there are 120 large IT companies providing ICT services which have around 300 employees each. There are also many self-employed IT workers in Ghana (Trade Ghana, 2018). Many IT workers hold professional certifications and undergraduate degrees.

About 78% of IT companies in Ghana focus on hardware and software services, and 22% concentrate on retailing of IT products and communication services and various IT peripherals. Most of the IT products are imported into the country instead of being locally manufactured (IT News Africa, 2017). A unique aspect of IT companies in Ghana is that many of them, especially the leading firms, are located in the city of Accra and have smaller satellite stations in other regions in the country. Ghana can also boast a number of multinational top IT firms, such as Samsung, Dell, and Vodafone that have established their presence and partnered with local firms to provide services on their behalf.

The IT sector in Ghana only contributes about 2.2% to the national GDP, and this represents a minute fraction of Africa's IT market. The IT sector in Ghana is growing at its slow pace (National Communications Authority, 2017). Nevertheless, IT is at the very foundation of the country's economy. Currently, the competitive nature of firms in Ghana, other African countries, and the western world determines the success of the Ghana's IT sector.

The number of IT firms in Ghana is steadily growing. For example, the government has developed a massive data center and is using it extensively throughout the country. The number of IT employees in the health sector in the country has recently quadrupled. The presence of consulting IT firms has drastically increased. Many multinational companies have entered the country by establishing in-house facilities in the country's capital. These in-house centers are highly connected or linked to their Global In-house Centers where they handle the core aspect of IT operations and innovations. Nevertheless, this represents a major achievement for the local economy because it attracts both financial and intellectual capital into the country. The workers in the IT sector are presumed to enjoy a high pay as compared to other employment sectors. The number of IT service firms in Ghana has been steadily growing for the last 10 years (Statista, 2017).

Currently, a need for IT professionals has led many colleges and universities to offer IT programs with various options to choose from. However, due to a recent trend of Ghana's analytics economy, the IT courses introduced into the system are relatively new to most universities and most of the professionals depend on foreign courses for their certification. The general IT workforce in Ghana is made of certificate holders and those with undergraduate degrees.

11.4 Methodology

In Ghana, the original instrument developed for the World IT Project (Palvia *et al.*, 2017; Palvia *et al.*, 2018) was used with no translation or modifications. Data collection was done by including almost all of the industry sectors in Ghana. The researcher (i.e., the first author who did data collection) encountered many difficulties in the course of the distribution of questionnaires to respondents, as well as in collecting back the questionnaires. The response rate was not quite encouraging, and the turnout was very low because many respondents refused sharing information with outsiders. To overcome these obstacles, the researcher approached top executives directly by presenting an official introduction letter and a reference which explained the purpose of the study and the strategy for data collection. The respondents were ensured that the data collected would be for research purposes only, and their identities would not be disclosed or mentioned in publications. In many cases, most of the large corporations gave their permission to collect data from their employees.

Ghana is noted for adopting English as the official and national language for communication. For all the employees of the firms visited, English was used for internal and external communication, and in all schools in Ghana, the language of instruction is strictly English. The institutions were mostly based in the metropolitan. The questionnaires were given to the executives who forwarded them to their IT workers. They were given ample time to

Table 11.1: Descriptive Statistics

Characteristics	N	%	Characteristics	N	%
Education:			Years of work experience:		
High school or less	31	10.20	0–4 years	146	47.71
Associate degree	60	19.74	5–9 years	95	31.05
Bachelor's degree	126	41.45	10–19 years	38	12.42
Master's degree	87	28.62	20–29 years	14	4.58
Ph.D.	0	0.00	30+ years	11	3.59
Years of IT experience:			Organizational location:		
0–4 years	134	44.08	IT department employee	124	40.79
5–9 years	120	39.47	IT worker in non-IT department	58	19.08
10–19 years	35	11.51	Contract employee	97	31.91
20–29 years	15	4.93	Consultant	12	3.95
30+ years	0	0.00	Vendor employee	13	4.28
Work as:			Managerial description:		
Mostly full time	242	79.61	Not part of management	107	35.20
Mostly part time	48	15.79	In lower management	72	23.68
Mostly over time	14	4.61	In middle management	70	23.03
Been laid off from IT job:			In senior management	55	18.09
Yes	33	10.86			
No	271	89.14			

complete the survey instruments and return them to the researcher, and this was monitored by the head of IT. Furthermore, at certain times, IT departments' heads were contacted by phone or email to facilitate the completion of the questionnaires. No gifts or incentives were offered to coerce any respondents or to influence their decision to complete the survey. For the period from 2014 to 2015, out of 400 distributed questionnaires, 304 were returned fully completed and used for analysis. The data were later entered into an Excel spreadsheet, and all the entered data were validated for consistency and correctness.

Table 11.1 presents the descriptive statistics of IT professionals who partook in the study. It describes the education of respondents, years of IT experience, and the nature and employment status in their respective organizations.

11.5   Organizational IT Issues

Table 11.2 indicates organizational IT issues in Ghana, and these issues are ranked on a Likert-type scale in order of importance. It must be noted that

Table 11.2: Organization Issues and IT in Ghana

IT Issues in Organization	Ranking	Mean Rating*	Std. Deviation
Security and privacy	1	1.80	1.09
Business productivity and cost reduction	2	1.83	0.76
Project management	2	1.83	0.78
IT reliability and efficiency	4	1.95	1.23
Business agility and speed to market	5	1.99	0.94
IT strategic planning	6	2.00	0.88
Knowledge management	7	2.01	0.86
Attracting and retaining IT professionals	7	2.01	0.89
Alignment between IT and business	7	2.01	0.79
Business process reengineering	10	2.11	0.84
IT service management (e.g., ITIL)	11	2.15	1.04
Revenue-generating IT innovations	12	2.18	0.99
Globalization	13	2.21	1.12
IT cost reduction	14	2.21	0.84
Continuity planning and disaster recovery	15	2.26	1.04
Enterprise Architecture	16	2.62	0.84
Outsourcing	17	2.73	1.09
Bring your own computing device (BYOD)	18	3.03	1.1

*Rating scale ranges from 1 to 5: 1 as most important and 5 as no importance.

lower average scores stand for higher agreement with the statements. The top issue includes Security and Privacy, similar to the issue in the top five of the 2017 SIM IT Key Issues and Trends Study (Johnson *et al.*, 2018). This means that the entire world is concerned about security and privacy when it comes to the use of IT. In Ghana, ensuring security and privacy is a major issue when dealing with users or customers, as people demand the assurance and high level of security measures on the work being done and carried across the dispersed locations in the country. Surprisingly, only three issues are included in both lists of top 10 items.

Security and privacy is followed by business productivity and cost reduction as the second most important issue. Project management and IT reliability and efficiency followed third and fourth, respectively. The next two issues were made up of IT strategic planning and knowledge management. The result obtained from the table shows that there was a tie among attracting and retaining IT professional, alignment between IT and business, and knowledge management.

The second issue is business productivity and cost reduction and project management. This means that the introduction of IT into the domain of the Ghanaian market system and all other sectors of the country has really brought about an increase in the work output. The use of IT has made work easier and more comfortable. IT has also helped businesses to operate more efficiently and effectively. Currently, most of the companies in Ghana are conducting their business online and this has brought about the growth of IT usage in the business circles in the country.

The use of IT in organizations and businesses in the country has really brought about reduction of cost in the operational aspects of business. Currently, the cost of operation has significantly decreased, and the realized benefits are channeled to highly profitable ventures. For instance, the Government of Ghana has embarked on large usage of electronic registration of businesses.

Project management shows that the Ghanaian IT professionals are concerned about project management practices in their organizations, and they are looking for ways to improve this activity. This finding is not surprising given a high level of project failures in the IT sector in general, and this study shows that Ghana is not an exception. Nevertheless, the Ghanaian IT workers recognize this issue and are likely looking for ways to improve their project management practices. The fourth highly ranked outcome was IT reliability and efficiency. The least important on the table were bring your own computing device (BYOD), outsourcing, enterprise architecture, continuity planning and disaster recovery, IT cost reduction, and globalization.

11.6　Technology and Infrastructure Issues

Table 11.3 displays technology and infrastructure concerns about IT in Ghana. The top issue was business intelligence/analytics (BI/BA) because of a high demand for BI/BA professionals. Both small and large IT firms are providing BI/BA services in Ghana, and the demand is very strong. The related services are notably artificial intelligence (AI)-enabled BI/BA applications, and the use of such technologies assists in decision-making in the areas of customer involvement, information sharing, marketing, and sales. To address the shortage of skilled BI/BA professionals in Ghana, educational institutions have gone a long way to provide very important support to the IT industry through the revision of their IT curriculum to meet the ever changing needs of the industry.

Customer relationship management (CRM) systems were ranked second highest. Enterprise application integration, mobile and wireless applications, and networks/telecommunications are among the top five issues.

Table 11.3:   Technology Issues

IT Related Issues	Rating	Mean Rating*	Std. Deviation
Business intelligence/analytics	1	1.92	0.91
Customer relationship management (CRM) systems	2	2.05	1.21
Enterprise application integration	3	2.07	0.83
Mobile and wireless applications	4	2.15	1.04
Networks/telecommunications	5	2.22	1.04
Software as a service	6	2.32	1.05
Business process management systems	6	2.32	0.91
Mobile apps development	8	2.39	1.11
Data mining	9	2.41	1.05
Enterprise resource planning (ERP) systems	9	2.41	1.05
Big data systems	11	2.41	0.97
Virtualization (desktop or server)	12	2.42	1.11
Collaborative and workflow tools	13	2.43	1.24
Social networking/media	14	2.46	1.13
Service-oriented architecture (SOA)	15	2.57	1.19
Cloud computing	16	2.58	0.97

*Rating scale ranges from 1 to 5: 1 as most important and 5 as no importance.

The related systems provide a great contribution to the development and safeguarding of services rendered by companies in Ghana. The importance of mobile application development results from the rise in the use of mobile phones and services both domestically and globally. The cost of mobile Internet has begun to decrease due to internal competition among mobile service providers. A sizable number of consumers are able to access the Internet only through mobile channels, and this has led to the emergence of IT companies that provide products and services primarily for the mobile market. Software as a service (SaaS) and cloud computing are among the top 10 issues. This reflects the global trend towards outsourcing software and hardware platforms and applications.

11.7   Individual IT Employee Issues

Table 11.4 offers a summary of individual IT employee issues.

First of all, IT workers in Ghana are only marginally satisfied with their jobs. The outcome is consistent with the recent Statista report indicating that about 75% of employees in the country are somewhat satisfied with their jobs (Statista, 2017). Work pressure of the IT employees showed an acceptable level. The IT employees were not perturbed about any issues.

Table 11.4:   Individual IT Employee Issues in Ghana

Individual Issues	Mean Rating*	Std. Deviation
Job satisfaction		
In general, I like working here.	2.08	1.09
All in all, I am satisfied with my current job.	2.67	1.17
In general, I don't like my current job.	3.50	1.23
Work pressure		
I feel that the number of requests, problems or complaints that I deal with at work is more than expected.	2.91	1.32
I feel that the amount of work I do interferes with how well it is done.	3.05	0.89
I feel busy or rushed at work.	3.22	1.04
I feel pressured at work.	3.05	1.17
Work–life balance		
There is a blurring of boundaries between my job and my home life.	3.11	0.99
My work-related responsibilities create conflicts with my home responsibilities.	3.47	0.96
I feel drained from activities at work.	3.31	0.97
I feel tired from my work activities.	3.15	1.09
Working all day is a strain for me.	3.16	1.15
I feel burned out from my work activities.	3.06	0.89
Sense of accomplishment		
I feel I am making an effective contribution to what this organization does.	2.00	0.81
In my opinion, I do a good job.	1.95	0.97
I have accomplished many worthwhile things in this job.	2.16	1.01
At my work, I feel confident that I am effective at getting things done.	2.14	1.19
Threats to one's job		
I am worried that future technology advancements may pose a threat to my job.	3.63	1.06
I believe that other people may be able to perform my work activities.	2.55	0.97
I am concerned that my job may be eliminated soon.	3.82	1.16
I am concerned that my job may be outsourced soon.	3.86	1.05
Career plans		
I will be with this organization 1 year from now.	2.44	1.24
I will take steps during the next year to secure a job at a different organization.	3.08	1.19
I will be with this organization 5 years from now.	2.95	1.28
I will be working in the IT field 1 year from now.	2.99	1.21
I will take steps during the next year to secure a job outside the IT field.	3.37	0.95
I will be working in the IT field 5 years from now.	3.16	1.08

*Rating scale ranges from 1 to 5: 1 as strongly agree to 5 as strongly disagree.

They do not compromise on the quality of their work and they are not in any way rushed or put under severe pressure at work.

Furthermore, the work–life balance among Ghanaian IT workers is at a healthy level. IT employees in Ghana are able to strike a balance on the nature of work they do at their organizations. There were some boundaries between the personal lives and work responsibilities. Achieving a work–life balance is vital for employees' mental and physical well-being, as this results in higher productivity and satisfaction as well as in the reduction in voluntary turnover. The IT workers experience a good and reasonable level of workload and have an acceptable intensity of burnout. The modern fast-paced workplace environment is accompanied by stress, fatigue, and pressure from the work activities, but this does not seem to be an issue in Ghana.

Most importantly, the sense of accomplishment of the IT workers in Ghana is good. As such, many believe that they are making an effective contribution to their organization, doing a good job, achieving substantial accomplishments, and expressing some degree of confidence in their ability to be effective employees.

The majority of IT professionals in Ghana are not worried about losing their employment due to the development of new technologies and outsourcing trends. They are moderately loyal to their organization but are more loyal to the IT profession. In addition to the other factors, such as high job satisfaction as well as low pressure, work–home conflict, and burnout, Ghana has a relatively high unemployment rate which prevents IT workers from moving between jobs or industries.

11.8 Conclusion

The most important organizational issue was security and privacy, which was consistent with the 2017 SIM IT Issues and Trends Study (Johnson *et al.*, 2018). This was followed by business productivity and cost reduction. Business intelligence/business analytics is the top technology issue. Other important issues pertained to enterprise application integration, mobile and wireless applications and networks/telecommunications.

This shows that the IT industry looks very attractive as its employees earn more than those in other sectors and have more opportunities to improve their skills. These systems provide a great contribution to the development of Ghana, and Ghana is a developing country that is trying to put itself into the limelight. The government of Ghana has seen a need for IT and created the enabling environment to boost the intra- and inter-communication among organizations in the country. This is in consonance with the findings reported for the other countries.

With respect to individual issues, it was observed that IT workers in the country are able to strike a balance between work and life, manage their workload, and avoid burnout from pressure at work. The IT employees see that their contributions to the organization are noticed, and their organizations help them realize their full potential to achieve professional accomplishment.

References

countryipblocks.net. (2016). *Country IP Blocks*. Retrieved August 08, 2018, from Select Formats, www.countryipblocks.net: http://www.countryipblocks.net/country-blocks/select-formats/.

Fosu Ignatius. (2016). "Exploring the potential of wireless technologies to accelerate universal Internet access in Ghana". In F. Ignatius, *Telecommunications Policy*, 35(6), 494–504.

Johnson *et al.* (2018). The 2017 SIM IT Issues and Trends Study. *MIS Quarterly Executive*, 53.

Palvia, P., Jacks, T., Ghosh, J., Licker, P., Romm-Livermore, C., Serenko, A., & Turan, A. H. (2017). The World IT Project: History, trials, tribulations, lessons, and recommendations. *Communications of the Association for Information Systems*, 41(18), 389–413.

Palvia, P., Ghosh, J., Jacks, T., Serenko, A., & Turan, A. (2018). Trekking the globe with the World IT Project. *Journal of Information Technology Case and Application Research*, 20(1), 3-8.

Shemcy Shem (2018). *Internet Service Providers in Ghana 2018*. Retrieved August 08, 2018, from yen.com.gh: https://yen.com.gh/103828

Statista (2017). *Statista*. Retrieved September 22, 2018, from Statista: https://www.statista.com/statistics/447530/employment-by-economic-sector-in-ghana/.

Trade Ghana (2018). *Ghana Trade*. Retrieved September 21, 2018, from Ghana Trade: http://ghanatrade.com.gh/Trading-Tips/internet-penetration-in-ghana.html.

Zurek, K. (2018). Retrieved September 22, 2018, from https://www.graphic.com.gh/news/general-news/over-10-million-ghanaians-using-the-internet-report.html.

Chapter 12

Information Technology Issues in Greece

Alexander Serenko[*,†,§] and Gokul Bhandari[‡,¶]

*University of Toronto, Toronto, Canada
†University of Ontario Institute of Technology, Oshawa, Canada
‡University of Windsor, Windsor, Canada
§a.serenko@utoronto.ca
¶gokul@uwindsor.ca

Summary

This chapter covers the organizational, technological, and individual information technology (IT) issues among IT workers in Greece. The results were obtained from a survey of 106 IT workers in Greece. Top organizational issues identified by IT workers include IT reliability and efficiency, security and privacy, and IT strategic planning. Business intelligence/analytics is the top-ranked technology and infrastructure issue whereas social networking/media is viewed as the least important one. IT workers in Greece seem to be generally satisfied with their jobs and experience moderate work pressure. They also appear to be marginally concerned about job outsourcing or elimination.

12.1   Introduction

In 2010, the Greek economy was almost at the brink of bankruptcy, but recently the country has started showing signs of recovery. Greece is comparable to the UK and Denmark in terms of its investment in business R&D. Cloud computing has become a part of the national e-government strategy (CIA, n.d.). While the future of IT in Greece appears promising, there are few studies exploring the IT issues from organizational, infrastructure, and employee perspectives.

12.2   Country and People Background

Greece is a country with a capitalist economy with 18% of its GDP coming from tourism, and the public sector accounting for 40% of GDP. In April 2010, Greece received the lowest possible credit rating which prompted the IMF and Euro-zone governments to provide Greece with short- and medium-term emergence loans worth of US$147 billion. In 2016, the unemployment rate for youth (15–24 year-olds) was 44.3% and 50.7% for male and female citizens, respectively (CIA, 2018). In 2017, the country's GDP improved, and the unemployment rate decreased.

Greece belongs to the high income category of countries with GNI per capita of US$20,290. As of 2017, it had a population of 10,823,732. To start a business in Greece, an entrepreneur has to spend about 2.2% of per capita income and needs to go through five different procedures. The whole process, on average, takes about 13 days which puts Greece in the 56th position (out of 190) for the ease of starting a business. On the ease of enforcing a business contract, Greece is far behind and ranks 133rd (out of 190 countries). In 2012, an electronic platform was implemented to interconnect government agencies with the goal of simplifying the process of business registration (World Bank, 2018).

Hofstede's 6D model assesses a country's culture relative to other world cultures on six dimensions. The scores for Greece are as follows: power distance (60), individualism (35), masculinity (57), uncertainty avoidance (100), long-term orientation (45), and indulgence (50).

Power distance is defined as "the extent to which the less powerful members of institutions and organizations within a country expect and accept that power is distributed unequally". A score of 60 indicates that the Greeks believe, although marginally, that hierarchy should be respected. *Individualism* refers to "the degree of interdependence a society maintains among its members". A score of 35 suggests that Greece is a collectivist culture. *Masculinity* is a measure of competition, achievement and success. A score of 57 suggests that the Greek society is marginally competitive and success-driven. *Uncertainty avoidance* refers to "the extent to which the members of a culture feel threatened by ambiguous or unknown situations and have created beliefs and institutions that try to avoid these". A score of 100 indicates that the Greeks as a nation are not comfortable with ambiguity and uncertainty at all. *Long-term orientation* is a measure of how every society maintains some links with its own past while dealing with the challenges that lie ahead. Greece scores 45 on this dimension indicating that the country prefers to maintain its culture and tradition and views contemporary changes with scepticism. The dimension of indulgence measures individuals' restraints on their desires and impulses. A score of 50 suggests that the Greeks stand at the middle of indulgence and restraint (Hofstede, 2018).

12.3 Information Technology in Greece

According to the 2016 Global Information Technology Report, Greece ranked 70 out of 139 countries in the Networked Readiness Index with a score of 4.1 with 7 being the highest and 1 being the lowest. The Networked Readiness Index aggregates the scores from the following 10 areas: political and regulatory environment; business and innovation environment; infrastructure; affordability; skills; individual usage; business usage; government usage; economic impacts; and social impacts (Baller *et al.*, 2016).

While the performance of ICT in Greece was negative in 2014–2016, it showed signs of recovery in 2017 with an expected growth of 0.35% in 2018. According to the European IT Observatory's sales figures in 2017, Greece's IT sector accounted for 31% of the ICT market with the remaining 69% coming from the telecommunications sector (Export, 2018).

In Greece, cloud computing is a part of the national e-government strategy with the OpenGov Private Cloud, Greek National Network for Research and Technology (GRNET) Cloud Computing Services, G-cloud, Storm Cloud, and Novoville. Greece belongs to a group of European countries that are considered "well informed" in the adoption of cloud computing by the government. These countries have a well-developed cloud strategy and are planning to implement massive cloud computing in the immediate future (Nanos *et al.*, 2019).

In Greece, about 18% of business R&D is in the ICT sector, which is comparable to the UK and Denmark. Labor productivity in information industries is almost double compared to all other non-agriculture business sectors in Greece (OECD, 2017).

12.4   Methodology

Data was collected by Dr. Spyros Kokolakis and Dr. Maria Karyda, both are with the University of the Aegean, Greece, who served as country investigators. An unmodified version of an online survey of the World IT Project (Palvia *et al.*, 2017; Palvia *et al.*, 2018) was administered in English in about 30 organizations. These organizations were chosen from both the private (approximately 70%) and public (approximately 30%) sectors of the economy. This was done because Greece has a very large, well-developed public sector, especially in the service industry. The following industries were included in data collection: transportation, electronic commerce, education, public administration, consulting, health, software, finance/insurance, etc. A number of senior executives at these organizations were directly contacted by the country investigators who, in turn, forwarded the survey link to their subordinate IT workers and encouraged them to complete the survey instrument. To establish initial contacts with these executives, the country investigators relied on their university's alumni network, postgraduate students, and personal connections. No incentives were offered because most senior executives in Greece favor collaboration with local academics. To ensure a high response rate, an initial invitation was followed by several weekly reminders.

Around 106 valid responses were collected and used in data analysis. Table 12.1 outlines descriptive statistics.

A large majority of the respondents (88.7%) had at least a bachelor's degree. This is not surprising because IT jobs in Greece are likely to be white

Table 12.1: Descriptive Statistics

Characteristics	N	%	Characteristics	N	%
Education:			Years of work experience:		
High school or less	2	1.9	0–4 years	11	10.4
Associate degree	10	9.4	5–9 years	16	15.1
Bachelor's degree	37	34.9	10–19 years	51	48.1
Master's degree	53	50.0	20–29 years	20	18.9
Ph.D.	4	3.8	30+ years	8	7.5
Years of IT experience:			Organization location:		
0–4 years	18	17.0	IT department employee	78	73.6
5–9 years	18	17.0	IT worker in non-IT department	4	3.8
10–19 years	47	44.3	Contract employee	8	7.5
20–29 years	17	16.0	Consultant	14	13.2
30+ years	6	5.7	Vendor employee	2	1.9
Work as:			Work position:		
Mostly full time	85	80.2	Not part of management	43	40.6
Mostly part time	21	19.8	In lower management	25	23.6
Been laid off from a job:			In middle management	24	22.6
			In senior management	14	13.2
Yes	14	13.2			
No	92	86.8			

collar jobs. Although there appears to be a mismatch between university learning and job expectations (Chang, 2015), the IT field changes rapidly and IT workers are often expected to update their skills while on the job. However, performing more ICT-related tasks does not necessarily increase wages significantly in Greece (OECD, 2017). In terms of work experience, 89.6% of them worked for 5 years or more, while slightly less (83%) have worked as IT professionals for at least 5 years.

12.5 Organizational IT Issues

The organizational IT issues perceived by respondents are given in Table 12.2. Top-five organizational IT issues in Greece are: IT reliability and efficiency, security and privacy, IT strategic planning, alignment between IT and business, and knowledge management. For 2016, top-five IT management issues identified by the Society for Information Management (SIM) were alignment of IT with the business, security/privacy, innovation, agility/flexibility (IT), and agility/flexibility

Table 12.2:   Organizational IT issues in Greece

Organizational IT Issues	Rank	Mean Rating*	Std. Deviation
IT reliability and efficiency	1	1.76	0.86
Security and privacy	2	1.82	0.87
IT strategic planning	3	1.88	0.90
Alignment between IT and business	4	1.92	0.85
Knowledge management	5	1.95	1.00
Project management	6	2.04	0.98
IT service management (e.g., ITIL)	7	2.09	0.83
Business agility and speed to market	8	2.11	0.87
Continuity planning and disaster recovery	9	2.11	1.00
Business productivity and cost reduction	10	2.12	0.81
Attracting and retaining IT professionals	11	2.17	0.90
Enterprise architecture	12	2.19	0.83
Revenue-generating IT innovations	13	2.27	0.96
Business process reengineering	14	2.29	0.84
IT cost reduction	15	2.51	0.92
Globalization	16	2.73	1.01

*Rating scale ranges from 1 to 5: 1 as most important and 5 as no importance.

(business) (Kappelman et al., 2017). Therefore, it appears that the critical IT issues in Greece are not very different from those identified by the SIM.

IT cost reduction and globalization are the least pressing organizational IT issues in Greece. Because ICT-related jobs in Greece do not demand significantly higher wages than those in the non-ICT sector, IT cost reduction may not have been a critical concern at this time.

12.6   Technology and Infrastructure Issues

Table 12.3 summarizes the technology and infrastructure issues ranked by the IT workers in Greece. The fairly low standard deviations suggest that there is not much variation in their responses. Business intelligence/analytics, enterprise application integration, and networks/telecommunications are viewed as the top three infrastructure issues in Greece whereas cloud computing, mobile apps development, and social networking/media are considered the three least important issues. One reason why cloud computing was not ranked very high could be the fact that there are already several cloud platforms initiated by the government as a part of the national strategy. Similar to Greece, analytics/business intelligence was also ranked as the most important area that should get more investments in the US in 2017 (Kappelman et al., 2017).

Table 12.3:   Technology and Infrastructure Issues in Greece

IT Related Issues	Rank	Mean Rating*	Std. Deviation
Business intelligence/analytics	1	2.06	0.89
Enterprise application integration	2	2.13	0.95
Networks/telecommunications	3	2.16	1.19
Mobile and wireless applications	4	2.22	1.17
Business process management systems	5	2.26	1.02
Virtualization (desktop or server)	6	2.27	1.04
Software as a service	6	2.27	0.95
Collaborative and workflow tools	8	2.28	0.99
Data mining	9	2.31	1.03
Enterprise resource planning (ERP) systems	10	2.32	1.07
Customer relationship management (CRM) systems	11	2.37	1.12
Big data systems	11	2.37	1.11
Service-oriented architecture (SOA)	13	2.42	1.07
Cloud computing	14	2.44	1.03
Mobile apps development	15	2.49	1.13
Social networking/media	16	2.60	1.06

*Rating scale ranges from 1 to 5: 1 as most important and 5 as no importance.

12.7   Individual IT Employee Issues

Table 12.4 summarizes the issues faced by individual IT employees in Greece. They are categorized into job satisfaction, work pressure, work–life balance, workload and burnout, sense of accomplishment, threat to one's job, and career plans. Overall, IT workers in Greece appear to be happy with their jobs and their industry, and they are successful in managing job expectations, among other commitments, at least in the short term.

As mentioned in the previous section, Greece is a collectivist culture with a marginal drive for competition and success. This could be the reason why the IT workers in Greece appear to be generally satisfied. In our survey, 83% of the participants have experience of working in the IT industry for 5 years or more. Individuals with significant experience in their jobs tend to be satisfied because they move up in their career ladder commanding more respect and salary increase for their knowledge and experience.

IT workers in Greece seem to manage the work-related pressure reasonably well. They are less inclined to believe that the amount of problems or complaints they encounter in their work is more than expected. They also do not necessarily feel busy or rushed at work. One explanation for their stoicism could be that Greece had been the cradle of western philosophy.

Table 12.4: Individual IT Employee Issues in Greece

Individual Issues	Mean Rating*	Std. Deviation
Job satisfaction		
In general, I like working here.	2.04	0.93
All in all, I am satisfied with my current job.	2.25	0.94
In general, I don't like my current job.	4.08	1.07
Work pressure		
I feel that the number of requests, problems or complaints that I deal with at work is more than expected.	3.04	1.14
I feel that the amount of work I do interferes with how well it is done.	2.59	0.99
I feel busy or rushed at work.	2.71	1.05
I feel pressured at work.	2.97	1.04
Work–life balance		
There is a blurring of boundaries between my job and my home life.	3.22	1.12
My work-related responsibilities create conflicts with my home responsibilities.	3.52	1.07
I do not get everything done at home because I find myself completing job-related work.	3.26	1.17
Workload and burnout		
I feel drained from activities at work.	3.10	1.07
I feel tired from my work activities.	3.13	1.11
Working all day is a strain for me.	2.93	1.11
I feel burned out from my work activities.	3.52	1.03
Sense of accomplishment		
I feel I'm making an effective contribution to what this organization does.	2.25	0.90
In my opinion, I do a good job.	1.97	0.88
I have accomplished many worthwhile things in this job.	2.11	0.95
At my work, I feel confident that I am effective at getting things done.	2.03	0.91
Threat to one's Job		
I am worried that future technology advancements may pose a threat to my job.	3.85	0.99
I believe that other people may be able to perform my work activities.	2.85	0.97
I am concerned that my job may be eliminated soon.	3.92	1.09
I am concerned that my job may be outsourced soon.	3.89	1.01
Career plans		
I will be with this organization 1 year from now.	2.11	1.18
I will take steps during the next year to secure a job at a different organization.	3.61	1.25

(*Continued*)

Table 12.4: (*Continued*)

Individual Issues	Mean Rating*	Std. Deviation
I will be with this organization 5 years from now.	2.55	1.12
I will be working in the IT field 1 year from now.	2.05	1.21
I will take steps during the next year to secure a job outside the IT field.	3.83	1.23
I will be working in the IT field 5 years from now.	2.22	1.19

*Rating scale ranges from 1 to 5: 1 as strongly agree, to 5 as strongly disagree.

The thoughts and wisdom of ancient philosophers such as Socrates, Plato, Aristotle, Pythagoras, and many others may still be shaping and influencing the temperament of the Greeks today.

In terms of work–life balance, the Greeks seem to have done phenomenally well. This is not surprising given that Greece has a score of 50 in Hofstede's dimension of indulgence which is exactly at the middle in the indulgence-restraint spectrum. IT workers in Greece are not inclined to feel drained or tired from their job-related activities. This conclusion is consistent with the fact that they have balanced their work and life reasonably well and that they don't feel any pressure in their work.

On average, IT workers in Greece have positive self-esteem and a sense of accomplishment. The country scores 35 on the Hofstede's masculinity dimension suggesting that the Greeks are generally content people and do not exhibit a high competitive drive.

IT workers in Greece do not seem to be worried about the prospect of losing their jobs either because of the advent of new technologies or outsourcing. This complacence is explained by the fact that the Greeks tend to maintain their culture and tradition, and view any changes (real or projected) with scepticism (Hofstede's score of 45 in the dimension of long-term orientation). This outlook is consistent with the fact that the IT workers in Greece would not be looking for a new job or are not looking to join a new organization within the next 5 years of their career.

12.8 Conclusion

Greece is expected to move forward in IT sector in a sustainable manner as evidenced by its workers who feel satisfied with their jobs. Their satisfaction arises from the fact that they are able to maintain a balance between their work and personal life, and that they enjoy some level of job security without losing a sense of accomplishment. In Greece, IT reliability and efficiency are the most important organizational issues, while business intelligence/analytics is ranked as number one in the technology sector.

References

Baller, S., Dutta, S., & Lanvin, B. (2016). *Global Information Technology Report 2016*. Ouranos.

Central Intelligence Agency (CIA) (n.d.). World Factbook: Europe Greece. Retrieved December 02, 2018, from https://www.cia.gov/library/publica tions/the-world-factbook/geos/gr.html.

Chang, F. (2015). Educational Resources, Job Match, and Employment Outcomes in Greece. *TEPS-B Working Paper*, (2015–02). Retrieved from http://teps b.nccu.edu.tw/download/TEPS-B_working_paper_201502.

Export (n.d.) Greece — Information and Communications Technology. Retrieved November 05, 2018, from https://www.export.gov/article?id=Greece-Info rmation-and-communications-Technology.

Hofstede Insights (n.d.). Country Comparison. Retrieved November 2, 2018, from https://www.hofstede-insights.com/country-comparison/greece/.

Kappelman, L., Nguyen, Q., McLean, E., Maurer, C., Johnson, V., Snyder, M., & Torres, R. (2017). The 2016 SIM IT Issues and Trends Study. *MIS Quarterly Executive*, 16(1).

Nanos, I., Manthou, V., & Androutsou, E. (2019). Cloud Computing Adoption Decision in E-Government. In *Operational Research in the Digital Era–ICT Challenges*. Springer, Cham, pp. 125–145.

OECD (Nov, 2017). Highlights from the OECD Science, Technology and Industry Scoreboard 2017 — The Digital Transformation: Greece. Retrieved November 10, 2018, from https://www.oecd.org/greece/sti-scoreboard-20 17-greece.pdf.

Palvia, P., Jacks, T., Ghosh, J., Licker, P., Romm-Livermore, C., Serenko, A., & Turan, A. H. (2017). The World IT Project: History, trials, tribulations, lessons, and recommendations. *Communications of the Association for Information Systems*, 41(18), 389–413.

Palvia, P., Ghosh, J., Jacks, T., Serenko, A., & Turan, A. (2018). Trekking the globe with the World IT Project. *Journal of Information Technology Case and Application Research*, 20(1), 3–8.

World Bank (n.d.) Doing Business 2017: Economy Profile 2017 Greece. Retrieved November 5, 2018, from https://openknowledge.worldbank.org/bitstream/ handle/10986/25530/WP-DB17-PUBLIC-Greece.pdf.

Chapter 13

Information Technology Issues in Hungary

Tamara Keszey*,‡, Ádám Katona*,§, and Prashant Palvia†,¶

*Corvinus University of Budapest,
Budapest, Hungary
† University of North Carolina at Greensboro,
Greensboro, NC, USA
‡ tamara.keszey@uni-corvinus.hu
§ adam.katona@appsite.hu
¶ pcpalvia@uncg.edu

Summary

The central theme that arises from the study of the information technology (IT) industry in Hungary is the current and future potential of the sector. The IT industry has already been a prominent contributor to the country's GDP, has the highest employment share within the economy and offers one of the highest starting salaries. Our study however highlights that IT is still regarded as a function dedicated to support day-by-day effective operations of the firm. Hence, establishment of a novel view regarding IT as a future-oriented, cutting edge strategic tool to manage not only in-firm but also firm-external stakeholder relations (i.e., big data analysis, managing customers by means of mobile apps, data mining, cloud computing, etc.) may open up new avenues of development for IT in Hungary.

13.1 Introduction

Hungary, being a former "Soviet Bloc" country, moved from a communist style central planning system to a free market system (Roztocki and Weistroffer, 2008). The information technology (IT) sector has been a catalyst in this process of transition and joining the bloodstream of international economy. Nowadays, the IT sector in Hungary plays an important role in the economy in terms of adding value to the GDP and employment rates. The residues of 40 years of central economic planning engendered significant hurdles to the diffusion of IT knowledge. For example, survey results at the time of early transition showed that the least important managerial capability contributing to managerial success were IT skills (Zoltay-Paprika, 2002).

Managers and firms operating in Hungary have undergone positive significant changes in terms of IT capabilities. However, compared to economically more developed countries, there are still lags (Roztocki and Weistroffer, 2008). Recent studies show that the number of firms using modern enterprise systems is increasing, with 50% of the companies using modular enterprise systems in 2011, and 37% using social networks for business in 2017, compared to EU average of 45% (Drótos and Móricz, 2012; Eurostat Statistics, 2017a). According to a representative study of firms with more than 50 employees, managers do not consider IT to be a central driver of competitiveness, hence, compared to firms operating in more developed economies, they spend less on IT (Drótos and Móricz, 2012). Hungarian firms are aware of forthcoming IT-related major challenges that will affect their firms' competitiveness, but their current practices are not in line with these future expectations, their attitudes are yet not translated down to behavior.

13.2 Country Background and History

Hungary is a country in Central Europe with 10 million inhabitants. Its official language is Hungarian, a Finno-Ugric one. Hungary joined the European Union in 2004, and became a member of the United Nations, NATO, World Bank, World Trade Organization, the Asian Infrastructure Investment Bank, the Council of Europe, and the Visegrád Group.

In the first quarter of 2018, Hungary's GDP growth rate was 4.4%, and the most important sectors were industry (including food processing, pharmaceuticals, motor vehicles, IT, and chemicals), wholesale and retail trade, transport, accommodation and food services. Both in terms of size and population, Hungary is a medium-sized country of the European Union and in terms of economic power, it is among the laggards in Europe (European Union, 2018; Eurostat Statistics, 2017b).

The history of Hungary dates back a long time. As a federation of united tribes, Hungary was established in 895. Hungary reached one of its greatest extents during the Árpádian kings, yet royal power was weakened at the end of their rule in 1301. After the last strong king of medieval Hungary, its international role declined and political stability was shaken. The resulting degradation of order opened avenues for the Turkish occupation. With the conquest of Buda by the Turks in mid-16th century, 1541, Hungary was divided into three parts and remained so for 150 years, until the end of the 17th century when it was freed from Turks by the Habsburgs, who later ascended the throne in Hungary. Later on, Hungary became a partner in the Austro-Hungarian empire in the mid-19th century. After a period of turmoil following World War I, an independent kingdom of Hungary was established. The redrawing of European borders that took place after World War I left about 5 million ethnic Hungarians living in neighboring countries. Following World War II, the country found itself under communist rule for four decades. An uprising against Soviet domination in 1956 was crushed by Soviet Red Army forces, but Hungary did later become the first Eastern European country to gain some economic freedom. Hungary played a pivotal role in the collapse of communism across Eastern Europe when it opened its border with Austria. On 23rd October 1989, the Hungarian Republic was proclaimed and in March 1990 the first democratic elections were held. Hungary became a free, democratic European republic. In 1999, Hungary joined NATO and the European Union in 2004. It has been part of the Schengen Area since 2007. Hungary is well known for its rich cultural history and has contributed significantly to arts, music, literature, sports, science and technology.

13.3 Information Technology in the Country

The Information and Communication Technology (ICT) sector in Hungary is an important driver of the economy. Compared to OECD countries, the sector's value added to the GDP is remarkably high with 7.3% of total value added, that exceeds the OECD average of 6.0% (OECD, 2017) and countries such as the UK, Germany, France, Ireland or the Netherlands. The productivity of the ICT sector compared to the total economy is 1.3% in Hungary, that is slightly below EU average of 1.5% (EU Science Hub, 2018). These results indicate that this industry is very important for the Hungarian economy, more productive than other sectors in the country, but its overall productivity is below the EU average. The fact that the growth rate of value added by the ICT sector is slightly higher (4.7%) than the EU average of 3.1% signals that slowly but steadily it is catching up in terms of productivity. Majority of the value added by the sector is produced by foreign firms operating in Hungary (OECD, 2017). While foreign firms employ 59.5% of the sectors, they produce 77% of value added to the GDP compared to all (Hungarian, foreign, state-owned) firms (OECD, 2017).

In Europe, Hungary has one of the highest ICT sector employment share with 4.1% if we take all sectors (both for-profit and non-profit) and 6.2% if we count the for-profit sector only (Fernandez de Guevara *et al.*, 2017). IT specialists are the fourth best paid professionals in Hungary and IT jobs offer the highest starting salaries for fresh graduates. The average monthly salary of a senior manager in the ICT sector was about 3500 euros in 2015 (Bell Research, 2015). However, according to a recent survey, there are 22,000 vacant jobs available on the Hungarian IT market (Bell Research, 2015). Despite the favorable income situation, the sector is struggling with significant labor shortages due to large number of IT professionals working abroad (typically in Western-Europe where the average salaries are double the Hungarian salary) and to lower academic output than demand. This labor shortage inhibits growth in the sector and endangers competitiveness.

13.4   Methodology

The data for this study were collected in the first quarter of 2016 by means of a survey that was sent to a sample of companies in Hungary listed in a commercial database (Bisnode) via e-mail with the option to administer the questionnaire online. The original instrument (see Palvia *et al.*, 2017; Palvia *et al.*, 2018) was adopted from English version. To improve the response rate, follow-up phone calls were made to inquire whether the questionnaire had reached a competent key respondent and to gain information about the causes of non-response. Key respondents were IT employees in various positions. No incentives or gifts were offered to the respondents for completing the survey, but the objective, academic nature, and confidentiality of the

Table 13.1: Descriptive Statistics

Characteristics	N	%	Characteristics	N	%
Education:			Years of work experience:		
High school or less	12	4.4	0–4 years	11	4.0
Associate degree	58	21.2	5–9 years	22	8.1
Bachelor's degree	126	46.2	10–19 years	84	30.8
Master's degree	76	27.8	20–29 years	123	45.1
Ph.D.	1	.4	30+ years	33	12.1
Years of IT experience:			Organizational location:		
0–4 years	11	4.0	IT department employee	167	61.2
5–9 years	22	8.1	IT worker in non-IT department	10	3.7
10–19 years	84	30.8	Contract employee	77	28.2
20–29 years	123	45.1	Consultant	19	7.0
30+ years	33	12.1	Vendor employee	0	.0
Work as:			Work position:		
Mostly full time	265	97.1	Not part of management	167	61.2
Mostly part time	7	2.6	In lower management	10	3.7
Mostly over time	1	.4	In middle management	77	28.2
Been laid off from IT job:			In senior management	19	7.0
Yes	39	14.3			
No	234	85.7			

data were clearly explained in the accompanying letter, and an online support e-mail as well as a contact phone number were provided for subsequent questions related to data gathering. We sent out 2600 e mails, and the data collection resulted in 273 usable questionnaires and an effective response rate of 10.5%. Survey data were entered into an Excel spreadsheet directly from the questionnaires. The entered data were validated by checking every 10th questionnaire for accuracy.

Table 13.1 shows the descriptive statistics of the IT professionals who took part in the study from Hungary.

13.5 Organizational IT Issues

Table 13.2 shows the organizational IT issues from the respondents from Hungary ranked in order of importance. Two items in the top five, security and privacy (security, cybersecurity, privacy), and alignment between IT and business (alignment of IT with the business) match items (in parentheses) in the top five of IT Key Issues and Trends study from the 2017 US study (Kappelman *et al.*, 2018). The "missing" three of the top five items from the study of Kappelman *et al.* (2018) are innovation that ranks 12th,

Table 13.2: Organizational IT Issues in Hungary

Organizational IT Issues	Rank	Mean Rating*	Std. Deviation
IT reliability and efficiency	1	1.65	0.65
Security and privacy	2	1.70	0.67
Alignment between IT and business	3	1.79	0.71
Knowledge management	4	1.85	0.74
Attracting and retaining IT professionals	5	1.87	0.80
Continuity planning and disaster recovery	6	1.88	0.90
IT service management (e.g., ITIL)	7	1.92	0.73
Project management	8	1.92	0.76
Networks/telecommunications	9	1.93	0.77
Virtualization (desktop or server)	10	1.95	0.84
Enterprise application integration	11	1.97	0.80
Revenue-generating IT innovations	12	1.97	0.67
IT strategic planning	13	2.06	0.80
Business agility and speed to market	14	2.08	0.77
Software as a service	15	2.09	0.88
Business process reengineering	16	2.12	0.74
Mobile and wireless applications	17	2.14	0.97
Collaborative and workflow tools	18	2.17	0.88

*Rating scale ranges from 1 to 5: 1 as most important and 5 as no importance.

and business/IT flexibility and agility that ranks 14th in the Hungarian sample. Surprisingly, cost reduction that has become an increasingly important IT management issue over the past 3 years in the US according to the members of Society for Information Management, does not even fit into the top-18 issues in Hungary.

The Hungarian survey depicts that operational, every-day management of organizational IT issues are the top priorities, with IT reliability and efficiency being the most important one, followed by issues such as alignment between IT and business, knowledge and project management, and IT service management among the top 10.

Data security-related issues are acknowledged as being important, with security and privacy ranked among top three, and continuity planning and disaster recovery being among the top-10 ranked issues. In harmony with employment issues in the IT sector in Hungary, attracting and retaining IT professionals ranks as one of the five most important issues.

Issues related to IT as a future-oriented strategic enabler are regarded as being less important compared to operational IT issues. For example, IT strategic planning ranks 13th, with business process reengineering being the 16th only. While IT can be an important catalyst of innovation, items related to IT innovations are laggards in the list. Revenue-generating IT innovations and business agility and speed to market ranks 12 and 14, subsequently.

Technology-intensive IT issues — such as mobile and wireless applications, collaborative workflow tools, virtualization — can also be found towards the bottom of Table 13.2. The survey also measured challenges related to big data, mobile apps development, globalizations, but these do not qualify in Hungary among the top-18 issues.

In summary, the most important IT organizational challenges in Hungary are related to the short-term, day-to-day, operational, effective and efficient management of IT, providing data security and solving human resources issues. More strategic, future-oriented goals, such as innovation, rethinking the role of IT as a service function and business enabler receive less attention. In a similar vein, Hungarian organizations seem to pay less attention to newer technologies such as mobile apps, virtualization, etc., hence, they may be laggards in adopting technological innovations

13.6   Technology and Infrastructure Issues

Taking a closer look into the technology and infrastructure issues in Hungary (Table 13.3), it can be seen that compared to organizational issues presented in Table 13.2, these are somewhat perceived as being less important. While survey results based mainly on data from the US claim the largest IT investments in 2017 are related to data analytics, business intelligence, data mining forecasting and big data (Kappelman *et al.*, 2018), results from Hungary show that the role of IT is perceived to be the most important in enabling intra-organizational communication and networking. Each of the six more important technology and infrastructure issues are related to this broader domain, with the most important being networks and telecommunication, followed by items such as collaborative workflow tools, mobile applications, and application integration. This result is very interesting in the light of international results, highlighting that only 2.7% of IT professionals claim that investments related to networks and telecommunications should get top investment priority (Kappelman *et al.*, 2018).

Issues that follow the first block of items related to network and communication in Hungary are related to supporting business functions (i.e., business process management systems, customer-relationship management, service-oriented architecture). These items suggest that Hungarian

Table 13.3:   Technology and Infrastructure Issues in Hungary

IT Related Issues	Rank	Mean Rating*	Std. Deviation
Networks/telecommunications	1	1.93	0.77
Virtualization (desktop or server)	2	1.95	0.84
Enterprise application integration	3	1.97	0.80
Software as a service	4	2.09	0.88
Mobile and wireless applications	5	2.14	0.97
Collaborative and workflow tools	6	2.17	0.88
Business process management systems	7	2.19	0.91
Service-oriented architecture (SOA)	8	2.19	0.90
Customer relationship management (CRM) systems	9	2.27	0.93
Business intelligence/analytics	10	2.32	0.90
Mobile apps development	11	2.40	1.04
Enterprise resource planning (ERP) systems	12	2.41	1.00
Data mining	13	2.61	0.97
Social networking/media	14	2.64	1.08
Cloud computing	15	2.67	1.09
Big data systems	16	2.75	1.03

*Rating scale ranges from 1 to 5: 1 as most important and 5 as no importance.

managers perceive IT as an important technology and infrastructure tool for business process management.

International top investment priorities of big data, cloud computing only appear at the end of our list (worth noting that the hot topic of big data is considered as a less important technology and infrastructure issue in Hungary). Items related to the role of IT as a pivotal element of firms' capabilities to manage customers (i.e., social networking, media, mobile apps development) are also ranked among the least important technology and infrastructure issues.

In summary, the technology and infrastructure issues ranked highly by the respondents in Hungary reflect that managers look at IT in an old-school manner, primarily as an enabler of communication and networking within the firm, while value-added and innovative modes (i.e., big data analysis, managing customers, etc.) of IT are regarded as being less important.

13.7   Individual IT Employee Issues

Table 13.4 shows the summary of the responses about individual issues. Note that lower average score means higher agreement with each statement. The results highlight that IT employees are typically satisfied with

Table 13.4:   Individual IT Employee Issues in Hungary

Individual Issues	Mean Rating*	Std. Deviation
Job satisfaction		
In general, I like working here.	1.87	0.86
All in all, I am satisfied with my current job.	1.99	0.92
In general, I don't like my current job.	3.96	1.04
Work pressure		
I feel that the number of requests, problems or complaints that I deal with at work is more than expected.	3.37	1.12
I feel that the amount of work I do interferes with how well it is done.	3.21	1.20
I feel busy or rushed at work.	3.40	1.08
I feel pressured at work.	3.07	1.17
Work–life balance		
There is a blurring of boundaries between my job and my home life.	3.29	1.18
My work-related responsibilities create conflicts with my home responsibilities.	3.25	1.17
I do not get everything done at home because I find myself completing job-related work.	3.39	1.18
Workload and burnout		
I feel drained from activities at work.	3.10	1.19
I feel tired from my work activities.	3.08	1.19
Working all day is a strain for me.	3.07	1.26
I feel burned out from my work activities.	3.75	1.15
Sense of accomplishment		
I feel I'm making an effective contribution to what this organization does.	1.87	0.76
In my opinion, I do a good job.	1.92	0.78
I have accomplished many worthwhile things in this job.	1.98	0.89
At my work, I feel confident that I am effective at getting things done.	1.89	0.75
Threats to one's job		
I am worried that future technology advancements may pose a threat to my job.	3.85	1.05
I believe that other people may be able to perform my work activities.	2.75	1.13
I am concerned that my job may be eliminated soon.	3.84	1.11
I am concerned that my job may be outsourced soon.	4.32	0.89
Career plans		
I will be with this organization 1 year from now.	2.00	0.98
I will take steps during the next year to secure a job at a different organization.	4.07	1.00

(*Continued*)

Table 13.4:   (*Continued*)

Individual Issues	Mean Rating*	Std. Deviation
I will be with this organization 5 years from now.	2.53	1.24
I will be working in the IT field 1 year from now.	1.87	1.02
I will take steps during the next year to secure a job outside the IT field.	4.26	0.99
I will be working in the IT field 5 years from now.	2.02	0.99

*Rating scale ranges from 1 to 5: 1 as strongly agree and 5 as strongly disagree.

their jobs and enjoy working in their current positions. Work pressure seems to be manageable; IT professionals rather disagree than agree with being busy, rushed, and pressured at work. Level of work–life balance is comparable to work pressure; Hungarian IT employees moderately disagree with the notion that boundaries are blurring between job and home life. Work load also does not seem to be a major issue in IT-related jobs. Employees moderately disagree with becoming tired of work activities, but the item measuring burnout scores higher (lower agreement) than items referring to volume of workload. This survey of Hungarian employees depicts an optimistic view of self-accomplishment, with typical positive evaluation of effective organizational contribution. Items related to profession-related threats receive low agreement scores. Thus, IT professionals do not worry about their job being eliminated or outsourced. Most respondents consider IT career as a long-term employment goal.

Employment in the IT industry is an appealing career path in Hungary. This sector offers one of the highest starting salaries for young professionals, with the opportunity of fast progress. IT is one of the five best paid professions. Although, compared to other career paths, salaries in the IT sector are comparable to Western-European countries and developed economies, many professionals choose to pursue a career abroad. Hence, the number of vacancies in the sector is relatively high. This is reflected in our survey's scores showing that IT professionals in Hungary are not very concerned about their job being eliminated.

While many IT professionals work abroad, paradoxically, internal work-related mobility within Hungary is traditionally low. Most people tend to work in their vicinity and professionals migrate rarely from one region to another for the sake of a job. This might also explain our results in terms of IT professionals planning to keep on working at their current organization as well as remaining within the sector.

13.8 Conclusion

The top organizational IT concerns in Hungary are effectively and securely managing IT systems and ensuring adequate levels of employment to enable operations. The top technology and infrastructure issues are related to telecommunication and networking, integration of enterprise applications, virtualization and wireless applications. Regarding both organizational IT concerns and technology-related issues, the emphasis is on the provision of day-to-day operations, while long-term and innovative issues are perceived as being less important. Employees indicate that the IT career continues to be very attractive. IT professionals are in general satisfied with their jobs, seem to contribute and add value to the organization, consider both workload and work–life balance manageable, and are optimistic about their continued role in the sector.

References

Bell Research (2015). Hungarian information and IT engineering situation, problems and gazior factors' examination - ivsz.hu. (January 1, 2015). Retrieved August 5, 2019, from http://ivsz.hu/wp-content/uploads/2016/03/a-haza i-informatikus-es-it-mernokkepzes-helyzetenek-probemainak-gatlo-tenyez oinek-vizsgalata.pdf.

Drótos, G., & Móricz, P. (2012). An information technology and the companies' competitiveness during the financial and economic crisis. *Management Review*, 42(2), 80–89.

EU Science Hub (2018). ICT sector analisys 2017. Retrieved August 5, 2019, from https://ec.europa.eu/jrc/en/predict/editions/2018.

European Union (2018). Living in the EU. Retrieved August 5, 2019, from https://europa.eu/european-union/about-eu/figures/living_en.

Eurostat Statistics (2017a). Digital economy and society statistics — Enterprises. Retrieved August 5, 2019, from https://ec.europa.eu/eurostat/statistics-e xplained/index.php/Digital_economy_and_society_statistics_-_enterprises.

Eurostat Statistics (2017b). Volume indices of GDP per capita, 2017 (EU-28=100).

Fernandez de Guevara, J., Robledo, J. C., & López-Cobo, M. (2017). An Analysis of ICT R&D in the EU and Beyond: European Comission — JRC Science for Policy Report.

Kappelman, L., Johnson, V., McLean, E., & Maurer, C. (2018). The 2017 SIM IT Issues and Trends Study. *MISQ Exec*, 17(1), 53–88.

OECD (2017). Digital Economy Outlook, Retrieved August 5, 2019, from https://www.oecd-ilibrary.org/science-and-technology/oecd-digital-economy-outlook-2017_9789264276284-en.

Palvia, P., Jacks, T., Ghosh, J., Licker, P., Romm-Livermore, C., Serenko, A., & Turan, A. H. (2017). The World IT Project: History, trials, tribulations, lessons, and recommendations. *Communications of the Association for Information Systems*, 41(18), 389–413.

Palvia, P., Ghosh, J., Jacks, T., Serenko, A., & Turan, A. (2018). Trekking the globe with the World IT Project. *Journal of Information Technology Case and Application Research*, 20(1), 3–8.

Roztocki, N., & Weistroffer, H. R. (2008). Information technology in transition economies. *Journal of Global Information Technology Management*, 11(4), 1–8.

Zoltay-Paprika, Z. (2002). The competitiveness of Hungarian managers. In A. Chikán, E. Czakó & Z. Zoltay-Paprika (Eds.), *National Competitiveness in Global Economy — The Case of Hungary*, Vol. 4. Budapest: Akadémiai Kiadó.

Chapter 14

Information Technology Issues in India

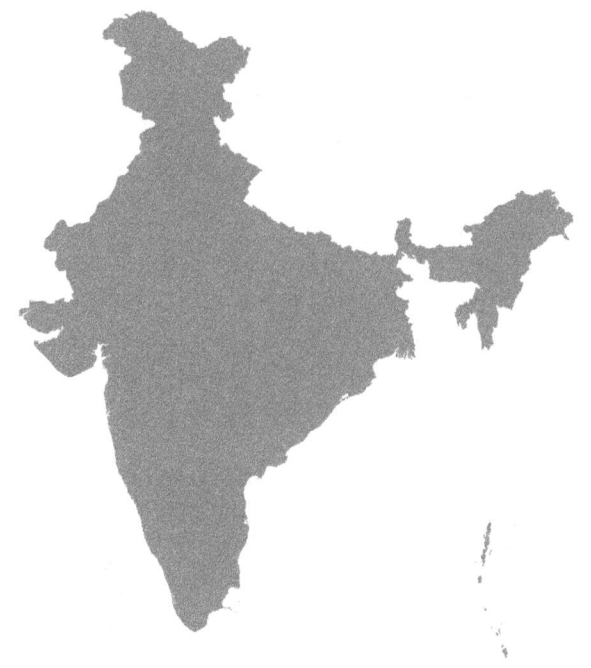

Jaideep Ghosh[*,§], Dolphy Abraham[†,¶], and Prashant Palvia[‡,||]

[*]*Shiv Nadar University, Uttar Pradesh, India*
[†]*Independent Researcher, Bengaluru, India*
[‡]*University of North Carolina at Greensboro, Greensboro, NC, USA*
[§]*jghosh20770@gmail.com*
[¶]*dolphyabraham@gmail.com*
[||]*pcpalvia@uncg.edu*

Summary

The central theme that arises from the study of the information technology (IT) industry in India is the importance of the size and adaptability of the IT workforce in the country. The prominence that Indian companies receive globally is due primarily to their ability to provide knowledgeable and skilled IT workers to clients who have a lot of projects that need to be carried out. The IT industry in India has also been able to adapt fairly well to the changing needs of their global clients. Global businesses have not only been outsourcing their IT work to Indian IT service companies over the past three decades, but they have also started to locate in-house units in India in order to make appropriate use of the availability of this sizable and skilled workforce.

14.1 Introduction

The majority of the mature Indian information technology (IT)-service companies were incorporated in the mid-1970s. This industry experienced a major boom in the early-1990s, because of skyrocketing demands for IT services in the developed economies, where companies exploited opportunities to leverage the services offered by the Indian firms with their cost-effective, skilled human capital and improved IT-related infrastructure. Although the leading Indian IT vendor firms initially lacked a good understanding of clients' business requirements and technical needs, they quickly acquired maturity and started to get more involved in diverse high-level operations, including analysis, design, and business processes (Arora *et al.*, 2001; Kaiser and Hawk, 2004). Today, Indian IT vendors have become leaders in the international sourcing market, providing end-to-end, diversified service solutions to their clients all over the world (Lewin *et al.*, 2009; Palvia and Palvia, 2017). In numerous large-scale ventures, they work in full collaboration with their clients in Europe and the US (Ethiraj *et al.*, 2005). Recent NASSCOM (2018) and IBEF (2018) reports estimate that the IT-BPM sector in India has expanded at a compound annual growth rate of nearly 14% over the period 2010–2016, which is about three times higher than the global growth rate. Besides, India is still the most attractive sourcing destination in the world, with approximately 56% of the market share in the global sourcing services and equipped with over 5 million digitally skilled employees, a large number of innovation centers, technology parks, incubation labs, and design studios (NASSCOM, 2017; IBEF, 2018).

14.2 Country Background and History

Today, India is the largest democracy and the second-most populous nation in the world (with a population close to 1.4 billion). It is also the seventh-largest country by area. It has a very ancient history, with the highly urbanized Harappan or Indus Valley Civilization as its earliest known civilization that flourished around the period 2500–1700 BC. The British arrived in India in the early 17th century and initially established a number of trading outposts under the British East India Company. The company rule continued until 1858, when direct administrative control was transferred to the British Crown in the name of the British Raj. India earned independence from the British in 1947, when it joined the Commonwealth. It earned the status of a federal republic in 1950 with a parliamentary system of government. India went through a post-independence government-controlled, closed-market economic phase, and severely restricted licensing policy (jocularly called the "License Raj"), internal long-term planning of core industries (the "5-year Plans"), aggressive market liberalization in the early 1990s, and the current phase of economic globalization.

English is a widely used lingua franca in the country, and Hindi is the national language. In the first quarter of 2018, India's GDP growth rate has amounted to 7.7% on an annual basis, and the rate averaged 6.2% over a period of approximately 66 years, starting from 1951. The rapid growth rate is sustained by high demands in the domestic and global service sectors as well as by manufacturing and investments growth at an accelerating pace in recent years.

14.3 Information Technology in India

The IT industry in India gained its global exposure in the mid-late 1990s by being able to provide the talent needed to handle the Y2K problem. The closed nature of the centrally-planned Indian economy till 1991 made it necessary for Indian IT companies to use legacy software and systems. That situation also required the Indian IT companies to develop talent to maintain and modify those legacy systems. When the Y2K problem created the demand for IT professionals who could work on legacy applications, the Indian IT companies were able to meet that demand. The successful completion of those Y2K projects subsequently provided the Indian IT companies numerous contracts for newer technologies, specifically, systems integration and Internet-based systems during the early 2000s.

The Indian IT companies utilized a global or distributed delivery model that leveraged the lower cost of IT professionals working in India. A small proportion of the employees were based at client sites dealing primarily with sales (client acquisition and management) and systems analysis work. The bulk of the work was handled by employees based at various locations in India. The contracts were priced based on the volume of work to be done. The larger clients had large scale projects requiring a lot of people; as a result, the value of the projects became tied to the number of people working on the project. This led the Indian IT companies on a path of linear growth, where growth in revenues was in direct proportion to the growth in the headcount at these companies. The larger IT companies in India grew rapidly as a result. For instance, the number of employees at Tata Consultancy Services (TCS) more than tripled from about 45,000 employees in 2005 to over 143,000 in 2009 and to about 387,000 by the end of 2017 (Statista, 2018). Indian companies providing Business Process Outsourcing (BPO) services have also been growing rapidly since the turn of the century.

More recently, there has been a rapid growth in the number of multi-national companies setting up in-house centers or captive units in India. These Global In-house Centers (GICs) are more likely to focus on core IT operations and IT-based innovations compared to the back-office operations that were handled by the IT service companies. The growth in the number of GICs has resulted in the formation of clusters with a variety of companies providing complementary IT talent to the GICs. For instance, automotive GICs are clustered in Pune and GICs of financial companies are located primarily in Mumbai. Bengaluru has the largest concentration of GICs representing a variety of industries.

14.4   Methodology

Collecting primary data on IT workers in India for the World IT Project (Palvia *et al.*, 2017; Palvia *et al.*, 2018) was not an easy task. Some of the common problems encountered at the initial stages of data collection included low response rates, distrust of research objectives, and a general reluctance to share information with outsiders. To overcome these limitations, the investigators directly approached a few of the high-level executives in a number of firms through official letters of introduction and reference, in which we clearly explained the research objectives and the data collection plan. In a large number of other firms, access was provide to collect data from the employees by upper level executives, to whom we were introduced by our personal contacts in the industry. Observations for a total of 350 respondents were collected over the period 2012–2014 inclusive. The participants were assured that the data would be used at an aggregate level for research purposes, and their identities would not be disclosed in any form whatsoever.

In India, Hindi is considered the national language, although not all people from all states speak Hindi, and the official language varies from one state to another. English has become the lingua franca, and a large number of schools in urban India use it as the medium of instruction. Since the company offices from where we collected the data were located in the major metropolitan cities, we administered the survey instrument in its original English language version. The questionnaires were given to the individual managers in the companies in sealed envelopes to be distributed to the IT employees working directly under them, with the arrangement that, when completed, they were to be returned to us in envelopes duly sealed and signed by the same managers themselves. In some instances, a few managers were contacted by phone and email to remind them to return the completed surveys to us on time. On occasion, phone calls were made to resolve ambiguities in the responses. No incentives or gifts were provided to the respondents for completing the survey, but the objective of the project was clearly explained to their managers, and with their help, subsequently to the IT employees.

The two country investigators (the first two co-authors) divided the data collection endeavor almost equally between them. They constantly kept in touch with each other through phone and email and shared their experiences on the progress of their data-collecting efforts. Each investigator entered the data into Excel spreadsheets directly from the completed questionnaires. The entered data were subsequently validated and checked for accuracy.

Table 14.1 shows the descriptive statistics of the IT professionals who took part in the study from India. Note that the totals may be less than 350 due to some missing data.

14.5   Organizational IT Issues

Table 14.2 shows the organizational IT issues from the respondents from India ranked in order of importance. Only one item in the top five, namely security and privacy, matches an item in the top five of the 2017 SIM IT

Table 14.1:   Descriptive Statistics

Characteristics	N	%	Characteristics	N	%
Education:			Years of work experience:		
High school or less	0	0.0	0–4 years	40	11.6
Associate degree	6	1.7	5–9 years	129	37.4
Bachelor's degree	159	46.2	10–19 years	130	37.7
Master's degree	177	51.5	20–29 years	33	9.6
Ph.D.	2	0.6	30+ Years	13	3.8
Years of IT experience:			Organizational location:		
0–4 years	63	18.3	IT department employee	260	75.8
5–9 years	138	40.1	IT worker in non-IT department	58	16.9
10–19 years	111	32.3	Contract employee	4	1.2
20–29 years	29	8.4	Consultant	18	5.2
30+ years	3	0.9	Vendor employee	3	0.9
Work as:			Work position:		
Mostly full time	339	98.3	Not part of management	235	68.3
Mostly part time	0	0.0	In lower management	30	8.7
Mostly over time	6	1.7	In middle management	58	16.9
Been laid off from IT job:			In senior management	21	6.1
Yes	14	4.1			
No	328	95.9			

Table 14.2:   Organizational IT Issues in India

Organizational IT Issues	Rank	Mean Rating*	Std. Deviation
Security and privacy	1	1.27	0.52
Attracting and retaining IT professionals	2	1.30	0.61
IT reliability and efficiency	3	1.31	0.55
Project management	4	1.31	0.58
Knowledge management	5	1.32	0.58
IT strategic planning	6	1.34	0.60
Revenue-generating IT innovations	7	1.34	0.61
Alignment between IT and business	8	1.37	0.62
Business agility and speed to market	9	1.38	0.56
Continuity planning and disaster recovery	10	1.41	0.71
Enterprise architecture	11	1.55	1.28
IT service management (e.g., ITIL)	12	1.57	0.72
Globalization	13	1.57	0.77
Business productivity and cost reduction	14	1.78	0.72
Outsourcing	15	2.02	0.89
Business process reengineering	16	2.03	0.85
IT cost reduction	17	2.12	0.93
BYOD (Bring Your Own Computing Device)	18	2.95	1.09

*Rating scale ranges from 1 to 5: 1 as most important and 5 as no importance.

Key Issues and Trends study (Kappelman *et al.*, 2018). Further only three items in the top 10 match items in the top 10 of the 2017 SIM issues list.

The highest-ranked item is security and privacy, which also appears first in the 2017 SIM issues list. This issue is faced by IT companies across the world and is not specific to any one country. For Indian IT organizations, this is a key issue because their clients demand a high level of security on the work that is being carried out in globally distributed locations.

The second highest-ranked issue is that of attracting and retaining IT professionals. In the larger IT service companies, the size of the projects drove the need to keep a large number of employees who were readily available to be deployed as new projects came on line. This required the larger Indian IT service companies to focus on managing human resources as a core competency. The larger companies were required to hire tens of thousands of employees each year while trying to control the number of employees leaving the firm. This demand for good employees meant that it was easy for employees to switch companies quickly. The larger companies resorted to hiring fresh college graduates in large numbers in anticipation of work being assigned to them once initial training was completed. This bulk hiring approach was the common practice till the 2009 recession and the change in business outlook during 2017–2018. For instance, TCS, the largest of the Indian IT companies, added fewer than 3600 employees during 2017–2018 compared to almost 25,000 in the previous year (Statista, 2018).

The reduction in the growth of the workforce during 2017–2018 was also due to the changes in the political climate in the US. A substantial portion of the revenues of the larger IT service companies are based on projects with US clients. The change in visa regulations has meant that these companies have a much harder time sending enough people to staff projects at client locations. These companies have resorted to hiring more employees locally to avoid the impact of such regulations.

For the smaller IT service companies, getting newer projects to maintain a continuous stream of work was the primary concern. The clients of these companies ranged from mid-sized to smaller firms. These client firms were more sensitive to changes in the business climate and were likely to cut IT budgets when the outlook was poor. The smaller companies had to ensure that there was a good pipeline of projects. As a result, the headcounts at these firms were more closely managed, as they could not pay employees who were not billable for a long time. The smaller companies were also more likely to be specialized in a specific set of IT tools.

The sustained demand for skilled IT workers has meant that the good employees could demand a higher rate of growth in their salaries. As a result, the growth in labor costs for the Indian IT companies was much higher than the increase in labor costs in other countries. In addition, the inflation rate in India is much higher than that of the developed economies across the world. Lower cost and lower margin back office work has not been sustainable due to the rise in costs and has moved out of India to lower cost locations such as the Philippines. The difference in labor costs is not the real driver for locating in India, rather it is the availability of skilled talent.

IT reliability and efficiency was ranked the third highest. For Indian IT firms, the number of employees who are on billable projects drives their efficiency. The larger IT companies whose clients are primarily the Fortune 500 size companies have projects or engagements that are large in scope. The smaller IT service companies, on the other hand, service numerous smaller clients and have smaller projects in comparison. For both types of IT service companies, the value being provided to clients is a combination of the programming skills being provided at a lower price point. For GICs, the need was to build teams with skills in the specific technologies required by a company and to transform those teams as the technologies continue to change.

Project management and knowledge management round out the top-five issues. Both issues relate to the ability of the IT organization to deliver services on time, within budget and at a high level of quality. As such, these two issues are closely linked to the IT reliability and efficiency issue which is ranked immediately higher. As client requirements change, the IT organization has to be able to respond to the changes rapidly; this creates a requirement for new knowledge to be disseminated effectively. An outcome of the demand for talent is that the mobility of employees is very high. Employee turnover rates typically exceed 20% for newer employees (first 3–4 years of their career) each year at the larger IT service companies. The ability to move between companies easily for better prospects is one reason that IT workers are general happy with their career choice. However, the easy mobility means that IT companies have to manage more carefully the competency of project teams and ensure that projects are completed on time and within budget despite changes in the composition of the team.

One reason for the significant difference between the SIM list and those identified in India is that a large proportion of IT workers in India is likely to work for IT service companies than internal IT departments. Following the 2009 recession, there has been an increase in global in-house centers (GICs) as companies across the world have started to set up internal IT units in India to leverage the availability of human capital to innovate their processes and products. Consequently, the establishment and growth of clusters of GICs have increased during the 2010s. As the employees of GICs increase in proportion to the total number of IT workers in India, the organization IT issues identified as being important are likely to change over time.

In summary, the IT services companies and global in-house centers have had to place a great emphasis on managing the HR. For all types of IT service companies and GICs, the need to control labor costs has been of great importance. In an industry that does not depend on physical capital, it is the human capital that determines the sustained success of the firm. Thus, four of the top organizational IT issues are related to managing human capital.

14.6 Technology and Infrastructure Issues

Table 14.3 shows the technology and infrastructure issues from India ranked in order of importance. Two of the top five responses from India match those

Table 14.3: Technology and Infrastructure Issues in India

IT Related Issues	Rank	Mean Rating*	Std. Deviation
Business intelligence/analytics	1	1.39	0.60
Networks/telecommunications	2	1.41	0.69
Customer relationship management (CRM)	3	1.41	0.62
Enterprise application integration	4	1.47	0.72
Enterprise resource planning (ERP) systems	5	1.47	0.74
Software as a service	6	1.48	0.72
Collaborative and workflow tools	7	1.57	0.69
Business process management systems	8	1.60	0.79
Social networking/media	9	1.64	0.96
Mobile and wireless applications	10	1.78	0.87
Big data systems	11	1.93	0.76
Virtualization (desktop or server)	12	1.95	0.79
Service-oriented architecture (SOA)	13	1.97	0.80
Data mining	14	1.97	0.85
Cloud computing	15	2.03	0.86
Mobile apps development	16	2.04	0.93

*Rating scale ranges from 1 to 5: 1 as most important and 5 as no importance.

in the top five of the 2017 SIM IT Key Issues and Trends study (Kappelman *et al.*, 2018). Five of the top 10 in the India list match the top 10 in the US list. Overall, it can be surmised that the technology and infrastructure issues identified in India show some level of similarity to the 2017 SIM list.

The top issue identified in both lists is analytics and business intelligence. This is an area of expertise where the demand for services is beginning to grow. For the Indian IT service companies, developing this expertise is an important goal and they have been able to ramp up their capabilities in the last few years. Large IT service companies as well as smaller boutique consulting firms provide analytics-related services. The related area of artificial intelligence (AI) applications has also been addressed. The ability to provide skills in analytics and AI are one reason for the recent growth in GICs in India. For instance, several of the large retailers from across the world have set up GICs to enable them to develop teams skilled in data analytics and AI. Their internal efforts are used to help decisions in merchandising, distribution and customer engagement.

To provide services in analytics, business intelligence and AI, it was necessary for the Indian IT companies to be able to hire individuals with such skills. Educational institutions have provided significant support to the IT industry through the continued revision of IT curricula to meet the changing needs of the industry. With every change in the technologies demanded by the IT industry and their clients, institutions of higher education have been fairly quick to respond to those changes. Analytics and AI are two areas in which the education sector has ramped up the ability to provide new employees with the requisite skills.

Networking and telecommunications was the second highest rated technology and infrastructure issue in India. This is an area of concern for the IT industry because the primary customer base is outside India, and reliable networks are critical to success. India as a whole has excellent connectivity through multiple undersea cables with the rest of the world. The redundancy in cables allows for very good reliability. This has allowed Indian companies to guarantee services to their clients even when the servers and project teams are located in India.

There are two sides to the IT infrastructure story. In most large cities, where the IT companies are located, the IT infrastructure is good and can support the demand. However, in smaller towns and in rural areas, the IT infrastructure is not as strong or reliable. This is true for both wired and wireless networks. The cost of mobile networks has been steadily declining especially since 2016 primarily due to the entry of a well-funded new competitor. Since 2016, there has been a substantial increase in consumers shifting to smartphones on 4G networks. However, access to wired broadband lags behind given the geography of the country.

The application areas of customer relationship management (CRM) systems, enterprise application integration and enterprise resource planning (ERP) systems round out the top-five issues. These areas constitute the majority of the development and maintenances services provided by the Indian companies. Software as a service (SaaS) and cloud computing also appear among the top-15 issues identified by the Indian respondents. These areas are not the key specializations of Indian service companies even though they do provide services in these areas. Indian IT companies are not major providers of software and hardware and are more likely to provide them from other vendors. Collaboration/workflow systems were ranked as seventh and business process management at eighth. These applications are important for Indian IT companies to complete projects successfully. Social networking/media was ranked ninth and reflects the demand for such applications from global customers.

Mobile and wireless applications appears at the tenth position while mobile application development appears in the fifteenth place. The importance of this area not only reflects the growth in mobile applications globally but also in the Indian market. The cost of mobile and internet access is dropping steadily due to increased domestic competition among mobile service providers. A substantial portion of the local consumer base tends to access the Internet only through mobile channels. This has led to the development of IT companies that provide products and services primarily for the mobile market.

In summary, the technology and infrastructure issues ranked highly by the respondents in India reflect the business model in use by the majority of the Indian IT industry, namely, providing IT services to global customers.

14.7 Individual IT Employee Issues

Table 14.4 shows the summary of the responses about individual issues. Note that once again a lower average score means higher agreement with each statement. The responses indicate that the group of workers agree that they are moderately happy with their work despite indicating that they do face significant work pressure although not excessively. They also seem to handle the workload without experiencing burnout. Furthermore, the IT employees felt a strong sense of accomplishment and did not feel threats

Table 14.4:   Individual IT Employee Issues in India

Individual Issues	Mean Rating*	Std. Deviation
Job satisfaction		
In general, I like working here.	1.48	0.66
All in all, I am satisfied with my current job.	1.60	0.79
In general, I don't like my current job.	4.48	0.84
Work pressure		
I feel that the number of requests, problems or complaints that I deal with at work is more than expected.	3.30	0.91
I feel that the amount of work I do interferes with how well it is done.	3.07	1.05
I feel busy or rushed at work.	3.43	1.00
I feel pressured at work.	3.61	0.96
Work–life balance		
There is a blurring of boundaries between my job and my home life.	3.71	1.08
My work-related responsibilities create conflicts with my home responsibilities.	3.82	1.00
I do not get everything done at home because I find myself completing job-related work.	3.76	1.05
Workload and burnout		
I feel drained from activities at work.	3.86	1.01
I feel tired from my work activities.	3.63	0.93
Working all day is a strain for me.	3.67	1.03
I feel burned out from my work activities.	3.81	0.94
Sense of accomplishment		
I feel I'm making an effective contribution to what this organization does.	1.50	0.76
In my opinion, I do a good job.	1.38	0.65
I have accomplished many worthwhile things in this job.	1.67	0.70
At my work, I feel confident that I am effective at getting things done.	1.49	0.59
Threats to one's job		
I am worried that future technology advancements may pose a threat to my job.	3.86	1.02
I believe that other people may be able to perform my work activities.	3.37	1.05
I am concerned that my job may be eliminated soon.	4.27	0.93
I am concerned that my job may be outsourced soon.	4.41	0.85
Career plans		
I will be with this organization 1 year from now.	1.56	0.99
I will take steps during the next year to secure a job at a different organization.	4.18	1.04
I will be with this organization 5 years from now.	2.73	0.95

(*Continued*)

Table 14.4: (*Continued*)

Individual Issues	Mean Rating*	Std. Deviation
I will be working in the IT field 1 year from now.	1.43	0.81
I will take steps during the next year to secure a job outside the IT field.	4.36	0.96
I will be working in the IT field 5 years from now.	1.89	0.97

*Rating scale ranges from 1 to 5: 1 as strongly agree, to 5 as strongly disagree.

to their jobs. While they do not foresee being with the same employer in 5 years time, they expect to continue in the IT industry.

Employment in the IT industry in India continues to be a very attractive career path for young people. The availability of well-paying jobs along with the opportunity to travel to client sites across the world and work with the latest information technologies all combine to attract large numbers of high school and college graduates into the profession. Prior to the early 1990s when India was a planned economy, the top students gravitated towards government (public-sector) jobs or opportunities outside India. The growth of the IT industry changed that pattern and it is the case that the IT industry gets a greater proportion of the brighter students.

One aspect that has proven very attractive for newer employees in the IT services companies is the opportunity to travel to client sites in other countries. Membership on globally distributed project teams gave those individuals the ability to travel outside India and, for some, the opportunity to settle in developed economies. As a large proportion of the employees in the IT service companies is relatively young, this international exposure is a significant draw.

The demand for international assignments has muted in recent years with the growth of domestic e-commerce and digital enterprises during the 2010s. Another reason for the increasing appeal of work locations in India is that the cohort of the workforce in the early part of the century wanted to have more stability as they started to have families. The strong inter-generational bonds that are common across Indian families draw employees to locate closer to their parents and extended family members. With children and families, the toll of travel to client sites becomes onerous and the stability of locating at an Indian site is more attractive.

For employees of IT service companies, the global standards of work to which they have to adhere has been a useful experience. The ideas of professional management and monitoring quality of work and output have spread from the IT industry to companies in other industries as well. At the present time, it is common that in Indian companies, the work practices that are followed match the practices in developed economies to a great extent.

While employment in the IT sector continues to be very attractive, there are some negative impacts faced by the employees. One of these is the need to work in time slots that match the client sites instead of the local time. Project teams have to align with the time zone of the client so that interaction between the developers in India, members of the team at the client site, as well as with the employees of the client is possible. For instance, employees at the Indian company would start work around 5:00 p.m. and end work around 2:30 a.m. as it matches with the eastern-time zone of the US client. This type of work pattern puts a lot of pressure on individuals and can lead to health problems. While employees could handle this work pattern when they were younger, it has a negative impact on those individuals as they grow older.

In recent years, the lure of start-ups has drawn employees away from IT service companies. The IT skills required for digital start-ups are in huge demand and a stake in the ownership of the start-up is very attractive to a lot of individuals. The combination of the work pressure typical in IT projects along with the availability of opportunities elsewhere has made the management of employee turnover in projects and companies a difficult task. In summary, employees of IT service companies and GICs in India find themselves among the privileged members of society. Their incomes are much higher than most other industries and the opportunities available to them in the IT industry are greater than in other industries.

14.8   Conclusion

The top organizational concerns in IT in India are security and privacy, attracting and retaining IT employees, and reliability and efficiency of IT systems. Top technology concerns include business intelligence and analytics, telecommunications, and customer relationship management systems. Employees indicate that the IT industry continues to be a very attractive career option. They are, in general, happy with their jobs and expect to continue in the industry though they are likely to move to other companies. Thus, while the IT industry is thriving, managing the rich human capital in India is among the most important issues confronting the industry.

References

Arora, A., Arunachalam, V. S., Asundi, J., & Fernandes R. (2001). The Indian software services industry. *Research Policy*, 30(8), 1267–1287.

Ethiraj, S. K., Kale, P., Krishnan, M. S., & Singh, J. V. (2005). Where do capabilities come from and how do they matter? A study in the software services industry. *Strategic Management Journal*, 26(1), 25–45.

IBEF (2018). https://www.ibef.org. Accessed April 2018.

Kaiser, K. M., & Hawk, S. (2004). Evolution of offshore software development: From outsourcing to cosourcing, *MIS Quarterly Executive*, 3(2), 69–81.

Kappelman, L., Johnson, V., McLean, E., & Maurer, C. (2018). The 2017 SIM IT Issues and Trends Study. *MISQ Exec*, 17(1), 53–88.

Lewin, A. Y., Massini, S., & Peeters, C. (2009). Why are companies offshoring innovation? The emerging global race for talent. *Journal of International Business Studies*, 40(6), 901–925.

NASSCOM (2018). http://www.nasscom.in. Accessed April 2018.

Palvia, S. C. J., & Palvia, P. (2017). *Global Sourcing of Services: Strategies, Issues and Challenges*. Singapore: World Scientific.

Palvia, P., Jacks, T., Ghosh, J., Licker, P., Romm-Livermore, C., Serenko, A., & Turan, A. H. (2017). The World IT Project: History, trials, tribulations, lessons, and recommendations. *Communications of the Association for Information Systems*, 41(18), 389–413.

Palvia, P., Ghosh, J., Jacks, T., Serenko, A., & Turan, A. (2018). Trekking the globe with the World IT Project. *Journal of Information Technology Case and Application Research*, 20(1), 3–8.

Statista (2018). https://www.statista.com/statistics/328244/tcs-employees-numbers. Accessed April 2018.

Chapter 15

Information Technology Issues in Iran

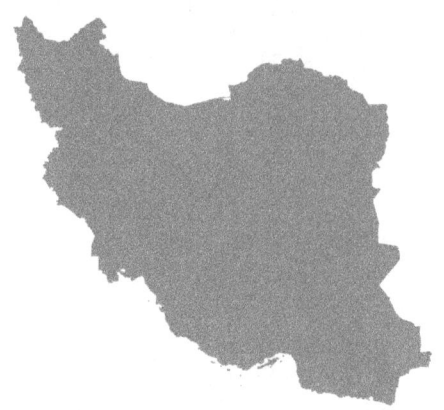

Jaideep Ghosh*,§, Dolphy Abraham†,¶, Prashant Palvia‡,||,
and Hamid Nemati‡,**

*Shiv Nadar University, Greater Noida,
Uttar Pradesh, India
†Independent Researcher, Bangalore, India
‡University of North Carolina at Greensboro,
Greensboro, NC, USA
§jghosh20770@gmail.com
¶dolphyabraham@gmail.com
||pcpalvia@uncg.edu
**hrnemati@uncg.edu

Summary

The importance of the information technology/information technology enabled services (IT/ITES) sector in modern Iran is indubitable. As IT has started to penetrate deeper into the country in recent years, IT and telecommunications infrastructure continue to be developed and enhanced throughout Iran. This expansion is creating many job prospects for employees in Iran, as well as providing greater availability of e-business, e-commerce, e-learning, and e-government in the country. Improving the quality of education and healthcare is also influenced by the growing importance of IT in the country. As demonstrated in this chapter, the three top IT-related organizational issues in Iran are: revenue-generating IT innovations, business agility and speed to market, and alignment between IT and business. The top three technology issues are: enterprise application integration, networks and telecommunications, and ERP systems. The IT employees indicate they are moderately happy with their work despite facing significant work pressure.

15.1   Introduction

Information technology (IT) and information technology enabled services (ITES) have played a major role in the development of modern Iran. The massive efforts in globalization have caused the Iranian government to contend with many technological, economical and infrastructural issues (Asemi, 2006). Among them, IT cost-cutting is a major concern both in the public and the private sectors. Increased employee training leads to more employee opportunities and job satisfaction. There is also evidence that IT/ITES has offered the financial and banking sectors many challenges in both B-2-B and B-2-C transactions, whose successful design and implementation will cause the economy to grow considerably in the near future (Dangolani, 2011). Besides, information and communications technology (ICT) has enriched the education and healthcare sectors as well as empowered (for example, with blogging, hosting personal sites, and so on) many marginalized groups in Iran, particularly women and the younger people (Shirazi, 2012).

Against this backdrop, this chapter explores organizational, technological and individual issues facing IT professionals in Iran.

15.2   Country Background and History

The Islamic Republic of Iran is located in the Middle East with an area of 1,648,195 km^2, constituting the 18th largest land mass in the world with a population of about 84 million making it the 70th most populous country in the world. Iran's neighbors include: Iraq and Turkey in the west; Afghanistan and Pakistan in the east; Armenia, Azerbaijan, Caspian Sea and Turkmenistan to the north; and the Persian Gulf in the south. Iran's population is diverse. Although Persians make up 51% of the population, other minorities include Azeris, with 24%, Gilaki and Mazandarani with 8%, Kurd with 7%, Arab with 3%, Lur with 2%, and Baloch and Turkmen with 2% each. About 68% of Iran's population is concentrated in urban areas, making it one of the most urbanized countries in the Middle East (CIA, 2019). It is estimated that the urban growth rate is about 2.1%, again one of the highest in the region with important consequences for information technology.

Iran's economy is the 18th largest in the world with a Gross Domestic Product (GDP) of $447 Billion (2017 estimate). Iran's economy is composed of 9% in agriculture, 35% in industry and 55% in service. With a GDP per capita of $20,100, Iran is the 89th richest country in the world. It is worth noting that the economy has shifted significantly from agricultural to industrial and service-based. This shift has had a major impact on IT growth and use in Iran. The major industrial sectors of the economy include petroleum, petrochemicals, textiles, and construction. However, the economy is dominated by the public sector with the state owning and administering several large industries, while relying on export earnings from oil and gas for more than 80% of its revenue. Iran is the fifth largest producer of oil and natural gas, respectively, in the world with third and second largest proven oil and natural gas reserves in the world. Iran's overreliance on the oil and gas revenues as a source of income has limited its growth potential and has made it susceptible to fluctuation in the price of oil. Although the economy is dominated by the public sector, the government has recently undertaken a massive privatization policy to divest itself from large portions of the economy. The private sector includes automobile, textile, metal manufacturing, and food-processing factories as well as thousands of small-scale enterprises such as workshops and farms (CIA, 2019).

15.3 Information Technology in Iran

Iran currently has over 31 million telephone land lines and 87 million mobile cellular lines, ranking it the ninth and 18th in the world, respectively. The World Telecommunication Development Conference (site accessed January 2019) estimates that Iran's penetration rate of mobile phones currently at 87% will reach 95% of the country's population by 2020. Recently, Iran's ministry of ICT has stated that a comprehensive development of the country's ICT infrastructure is its main priority and has pledged an ambitious goal that by the end 2020 an estimated 53 million people in Iran will have access to mobile communication and high-speed Internet services.

Demographic forces are the main drivers for the rapid growth of ICT in Iran. Iran's population of 84 million is young, educated and ethnically and linguistically diverse. The population of Iran is one of the youngest in the world with median age of 23 years, and an average age of 27 years. According to a 2017 census report, 24.3% of Iran's population is 14 years of age or younger and only 5.4% is 65 and older. It is estimated that 47% of Iran's 84 million population is under the age of 29, with 35% of population between 15–29 years of age and 14–19 years constituting about 9% of the total population. The youth of Iran is much better educated in comparison to other Middle Eastern countries. According to the World Bank (site accessed January 2019), Iran has a literacy rate of 86.8%. This is 5% age points higher than the average of the Middle East and North African region, making Iran the most educated nation in the Middle East, particularly in terms of the young student population attending universities. Consequently, Iran has a strong tradition of science and engineering education and entrepreneurship, as evident by Iranian technology entrepreneurs, who started companies such as eBay, Dropbox and Zoosk, among others.

Not only educated, the young population in Iran is also very politically active. According to United States Institute of Peace, Iran has the most politically active youth among the 57 nations of the Islamic world (World Bank). Another important contributing factor in the rapid growth of ICT in Iran can be found in the composition of its population. Unlike other Middle Eastern countries, Iran's population is mostly urban. According to the Asian Development Bank, the majority of population in the Middle East lives in rural areas with little or no access to most basic ICT, while in Iran, close to 70% of the population lives in cities and urban areas where access to Internet is more widely available. This factor alone gives Iran an edge over access and availability of ICT over other Middle Eastern countries since the infrastructure needed for ICT distribution and effective usage are more easily implemented when the majority of the population has a higher literacy rate and is located in urban areas.

Iran's ICT and telecommunications industry and infrastructure are almost entirely state-owned allowing the government to restrict access that hampers the free and open exchange of ideas. Despite the pervasive censorship and control imposed by the government, Iranians are very active online. The growth in the number of users of social networking sites such as Facebook, Instagram and instant messaging apps such as Telegram and WhatsApp in Iran is one of the highest in the world. For example, according to *Financial Times* (site accessed January 2019), Instagram has 24 million active users in Iran making the country the seventh largest market in the world. With 48 million Internet users, about 11 million of whom are broadband users, Iran has 61% Internet penetration making it the 17th most wired country in the world and first among the Middle Eastern counties. It is estimated that almost 63% of households own a computer. In addition, there are approximately 1,800 cybercafés mainly in the larger cities of Tehran, Mashhad and Isfahan. Internet-use in Iran has mushroomed over the past few years. The OpenNet Initiative Report (site accessed January 2019) on Iran states that there has been a sharp and perhaps an explosive growth in Internet usage among the Iranian population, much higher than any other Middle Eastern nation. Internet users have increased approximately 48% annually.

15.4 Methodology

The World IT Project (Palvia *et al.*, 2017; Palvia *et al.*, 2018) core team furnished the survey instrument for administering in Iran. While the goal was to obtain 300 responses from IT employees, we obtained 357 responses from IT employees representing various industries and job titles. Table 15.1 shows the descriptive statistics of the IT professionals who took part in the study.

15.5 Organizational IT Issues

Table 15.2 shows the organizational IT issues from the respondents from Iran ranked in order of importance (as per the measurement scale, a lower mean implies a higher rank). Only two of the five highest-ranked items match an item in the top five of the 2017 SIM IT Key Issues and Trends study (Kappelman *et al.*, 2018), i.e., alignment between IT and business, and security and privacy. Five of the top-10 items ranked by the respondents from Iran are common to the top 10 of the 2017 SIM IT Key Issues and Trends study (Kappelman *et al.*, 2018). The items that overlap between the two lists reflect the importance of the business value of IT. The item

Table 15.1:   Descriptive Statistics

Characteristics	N	%	Characteristics	N	%
Education:			Years of work experience:		
High school or less	31	8.7	0–4 years	48	13.4
Associate degree	70	9.6	5–9 years	109	30.5
Bachelor's degree	167	46.8	10–19 years	98	27.5
Master's degree	80	22.4	20–29 years	68	19.0
Ph.D.	9	2.5	30+ years	34	9.5
Years of IT experience:			Reporting relationships:		
0–4 years	66	18.5	IT department employee	246	68.9
5–9 years	103	28.9	IT worker in non-IT department	28	7.8
10–19 years	95	26.6	Contract employee	30	8.4
20–29 years	62	17.4	Consultant	28	7.8
30+ years	31	8.7	Vendor employee	25	7.0
Do you work as:			Work position:		
Mostly full time	301	84.3	Not part of management	229	64.1
Mostly part time	27	7.6	In lower management	84	23.5
Mostly over time	29	8.1	In middle management	30	8.4
Been laid off from IT job:			In senior management	14	3.9
Yes	56	15.7			
No	301	84.3			

Table 15.2:   Organizational IT Issues in Iran

Organizational IT Issues	Rank	Mean Rating*	Std. Deviation
Revenue-generating IT innovations	1	2.06	1.29
Business agility and speed to market	2	2.18	1.41
Alignment between IT and business	3	2.24	1.43
Security and privacy	4	2.37	1.06
Business productivity and cost reduction	5	2.39	1.13
Project management	6	2.42	1.09
IT service management (e.g., ITIL)	7	2.43	1.06
IT strategic planning	8	2.46	0.95
Continuity planning and disaster recovery	9	2.52	1.16
Attracting and retaining IT professionals	10	2.53	1.15
Outsourcing	11	2.54	1.15
Globalization	12	2.55	1.18
Knowledge management	13	2.66	1.03
Business process reengineering	14	2.72	1.14
IT reliability and efficiency	15	2.76	1.02
IT cost reduction	16	2.83	1.07
BYOD (Bring Your Own Computing Device)	17	2.99	1.01
Enterprise architecture	18	3.00	1.15

*Rating scale ranges from 1 to 5: 1 as most important and 5 as no importance.

that was ranked at the top in Iran was revenue generation through IT innovations. However, this item was ranked much lower on the 2017 SIM list. The second highest item indicated the importance of business agility and flexibility through the use of IT. The Iranian respondents indicate that their organizations value the impact of IT on business productivity and flexibility in the near term. IT issues internal to an organization, such as, cost reduction, BYOD or enterprise architecture, were ranked lower by the respondents in Iran. As was the case with many other countries, security and privacy was also ranked very high.

In summary, the organizational issues that were ranked at the top by the respondents from Iran primarily reflect a demand for benefits from the use of IT in the short run. The respondents from Iran indicate that their organizations place a greater importance on how IT improves business performance, while efficient use of IT resources and cost reduction are not prioritized as highly.

15.6 Technology and Infrastructure Issues

Table 15.3 shows the technology and infrastructure issues from the respondents from Iran ranked in order of importance. Two of the top five responses from Iran, i.e., business intelligence/analytics and ERP systems match those in the top five of the 2017 SIM IT Key Issues and

Table 15.3: Technology and Infrastructure Issues in Iran

IT Related Issues	Rank	Mean Rating*	Std. Deviation
Enterprise application integration	1	1.91	1.31
Networks/telecommunications	2	2.33	1.00
Enterprise resource planning (ERP) systems	3	2.45	1.10
Business intelligence/analytics	4	2.50	1.12
Service-oriented architecture (SOA)	5	2.62	1.20
Business process management systems	6	2.88	1.00
Collaborative and workflow tools	7	2.88	1.00
Virtualization (desktop or server)	8	2.88	1.02
Customer relationship management (CRM)	9	2.90	1.01
Cloud computing	10	3.10	1.28
Data mining	11	3.12	0.97
Software as a service	12	3.33	1.28
Mobile and wireless applications	13	3.39	1.19
Big data systems	14	3.40	1.22
Mobile apps development	15	3.48	1.10
Social networking/media	16	3.75	1.41

*Rating scale ranges from 1 to 5: 1 as most important and 5 as no importance.

Trends study (Kappelman *et al.*, 2018). Five out of the top 10 issues in the Iran list match the top 10. These include: business intelligence/analytics, ERP systems, networks/telecommunications, customer relationship management, and cloud computing. Issues important in Iran but not in the US include enterprise application integration, service oriented architecture, business process management systems, and collaborative tools. Overall, it is surmised that while there are similarities between Iran and the US, there are also important differences.

It is important to note that despite this phenomenal growth in the Internet usage, Iran is facing formidable challenges in maintaining such an explosive growth, mainly due to the recent economic embargo imposed on Iran by the US. However, there are signs that with the help of the European Union and China, Iran may be able to sustain this current rate of growth. The non-governmental IT industry in Iran is thriving and the number of technology startups has increased in recent years. Transitioning from governmental ownership of technology-based organization to private ownership, as well as dealing with restriction imposed on these companies as the result of the economic embargo, is evident in rankings shown in Tables 15.2 and 15.3. The rankings show that the salient organizational and technological issues are those considered as important to organizational viability and sustainability, in an environment that is dominated by the government and the restrictions imposed on private organizations.

15.7   Individual IT Employee Issues

Table 15.4 shows the summary of the responses from the Iranian respondents about employee-related issues. The responses indicate a group of workers who agree that they are moderately happy with their work despite indicating that they do face significant work pressure, though perhaps not

Table 15.4:   Individual IT Employee Issues in Iran

IT Employee Related Issues	Mean Rating*	Std. Deviation
Job satisfaction		
In general, I don't like my current job.	3.18	1.17
All in all, I am satisfied with my current job.	2.75	1.10
In general, I like working here.	2.58	1.05

(*Continued*)

Table 15.4:   (*Continued*)

IT Employee Related Issues	Mean Rating*	Std. Deviation
Work pressure		
I feel that the number of requests, problems or complaints that I deal with at work is more than expected.	3.12	1.15
I feel busy or rushed at work.	2.89	1.24
I feel pressured at work.	2.85	1.20
I feel that the amount of work I do interferes with how well it is done.	2.42	1.07
Work–life balance		
There is a blurring of boundaries between my job and my home life.	3.39	1.11
My work-related responsibilities create conflicts with my home responsibilities.	3.29	1.19
I do not get everything done at home because I find myself completing job-related work.	3.20	1.23
Workload and burnout		
Working all day is a strain for me.	2.95	1.22
I feel drained from activities at work.	2.91	1.25
I feel tired from my work activities.	2.88	1.22
I feel burned out from my work activities.	2.85	1.21
Sense of accomplishment		
I have accomplished many worthwhile things in this job.	2.96	1.08
I feel I'm making an effective contribution to what this organization does.	2.96	1.07
At my work, I feel confident that I am effective at getting things done.	2.93	1.17
In my opinion, I do a good job.	2.87	1.20
Threats to one's job		
I am worried that future technology advancements may pose a threat to my job.	3.57	1.11
I am concerned that my job may be eliminated soon.	3.43	1.26
I believe that other people may be able to perform my work activities.	3.42	1.28
I am concerned that my job may be outsourced soon.	3.37	1.29
Career plans		
I will be working in the IT field 1 year from now.	3.31	1.29
I will take steps during the next year to secure a job outside the IT field.	3.24	1.23
I will be with this organization 5 years from now.	3.12	1.18
I will be with this organization 1 year from now.	3.11	1.23
I will take steps during the next year to secure a job at a different organization.	3.04	1.24
I will be working in the IT field 5 years from now.	2.69	1.22

*Rating scale ranges from 1 to 5: 1 as strongly agree, to 5 as strongly disagree.

excessive. They also indicate heavy work load and burnout at the job. The responses also show that the IT employees surveyed felt a strong sense of accomplishment and did not feel any threats to their jobs. While they do not foresee continuing with the same employer over a 5-year horizon, they expect to continue in the IT industry.

15.8 Conclusion

The result of our survey shows that Iran's information technology is currently in a state of flux. In Iran, the primary organizational concerns relate to making the organization more productive and flexible. The top technology concerns mirror the organizational concern by the importance given to technologies that support operations across the organization. For most employees, IT jobs constitute a worthwhile career option. In general, employees seem to be satisfied with their jobs in the IT sector and are likely to continue in the industry in the near future, although there is a concern among some employees that their IT jobs may be outsourced from Iran to other countries.

A recent report by the International Telecommunication Union that analyzes the status of ICT developments around the world shows Iran as having one of biggest gains in terms of ICT Development Index (IDI) in 2017. The youthfulness of the population, its high level of education and urbanization are factors that contribute to this growth. Additionally, these factors collectively make Iran one of the most dynamic countries in the region in terms of high-tech entrepreneurship.

References

Asemi, A. (2006). Information technology and national development in Iran. *International Conference on Hybrid Information Technology (ICHIT'06)*. 0-7695-2674-8/06. IEEE Computer Society.

CIA, The World Facts, https://www.cia.gov/library/publications/the-world-factbook/geos/ir.html, last accessed January 2019.

Dangolani, S. K. (2011). The impact of information technology in banking system (a case study in Bank Keshavarzi Iran). *Procedia — Social and Behavioral Sciences*, 30, 13–16.

Financial Times, https://financialtribune.com/articles/economy-sci-tech/81384/iran-ranked-world-s-7th-instagram-user, last accessed January 2019.

Kappelman, L., Johnson, V., McLean, E., & Maurer, C. (2018). The 2017 SIM IT Issues and Trends Study. *MISQ Exec*, 17(1), 53–88.

OpenNet Initiative Report, https://opennet.net/research/profiles/iran, last accessed January 2019.

Palvia, P., Jacks, T., Ghosh, J., Licker, P., Romm-Livermore, C., Serenko, A., & Turan, A. H. (2017). The World IT Project: History, trials, tribulations, lessons, and recommendations. *Communications of the Association for Information Systems*, 41(18), 389–413.

Palvia, P., Ghosh, J., Jacks, T., Serenko, A., & Turan, A. (2018). Trekking the globe with the World IT Project. *Journal of Information Technology Case and Application Research*, 20(1), 3–8.

Shirazi, F. (2012). Information and communication technology and women empowerment in Iran. *Telematics and Informatics*, 29, 45–55.

The World Telecommunication Development Conference, https://www.itu.int/en /ITU-D/Conferences/WTDC/WTDC10/Pages/default.aspx, last accessed January 2019.

Chapter 16

Information Technology Issues in Italy

Chiara Frigerio*,§, Marcello Martinez†,¶, Mario Pezzillo Iacono†,||,
Federico Rajola*,**, and Tim Jacks‡,††

*Università Cattolica del Sacro Cuore, Milano, Italy
†Università della Campania Luigi Vanvitelli, Napoli, Italy
‡Southern Illinois University Edwardsville, Edwardsville, USA
§chiara.frigerio@unicatt.it
¶marcello.martinez@unicampania.it
||mario.pezzilloiacono@unicampania.it
**federico.rajola@unicatt.it
††tjacks@siue.edu

Summary

In this chapter, we report the organizational, technological, and individual information technology (IT) issues of Italian IT workers. The participants of our survey are mostly IT professionals, working full time in non-managerial positions. IT strategic planning, IT service management, knowledge management, alignment between IT and business were among the most pressing organizational issues. Top technology concerns include enterprise application integration, business intelligence/analytics, collaborative and workflow tools, and networks/telecommunications. Italian IT employees seem to be satisfied with their work and perceived their workloads to be meaningful. They exhibited low levels of turnover intention and felt quite secure in their jobs.

16.1 Introduction

Technology is a key competitive factor for firms and public administrations. This is true also in the Italian system since several investments are being made in the information and communication technology (ICT) field. The positive trend is driven by some specific governmental programs (Impresa 4.0; Agenda Digitale) and by independent regulatory agencies. It is related also to the digital market because it is aimed at increasing both the competitiveness and the productivity of the country, since they are both below the level of other European countries.

The main purpose of ICT projects is to boost innovation, both updating processes and services (especially considering public administrations and corporate firms) and developing new business models and new services based on the intensive use of information technology (IT). In many sectors, including banking, insurance, transport and logistics, the availability of big data is presented with an opportunity whose strategic and organizational effects are not yet fully realized. Even in public administration, thanks to relevant government investment plans, ICT is presenting itself as a priority to be faced with significant procedural and technical changes.

16.2 Country Background and History

Italy, officially the Italian Republic, is a country in Europe and at the beginning of year 2017, had an estimated population of 60.4 million. The economy of Italy is the third largest national economy in the Eurozone, the eighth largest by nominal gross domestic product (GDP) in the world, and the 12th largest by GDP (at purchasing power parity or PPP).

The country is a major advanced economy and is a founding member of the European Union, the Eurozone, the OECD, the G7 and the G20. Italy is the eighth largest exporter in the world with $514 billion exported in 2016. Its closest trade ties are with other countries of the European Union, with whom it conducts about 59% of its total trade.

The country is also well known for its influential and innovative business economic sector, an industrious manufacturing sector (Italy is the second largest manufacturer in Europe behind Germany) and competitive agricultural sector (Italy is the world's largest wine producer), and for its creative and high-quality products in categories such as automobiles, naval, industrial, appliance and fashion design. Italy is the largest hub for luxury goods in Europe and the third luxury hub globally.

Furthermore, the country's private wealth is one of the largest in the world. The stagnation in economic growth, and the political efforts to revive it with massive government spending from the 1980s onwards, eventually produced a severe rise in public debt. In addition, Italian living standards have a considerable North–South divide: the average GDP per capita in Northern and Central Italy significantly exceeds the EU average, while some regions and provinces in Southern Italy are dramatically below average.

16.3 Information Technology in Italy

Official data shows that, during 2017, the Italian digital market grew by 2.3% for a total of 68,722 million Euros and will continue to grow in the following years: 2.6% in 2018, 2.8% in 2019, 3.1% in 2020. Considering the pace of innovation, we observe an increase in the use of digital advertisement (+7.7%), ICT services (+4%), software and solutions (5.9%), whereas intangible systems and infrastructures are showing a continuous expense trend (Assinform, 2018).

According to IDC, 24% of investments are going to be allocated in four of the Third Platform Technologies, like cloud, mobility, big data, analytics and social media. The main stake of the investments is going to be addressed toward Innovation's Accelerators, that are technologies based on the Third Platform that are going to impact all the industrial sectors. Those new technologies are related to internet of things (IoT), robotics, artificial intelligence (AI), augmented and virtual reality, 3D printing, and blockchain.

Moreover, the Italian market is likely to appreciate the digital innovation component and it is going to be ready in testing and adopting all those technologies that lead to a change that involves both the firms' structures and their supply chain. As an example, IoT converts manufacturing objects in network system components with new functionalities that will generate new business value. Cloud and collaboration platforms allow firms to digitally reshape supply chains, evolving relations between client and supplier. Big data and AI are the fundamental elements of new activities based on knowledge management. New digital security platforms are the foundation for all these developments, especially in the area of mobile and payment frameworks.

The productivity trend shows that it is highly correlated to companies that invest more on innovation. Innovation also has a positive impact on salaries: firms that invest in innovation have reached a higher average wage (19.3% vs. 16.8% on average for the total number of employees).

IT investments are related to the innovation concept, considered as the capability to create new value, that should be pervasive and sustainable for corporate business. The strategic value of innovation is generated by the IT capability to generate value inside a dynamic and emerging environment. Value is often synonymous with efficiency, on one side, considered as the capability to increase IT productivity and lower the production factors used, and, on the other side, of effectiveness, considered as the capability to quickly meet customers' needs. Hence, managers and CIOs are constantly looking for mechanisms that can create value in both directions, even if they seem to be dichotomous and even opposed. Investment priority is typically directed towards solutions that can combine, at the same time, the exploration of new competitive fields and the exploitation of existing processes.

Even if the market shows an increase in the adoption of technologies aimed to boost innovation, there is a digital divide between large and small/medium enterprises (SMEs) which comprise 76% of Italian enterprises. SMEs suffer from limited capabilities that are both financial, considering investments in the medium and the long term, and organizational, that limit the implementation capability. Also, on the supply side, the Italian market is characterized by a lack of technological solutions suitable for the SME market, since they are often designed for large enterprises. Moreover, SMEs need specific solutions and approaches, based on their structural, cultural, and organizational peculiarities. It is rare for the transfer of solutions designed for large enterprises to be effective in SMEs. It is not appropriate to consider the SME environment as homogeneous, offering similar solutions for all the realties that constitute it. Hence, we can observe that the business environment is extremely heterogeneous, both in terms of size (some enterprises have 10 employees while others have 200 employees) and in terms of markets and sectors (Gregori *et al.*, 2014).

With regard to the geographical distribution of ICT spending, during 2018, the Northwest has been the biggest technological investor in Italy, with 35% of total spending on ICT, followed by Central Italy that represents 27% of the total, driven by the central public administration and by the headquarters of many large companies. The enterprises of southern Italy are expressing optimism on the development of their technological investment; almost one third forecast a growth rate of 5% during 2018. However, 60% of those find some difficulties both in terms of financial resources and in terms of managers' low risk appetite (Assinform, 2018).

More than 102,000 firms and 560,000 employees compose the supply side of the ICT market. Innovation also allows for the growth of new firms. Thanks to an incentivizing public policy, during the first half of 2017, there were 7,398 innovative Italian startups. In the first half of 2016, we observed an increase of 24.5%, equal to 1,456 new startups. Since their introduction in the jurisdictional Italian system, innovative startups and SMEs have grown significantly, doubling their number in the last 2 years. Today, they are no longer considered a niche reality, considering that they generate a turnover greater than 2 billion Euros and they offer around 50,000 jobs. However, even if those numbers continue to grow, they are still far from international benchmarks (MISE, Relazione 2017).

The pervasive role of IT for the firm's business processes, along with public administration, is leading to an increasing awareness concerning methodologies, tools, and knowledge that will allow for a wider culture of innovation inside Italian enterprises. But investments in ICT skills are still poor. Only 16.2% of small firms with at least 10 employees hire ICT experts (16.8% during the previous year) compared to 72.3% of large companies. Large companies tried to hire ICT experts during the previous year in 31% of the cases (29.8% in 2016 and 26.6% in 2015), versus only 4.2% of the cases for smaller firms. Both groups have faced some difficulties in finding ICT specialists: 12% of firms with at least 10 employees and 1.7% of large companies. This means that the ICT knowledge demand is not completely covered by university offerings (ISTAT, 2017).

In conclusion, the Italian ICT market is showing a positive investment trend focused on new technologies. This growth is mainly due to large companies that use innovation both with internal knowledge and with ecosystems that involve innovative startups (open innovation). Even if the overall environment is growing, it struggles with the expansion of innovation processes for SMEs across the country.

16.4   Methodology

The standard instrument from the World IT Project (Palvia *et al.*, 2017; Palvia *et al.*, 2018) was translated into Italian using the back-translation method. First, one of the authors translated the instrument into Italian and then another author translated it back into English. By involving another academician, the back-translation was compared with the original copy and discrepancies were resolved.

The completion of the survey by each participant took about 30 minutes. The surveys were administered by personal visits and were manually entered into Excel. There was no specific incentive or bonus offered to participants. Communication between the investigators and the core World IT Project team was mainly via emails and phone.

The respondents investigated came from the following sectors: ICT consulting (20%), financial and insurance services (19%), transport and logistics services (18%), public administration (17% including education and utilities), business consulting (12%), manufacturing, retail, and hospitality. They are typically employed in large companies (more the 50% of the participants work in organizations with more than 2,000 employees), with an ICT department located within their organizational structure.

In our sample, Italian IT employees are mostly in the middle age group (36.3% are between 40 and 49 years of age). The second largest group is between 30 and 39 years of age (29.3%), followed by the 50–59 years group (20.3%). The most common job roles are programming (12%), hardware or software maintenance (12%), consulting (10%), system analysis and design (10%), and project management (8%).

Table 16.1 shows the descriptive statistics of the IT professionals who took part in the study from Italy. With regard to educational qualifications, the sample of 310 respondents reported a high school diploma for 51.2% and a college degree for 41.3%. Most respondents have significant professional experience of more than 10 years related to IT. These are professionals with technical skills, working within the IT Department.

16.5   Organizational IT Issues

The survey participants were asked to rate 18 organizational IT-related issues in terms of their level of importance based on a 5-point Likert scale,

Table 16.1: Descriptive Statistics

Characteristics	N	%	Characteristics	N	%
Education:			Years of work experience:		
High school or less	161	51.2	0–4 years	32	10.3
Associate degree	2	0.6	5–9 years	40	12.9
Bachelor's degree	128	41.3	10–19 years	119	38.4
Master's degree	13	4.2	20–29 years	81	26.1
Ph.D.	6	1.9	30+ years	38	12.3
Years of IT experience:			Organizational location:		
0–4 years	46	14.8	IT department employee	212	68.4
5–9 years	67	21.6	IT worker in non-IT	56	18.1
10–19 years	115	37.1	department		
20–29 years	68	21.9	Contract employee	25	8.1
30+ years	14	4.5	Consultant	17	5.5
Work as:			Vendor employee	0	0
Mostly full time	259	83.5	Work position:		
Mostly part time	13	4.2	Not part of management	178	57.4
Mostly over time	38	12.3	In lower management	62	20
Been laid off from IT job:			In middle management	59	19
			In senior management	11	3.5
Yes	7	2.3			
No	303	97.7			

with 1 being the most important and 5 being the less important. The average rating for each issue was computed. Eighteen issues are ranked from 1 to 18 based on their average ratings and are shown in Table 16.2.

The analysis of the results highlights the growing importance of IT competencies for the strategic development of organizations according to a well-identified complementary approach doctrine (Martinez, 2011). Strategic alignment between IT and the business is in fact the main objective and it is what enables an increase in both the productivity of the business and its flexibility in a typical model based on ambidexterity (Marabelli *et al.*, 2012).

IT departments are expected, first of all, to be reliable and efficient. The integration of technologies and productive and administrative systems requires that there are no factors of delay or error in the execution of different strategic and organizational processes (Martinez *et al.*, 2017). Reliable and efficient information technologies play an important role in the compliance of companies that were objects of the survey. It was well recognized by the respondents that working with reliable information, organizations can make better decisions and achieve competitive advantage by improving their work performance. This basic assumption allows IT departments to play a conscious role of service toward corporate goals and support the

Table 16.2:   Organizational IT Issues in Italy

Organizational IT Issues	Rank	Mean Rating*	Std. Deviation
IT reliability and efficiency	1	1.57	0.71
IT strategic planning	2	1.63	0.71
IT service management (e.g., ITIL)	3	1.66	0.66
Knowledge management	4	1.69	0.62
Alignment between IT and business	5	1.70	0.61
Business productivity and cost reduction	6	1.72	0.59
Business agility and speed to market	7	1.76	0.71
Security and privacy	8	1.77	0.80
Project management	9	1.79	0.68
Continuity planning and disaster recovery	10	1.80	0.79
Business process reengineering	11	1.80	0.75
Attracting and retaining IT professionals	12	1.86	0.74
Revenue-generating IT innovations	13	1.87	0.76
Enterprise architecture	14	2.01	0.80
IT cost reduction	15	2.12	0.91
Globalization	16	2.29	0.88
BYOD (Bring Your Own Computing Device)	17	2.56	1.15
Outsourcing	18	2.60	0.97

*Rating scale ranges from 1 to 5: 1 as most important and 5 as no importance.

development of business growth strategies in the medium and long term (Venier, 2017).

In some sectors, relevant to the sample being investigated, such as the financial and insurance sector, the alignment between IT and business is driven by provisions of the regulatory authorities. The Bank of Italy requires the issuance of a Strategic Address Document in which the strategic IT initiatives must be illustrated in line with the company strategic plans and with an indication of the architectural model, sourcing strategies, and IT risk propensity. The approval of this document, the most important among the documents required for the governance and control of information systems, is the responsibility of the Body with the Strategic Supervision Function (OFSS), typically the Board of Directors. Furthermore, it is required that the document be promptly updated in the face of a new business plan, as business strategies change, and in the event of exceptional corporate changes.

Outsourcing is considered the least important issue, possibly because for some time this dynamic has already been addressed and does not represent a "new" challenge. The ICT outsourcing market in Italy has been fairly stable since 2013 (Venier, 2017).

16.6   Technology and Infrastructure Issues

Table 16.3 shows the technology and infrastructure issues from Italy ranked in order of importance. The respondents were asked to rate several IT issues in terms of their importance based on a 5-point Likert scale, with 1 being of most importance and 5 being of no importance. Sixteen technology related issues were rated on this scale.

The top seven most important issues rated by respondents were enterprise application integration, business intelligence/analytics, collaborative and workflow tools, networks/telecommunications, business process management systems, enterprise resource planning (ERP) systems, and mobile and wireless applications. This ranking fully reflects the challenges that Italian companies are called to face from a technological point of view. The development of smart organizational models requires an ever-greater pervasiveness of big data in particular (Caporarello *et al.*, 2015).

ICTs are used to sustain the managerial systems of evaluation and control the performance of different processes both in the case where they are fully automated and when they require a strong electronic interaction between professionals or groups with complementary skills (Martinez, Pezzillo Iacono 2013; Di Lauro *et al.*, 2018). As expected, from a technological point of view, ICT and mobile device networks are essential

Table 16.3:   Technology and Infrastructure Issues in Italy

IT Related Issues	Rank	Mean Rating*	Std. Deviation
Enterprise application integration	1	1.64	0.62
Business intelligence/analytics	2	1.71	0.69
Collaborative and workflow tools	3	1.74	0.66
Networks/telecommunications	4	1.75	0.72
Business process management systems	5	1.79	0.67
Enterprise resource planning (ERP) systems	6	1.79	0.69
Mobile and wireless applications	7	1.80	0.76
Customer relationship management (CRM)	8	1.94	0.69
Software as a service	9	1.96	0.76
Virtualization (desktop or server)	10	1.97	0.85
Service-oriented architecture (SOA)	11	2.04	0.71
Cloud computing	12	2.06	0.80
Mobile apps development	13	2.08	0.90
Data mining	14	2.13	0.77
Big data systems	15	2.23	0.89
Social networking/media	16	2.24	0.88

*Rating scale ranges from 1 to 5: 1 as most important and 5 as not important.

prerequisites for any analytical interpretation of big data (vom Brocke *et al.*, 2014).

16.7   Individual IT Employee Issues

Individual IT employee issues were investigated under seven broad themes. The detailed analysis of each theme is provided below in Table 16.4 that

Table 16.4:   Individual IT Employee Issues in Italy

Individual Issues	Mean Rating*	Std. Deviation
Job satisfaction		
In general, I like working here.	1.89	0.84
All in all, I am satisfied with my current job.	2.14	0.92
In general, I don't like my current job.	3.79	1.18
Work pressure		
I feel that the number of requests, problems or complaints that I deal with at work is more than expected.	3.12	1.04
I feel that the amount of work I do interferes with how well it is done.	2.40	1.18
I feel busy or rushed at work.	2.80	1.10
I feel pressured at work.	3.12	1.10
Work–life balance		
There is a blurring of boundaries between my job and my home life.	3.41	1.22
My work-related responsibilities create conflicts with my home responsibilities.	3.66	1.20
I do not get everything done at home because I find myself completing job-related work.	2.80	1.14
Workload and burnout		
I feel drained from activities at work.	2.74	1.40
I feel tired from my work activities.	3.07	1.02
Working all day is a strain for me.	3.53	1.11
I feel burned out from my work activities.	3.65	1.28
Sense of accomplishment		
I feel I'm making an effective contribution to what this organization does.	2.08	0.90
In my opinion, I do a good job.	1.95	0.87
I have accomplished many worthwhile things in this job.	1.97	0.85
At my work, I feel confident that I am effective at getting things done.	1.96	0.88

(*Continued*)

Table 16.4: (*Continued*)

Individual Issues	Mean Rating*	Std. Deviation
Threats to one's job		
I am worried that future technology advancements may pose a threat to my job.	3.74	1.17
I believe that other people may be able to perform my work activities.	2.83	1.07
I am concerned that my job may be eliminated soon.	3.88	1.04
I am concerned that my job may be outsourced soon.	3.79	1.03
Career plans		
I will be with this organization 1 year from now.	1.92	1.20
I will take steps during the next year to secure a job at a different organization.	3.99	1.13
I will be with this organization 5 years from now.	2.17	1.14
I will be working in the IT field 1 year from now.	2.30	1.51
I will take steps during the next year to secure a job outside the IT field.	4.03	1.33
I will be working in the IT field 5 years from now.	2.15	1.37

*Rating scale ranges from 1 to 5: 1 as strongly agree, to 5 as strongly disagree.

shows the summary of the responses about individual issues. Note that, once again, a lower average score means higher agreement with each statement.

In general, the survey shows a high level of job satisfaction. Clearly there is a need to perform one's duties and tasks with a certain rhythm or urgency, or even a temporary overload, but it does not translate into dissatisfaction due to excessively high expectations or unmanageable pressures.

In terms of work–life balance, once again the survey does not signal a critical tension. In fact, the boundaries between work carried out in the office and private life are not perceived as particularly unfair, and workload does not appear critical. Moreover, the attitude prevails that the competency in IT leads to results useful for the company and an ability to contribute to relevant results. The relevance of IT skills is consciously considered a high factor that supports future opportunities for most of the participants in the survey.

Finally, the future perspectives confirm the possibility of continuing to work within the current organization, and always within the IT field. However, the possibility of change of workplace is not excluded in the next 5 years, probably for career needs. It can therefore be assumed that the prevailing perception is of conscious optimism. Surely the basis of these emerging beliefs can be placed on the increasing importance of IT skills for the strategic and organizational development of companies, as evidenced also by the results of Tables 16.2 and 16.3.

16.8   Conclusion

The top organizational concerns in IT in Italy are IT reliability and efficiency, IT strategic planning, IT service management, knowledge management, and alignment between IT and business. Top technology concerns include enterprise application integration, business intelligence/analytics, collaborative and workflow tools, and networks/telecommunications. Employees indicate that they are, in general, satisfied with their jobs and that they perceive to be playing a relevant role for their companies. They expect to continue in the IT industry and they are confident they can reach new career opportunities.

References

Assinform (2018). *Il mercato digitale in Italia*, July 24th 2018.

Di Lauro, S., Antonelli, G., & Martinez, M. (2018). Understanding employees' perspectives on organizational identity change from their LinkedIn accounts. In: Cabitza, F., Lazazzara. A., Magni, M., & Za, S. (Eds.), *Organizing for Digital Economy: Societies, Communities and Individuals*. ROMA: Luiss University Press, ISBN: 9788868561291.

Caporarello, L., Di Martino, B., & Martinez, M., (2015). Composing and orchestrating the smart artifacts: Technological and organizational challenges. In: Caporarello, L., Di Martino, B., & Martinez, M. (Eds.), *Lecture Notes in Information Systems and Organisation*. Vol. 7. Heidelberg: Springer, pp. 1–8, ISBN: 978319-070391.

Gregori, G. L., Gigliarano, C., Cardinali, S., & Pascucci, F. (2014). Fattori influenti sul ricorso ad Internet nei processi gestionali delle micro-imprese. *Sinergie Italian Journal of Management*, 73–95.

ISTAT, *Report Imprese, Cittadini, ICT*, December 2017.

Marabelli, M., Frigerio, C., & Rajola, F. (2012). Ambidexterity in Service Organizations: Reference Models from the Banking Industry, *Industry and Innovation*, 19, 109–126.

Martinez, M., (2011). ICT, productivity and organizational complementarity. In: Rossignoli, C., & Carugati, A. (Eds.), *Emerging Themes in Information Systems and Organization Studies*. Berlin: Springer Verlag, pp. 271–281, ISBN: 9783790827385.

Martinez, M., Di Nauta, P., & Sarno D. (2017). Real and apparent changes of organizational processes in the era of big data analytics. *Studi Organizzativi*, 2.

Martinez, M., & Pezzillo Iacono, M., (2013). Dealing with Is critical research: Artifacts, drifts Electronic Panopticon and Illusions of Empowerment. In:

Baskerville, R., De Marco, M., Spagnoletti, P., *Designing Organisational Systems — An Interdisciplinary Discourse*. Berlin, Heidelberg: Springer-Verlag, pp. 83–102, ISBN: 9783642333712.

MISE Ministry of Economic Development (2017). *Relazione annuale al Parlamento sullo stato d'attuazione e l'impatto delle policy a sostegno di startup e PMI innovative*, anno 2017.

Palvia, P., Jacks, T., Ghosh, J., Licker, P., Romm-Livermore, C., Serenko, A., & Turan, A. H. (2017). The World IT Project: History, trials, tribulations, lessons, and recommendations. *Communications of the Association for Information Systems*, 41(18), 389–413.

Palvia, P., Ghosh, J., Jacks, T., Serenko, A., & Turan, A. (2018). Trekking the globe with the World IT Project. *Journal of Information Technology Case and Application Research*, 20(1), 3–8.

Venier, F., (2017). *Trasformazione digitale e capacità organizzativa. Le aziende italiane e la sfida del cambiamento*. EUT Edizioni Università di Trieste, ISBN 9788883038181.

vom Brocke, J., Braccini, A.M., Sonnenberg, C., & Spagnoletti, P. (2014). Living IT infrastructures — An ontology based approach to align IT infrastructure capacity and business needs. *International Journal of Accounting Information Systems*, 1, 1–29.

Chapter 17

Information Technology Issues in Japan

Hiroshi Sasaki*,§, Osam Sato†,¶, and Prashant Palvia‡,||

*Rikkyo University, Japan
† Tokyo Keizai University, Japan
‡ University of North Carolina at Greensboro, USA
§ sasaki-h@rikkyo.ac.jp
¶ osamsato@tku.ac.jp
|| pcpalvia@uncg.edu

Summary

This chapter explores current information technology (IT) issues in Japan from organizational, technological, and individual perspectives. Survey results included in the chapter reveal that security and privacy, IT reliability and efficiency, and business agility are the three top organizational IT issues in Japan. The top three concerns related to technology are business intelligence and analytics, networks and telecommunications, and enterprise application integration. Regarding individual issues, results indicate that IT workers are generally satisfied with their current jobs, experience a moderate level of stress, have a sense of accomplishment at work, and do not want to leave their jobs or the IT profession any time soon. These results are indicative of the Japan-specific situation in which the IT workplace is considered fairly stable.

17.1   Introduction

Japan is an island country located in East Asia. Since ancient times, the Japanese have eagerly adopted foreign customs and technologies. Recent IT products and services are no exception to this. However, when it comes to implementing new technologies into *gemba* (the place where manufacturing and office activities are conducted in Japanese companies), issues often arise because of unique business customs and processes (e.g., *gemba* work flows, decision-making processes, domestic financial reporting standards, and intra-organizational interfaces for electronic data sharing). To address these issues, frequent software customization or add-on application development is required. In addition, information systems (IS) development is also affected by history, culture and industry-specific factors. This chapter provides a quick overview of the country's background and history, followed by a description of the IT services industry in Japan, and finally presents the results of the World IT Project survey. These results include organizational, technological, and individual employee issues.

17.2   Country Background and History

It is widely recognized that Japan belongs to the Confucian group (House *et al.*, 2004). However, in reality, Japanese culture is a complete mixture of Eastern and Western cultures. The Japanese language is a typical example of this cultural mixture. Japanese is written using several different writing systems: two Japanese-original syllabic scripts (*Hiragana* and *Katakana*), Chinese characters (*Kanji*), and Latin script. Because sentence structure and pronunciation are vastly different from English, Japanese people generally experience difficulties when communicating with foreigners in English. The language barrier sometimes causes delays when implementing new technologies from overseas.

The Japanese have engaged in agriculture for more than 2,000 years. In order to sustain themselves, the Japanese people have had to work together in small village communities. According to Chow (2002), Japan has a stronger sense of collectivism compared with other Asian societies, yet its family relations are weaker. Nakane (1970) studied the importance of vertical relationships and hierarchies among people, groups, and organizations in Japan. The hierarchical structure and vertical relationship can be seen in the IT services industry in Japan. Abegglen (1958, 2006), who studied the Japanese management style from a Western perspective, found that the key characteristic of the Japanese employment system lies in "lifetime commitment." This is the practice of an employee entering a company directly after graduation from school, participating in continuous on-the-job training to become an experienced professional, and having guaranteed employment until retirement. While this system still exists, job hopping is becoming more common among talented, young IT professionals.

After the Second World War, the Japanese economy promptly recovered with the support of the US. In fact, a few decades after the war, Japan became the world's second largest economy, and today it has the world's third largest GDP. Unfortunately, the Japanese economy peaked in the latter half of the 1980s during the so-called "bubble economy", which was followed by an economic collapse in the 1990s, and a long-term recession which lasted for more than 20 years (the so-called "lost decades"). Following the global financial crisis in 2008 and the Great East Japan Earthquake in 2011, Japanese firms suffered serious damage. Fortunately, however, economic recovery began quickly thanks in part to a series of new economic policies which were adopted in 2012, popularly known as *Abenomics*. The survey reported in this chapter was conducted in this economic climate.

17.3 Information Technology in Japan

Historically, the rise of the Japanese IT industry can be traced back to the 1960s. At that time, IBM's release of the System/360 was epoch-making news in 1964 (IBM, 1964). In 1968, the Japan Productivity Center, a well-known non-profit think tank, sent a delegation comprising business leaders to the United States. The delegation published a report (Japan Productivity Center, 1968), which had a big impact on the Japanese business world. It was big news that IBM mainframes helped to support data aggregation for the Tokyo Olympic Games in 1964. Also in 1964, Japanese National Railways unveiled a computerized reservation system for purchasing bullet train tickets.

In the 1970s, the adoption of mainframe computers began picking up momentum. Mainframe manufacturers in the US tried to penetrate the Japanese market. In response, the government strategically urged the creation of three domestic computer manufacturing groups: Fujitsu and Hitachi, NEC and Toshiba, and finally, Mitsubishi and Oki, in 1971. The goal of these research groups was to compete with overseas manufacturers. Some manufacturers developed IBM-compatible machines, but some started developing proprietary machines including the *Ofukon* (a medium-sized office computer targeted at small- and medium-sized enterprises) that became popular in the 1980s (IPSJ, 2010).

Since the mid-1980s, IT business innovations coming from the US have caused booms in Japan, such as strategic information systems, enterprise resource planning, supply chain management, customer relationship management, cloud computing, big data, the internet of things, and most recently, artificial intelligence. For decades, the IT industry has taken a leadership role in Japanese IT innovations, closely cooperating with IT departments of companies in other industries. Along with these IT innovations, firms have also updated information systems (IS) several times. This is because Japanese business customs and processes are idiosyncratic, with numerous on-premises applications which fit Japanese business practices. However, after the bubble economy burst in the 1990s, these firm-specific applications caused much distress when organizations tried to integrate IT infrastructure across various organizational units.

The IT services industry plays a key role in Japan. The total sales in the IT services industry has been increasing since 1988, except for the period between 2008 and 2011 due to the global financial crisis and the Great Tohoku Earthquake. In addition, the total number of companies has been declining slightly after reaching its peak in 2004. Cusumano (2004, p. 11) uses the expression *software factories,* indicating that "Japan's computer manufacturers and software firms have shown significant skill in writing programs for all sorts of applications, ... However, most of these systems are custom-built for specific customers in Japan and contain relatively few innovative features by design." His image of IT workers in Japan is that of a craftsman who has a narrow range of skills and cares about making quality products with zero defects.

The IT industry in Japan has a rigid hierarchy, composed of a few large-scale firms at the top and many small and medium-sized enterprises at the bottom. Over time, each firm's hierarchical position in the industry has become fixed (Sasaki, 2012). Under such locked-in circumstances, mechanistic and bureaucratic management styles have become desirable. As a result, craftsmanship has been strengthened. Furthermore, the few large firms possess better business efficiency as measured by revenue per employee, meaning that employees in small-sized firms may be pushed to work harder in order to make up for lower revenue. In this particular case, "3K" situations are quite common (i.e., *Kitsui* — hard, *Kibishii* — demanding, and *Kaerenai* — having irregular working hours). Due to this pressure, employees at firms positioned at the bottom of the hierarchy typically display issues with work–life conflict, workaholism, stress, and burnout. It is safe to say that the same situation can happen in the small IT departments in non-IT industries.

While the hierarchical structure still functions generally well, web innovations since the 2000s have created opportunities for small ventures to grow outside of the traditional hierarchical structure. Further, some executives have changed their policies: not to use traditional outsourcing relationships, but to build new partnerships with overseas software developers.

17.4 Methodology

To conduct the World IT Project (Palvia *et al.*, 2017; Palvia *et al.*, 2018) survey in Japan, the original survey questions were first translated from English into Japanese. As part of the process, one team member translated the questionnaire into Japanese and another member translated it back into English. After repeating this task several times, the Japanese team (i.e., the first two co-authors) were able to finalize the Japanese version. Next, the questionnaire was posted online. In order to maximize the number of responses, several executives from the Japan Information Technology Service Industry Association and the Japan Institute of Information Technology were contacted and asked for their cooperation in approaching IT employees. The team also researched and selected the best marketing research company in terms of having a large number of IT workers to participate in the survey. The research company first built a Web gateway linking to a survey website, and second sent emails to groups of IT workers

Table 17.1:　Descriptive Statistics of Survey Respondents

Characteristics	N	%	Characteristics	N	%
Education:			Years of work experience:		
High school or less	55	17.7	0–4 years	14	4.5
Associate degree	35	11.3	5–9 years	19	6.1
Bachelor's degree	155	50.0	10–19 years	63	20.3
Master's degree	57	18.4	20–29 years	109	35.2
Ph.D.	8	2.6	30+ years	105	33.9
Employee reporting relationships:			Laid off experience:		
IT department employee	169	54.5	Yes	25	8.1
IT worker in a non-IT department	73	23.5	No	285	91.9
Contract employee	27	8.7			
Consultant	18	5.8			
Vendor employee	23	7.4			

asking them to respond to the questionnaire. As a result, 380 responses were obtained during the time period from September 2016 to November 2016. After excluding incomplete data or outliers, there were 310 valid responses.

Table 17.1 provides the descriptive statistics of the respondents. Our analysis found that 78.0% of the respondents were employees in non-IT industries (54.5% were IT department employees and 23.5% were IT workers in non-IT departments) and 7.4% were vendor employees in the IT industry. In addition, 91.9% of all the respondents had never been laid off from IT-related positions. Thus, most respondents are protected by Japanese lifetime employment systems.

17.5　Organizational IT Issues

In the survey, IT workers were asked to rate the importance of 18 IT-related organizational issues using a 5-point Likert scale ranging from 1 being most important to 5 being of no importance at all. Table 17.2 presents the ranking of each of the 18 questions based on the average score, sorted from most to least important.

Our results indicate that security and privacy is the most important organizational IT issue. This result is consistent with the results of the 2017 SIM IT Key Issues and Trends Study (Kappelman et al., 2018). Continuity planning and disaster recovery was not ranked at the top but somewhere in the middle (i.e., ranked sixth), even though Japan has faced several disasters in the 2010s. It appears that day to day issues are more important than preparing for unpredictable disasters. The results show that the

Table 17.2:   Organizational IT Issues in Japan

Organizational IT Issues	Rank	Mean*	Std. Deviation
Security and privacy	1	1.76	0.614
IT reliability and efficiency	2	1.90	0.554
Business agility and speed to market	2	1.90	0.517
Alignment between IT and business	4	1.91	0.526
Project management	4	1.91	0.593
Business productivity and cost reduction	6	1.94	0.485
Continuity planning and disaster recovery	6	1.94	0.694
Revenue-generating IT innovations	8	1.98	0.591
IT strategic planning	9	2.02	0.682
Enterprise architecture	10	2.04	0.622
Knowledge management	11	2.05	0.659
Business process reengineering	12	2.07	0.669
IT cost reduction	12	2.07	0.652
IT service management (ITIL)	14	2.09	0.655
Attracting and retaining IT professionals	15	2.13	0.779
Globalization	16	2.35	0.983
Bring your own device (BYOD)	17	2.58	1.072
Outsourcing	18	2.59	1.053

*Rating scale ranges from 1 to 5: 1 as most important and 5 as no importance.

second, third, and fourth most important issues were IT reliability and efficiency, business agility and speed to market, and alignment between IT and business, whereas issues related to financial performance, such as business productivity and cost reduction, revenue-generating IT innovations, and IT cost reduction, were ranked somewhat lower (ranked sixth, eighth, and 13th, respectively). It seems that the IT employees are more tuned to long term and strategic concerns than to short term and financial results.

The fourth highest ranked is project management (actually very close to the fourth ranked alignment). Project management is seen important in order to assure quality, cost, and delivery of key projects. Recently, many advanced developers have been keen to study more agile software development approaches and associated project management activities. In particular, they have paid attention to a framework called "scrum," which is self-organizing, with cross-functional teams taking the initiative (Takeuchi and Nonaka, 1986). The traditional model requires bureaucratic management style and the waterfall model is preferred for software development. The emergence of a scrum-based project management style may be an innovation that leads to new organizational capabilities.

IT innovation issues targeting organizational restructuring, such as enterprise architecture and business process reengineering, were ranked lower at 10th and 12th, respectively. Issues in attracting and retaining IT workers were also ranked low (ranked 15th). Three issues included in the survey: globalization, BYOD (Bring Your Own Device), and outsourcing were not seen as important (ranked 16th, 17th, and 18th, respectively). Interestingly, Japan has been facing difficulties related to globalization and outsourcing for a rather long time. However, these issues may be more familiar to top management instead of IT workers. The low rankings may reflect the fact that respondents do not face these issues on a daily basis.

17.6 Technology and Infrastructure Issues

IT workers were also asked to rate the importance of 16 areas of technology using the same 5-point Likert scale described previously. Table 17.3 summarizes the responses by ranking the issues from one to 16 from the most important to the least.

In order to interpret the results, the 16 issues were classified into four groups. The first and the highest-ranked group includes support for core business functions and decision-making (group A). The four topmost important issues are from group A and include business intelligence and analytics, networks/telecommunications, enterprise application integration and business process management systems. These are followed by collaborative

Table 17.3:   Technology and Infrastructure Issues in Japan

IT-Related Issues	Rank	Mean*	Std. Deviation	Group
Business intelligence and analytics	1	2.15	0.730	A
Networks/telecommunications	2	2.21	0.880	A
Enterprise application integration	3	2.25	0.847	A
Business process management systems	4	2.27	0.824	A
Virtualization (desktop or server)	5	2.28	0.942	S
Collaborative workflow tools	6	2.32	0.836	A
CRM	7	2.38	0.898	A
Big data systems	8	2.41	1.009	B
Data mining	8	2.41	1.026	B
Cloud computing	10	2.44	1.065	S
ERP systems	11	2.45	0.987	A
Mobile and wireless applications	11	2.45	1.032	C
Software as a service	13	2.47	1.038	B
Service-oriented architecture (SOA)	14	2.53	1.057	S
Mobile apps development	15	2.56	1.135	C
Social networking/media	16	2.88	1.226	S

*Rating scale ranges from 1 to 5: 1 as most important and 5 as no importance.

workflow tools, customer relationship management (CRM), and enterprise resource planning (ERP) systems. In the early days of computing, it was thought that an IS provides managers with necessary information anytime and anywhere (Japan Productivity Center, 1968). In this sense, the fundamental purpose of IS has remained unchanged. For IT workers, it can be considered as an endless challenge. The second group that is ranked lower than the first includes new technologies and trends (group B). Recently, the trend of big data has begun to be overtaken by IoT and AI, with both becoming more popular. Software as a service (SaaS) implementation has also become an attractive strategic option for firms considering shifting from on-premises applications to cloud-based ones. The third group issues are all related to mobile services (group C). These issues are less important because mobile technology and applications have now become widespread and no longer attract the special attention of IT workers. Finally, the fourth group includes an assortment of different technologies (group S). Included are service-oriented architecture, cloud computing, and social networking/media, which are all ranked lower.

17.7 Individual IT Employee Issues

Lastly, the IT workers were asked to rate individual issues related to their jobs under seven broad themes using a 5-point Likert scale ranging from 1 (for strongly agree) to 5 (for strongly disagree). Table 17.4 details the results. In the table, each statement is further evaluated as: A (overall agree: mean value being below 3.0), or D (overall disagree: mean value being above 3.0).

First, as for job satisfaction overall, IT workers said they were fond of their workplace and were satisfied with their current job. Second, although somewhat pressured, the respondents did not indicate much stress from the amount of work, requests, problems, or complaints. Third, regarding work–life balance, all three issues showed mean values above 3.0, meaning such issues were not critical. Fourth, although they did report being drained and tired at work, they did not feel burned out by such activities. Fifth, with regard to productivity and accomplishments, all four issues had mean values much lower than 3.0, suggesting that the workers are proud of their accomplishments and self-efficacy. Sixth, about threats to their jobs, on average, IT workers seemed to be assured of their jobs and did not see any serious threats although they acknowledged that others could do their jobs. Finally, regarding their career plans, all of the answers are revealing and

Table 17.4:   Individual IT Employee Issues in Japan

Individual Issues	Mean Rating*	Std. Deviation	A/D**
Job satisfaction			
In general, I like working here.	2.51	0.96	A
All in all, I am satisfied with my current job.	2.73	0.98	A
In general, I don't like my current job.	3.33	1.05	D
Work stress			
I feel that the number of requests, problems, or complaints that I deal with at work is more than I expected.	3.02	1.01	D
I feel that the amount of work I do interferes with how well it is done.	3.20	1.06	D
I feel busy or rushed at work.	3.00	1.11	A/D
I feel pressured at work.	2.81	1.04	A
Work–life balance			
There is blurring of boundaries between my job and my home life.	3.31	1.14	D
My work-related responsibilities create conflicts with my home responsibilities.	3.20	1.13	D
I do not get everything done at home because I find myself completing job-related work.	3.35	1.02	D
Workload and burnout			
I feel drained from activities at work.	2.93	1.13	A
I feel tired from my work activities.	2.92	1.18	A
Working all day is strain for me.	2.93	1.07	A
I feel burned out from my work activities.	3.42	1.09	D
Sense of accomplishment			
I feel I'm making an effective contribution to what this organization does.	2.49	0.85	A
In my opinion, I do a good job.	2.54	0.87	A
I have accomplished many worthwhile things in this job.	2.51	0.87	A
At my work, I feel confident that I am effective at getting things done.	2.34	0.79	A
Threats to one's job			
I am worried that future technology advancements may pose a threat to my job.	3.16	1.04	D
I believe that other people may be able to perform my work activities.	2.70	0.97	A
I am concerned that my job may be eliminated soon.	3.33	1.05	D
I am concerned that my job may be outsourced soon.	3.54	1.04	D

(Continued)

Table 17.4: (*Continued*)

Individual Issues	Mean Rating*	Std. Deviation	A/D**
Career plans			
I will be with this organization 1 year from now.	2.25	0.95	A
I will take steps during the next year to secure a job at a different organization.	3.39	0.99	D
I will be with this organization 5 years from now.	2.76	1.04	A
I will be working in the IT field 1 year from now.	2.27	0.94	A
I will take steps during the next year to secure a job outside the IT field.	3.54	0.98	D
I will be working in the IT field 5 years from now.	2.51	0.96	A

*Rating scale ranges from 1 to 5: 1 as strongly agree, to 5 as strongly disagree.
**A/D: A: agree (mean value below 3.0); D: disagree (mean value above 3.0).

perhaps unique to Japan. Overall, the respondents did not want to quit their current jobs, move to different fields, or get jobs in different organizations.

These results reflect the Japan-specific situation in which the IT workplace is relatively rewarding and stable. In fact, since IT workers are usually protected by lifetime employment systems, more than 90% of the respondents had not experienced lay-offs (see Table 17.1). Over the years, executives and human resource managers in both non-IT industries and the IT industry have made long-term efforts to improve the work environment, and the results seem to be paying off.

17.8 Conclusion

In this chapter, we have discussed numerous IT-related organizational, technical, and individual issues in Japan. Survey results reveal that security and privacy, IT reliability and efficiency, and business agility are the three top organizational IT issues in Japan. The top three concerns related to technology are business intelligence and analytics, networks and telecommunications, and enterprise application integration. Thus there is support for core business functions and decision-making. It implies that, under any technological circumstances, the fundamental purpose of IS development has remained unchanged.

There is no doubt that the IT workplace is influenced by the Japanese history and the country's unique cultural background. When evaluating individual issues, the key findings are that: the IT workplace is generally stable overall — in part due to being protected by lifetime employment systems, workers do not want to leave their current organization or the IT profession, and they are generally satisfied with their jobs and workplace without much impact on work–life balance. While it may be too early to predict, IT workers may have to learn to cope with a rapidly changing technological environment, and realize that there is the risk of their jobs being outsourced or eliminated.

Acknowledgment

The authors thank all those who participated in the World IT survey in Japan.

References

Abegglen, J. (1958). *The Japanese Factory: Aspects of its Social Organization.* Glencoe, IL: Free Press.

Abegglen, J. (2006). *21st Century Japanese Management: New Systems, Lasting Values.* Palgrave Macmillan.

Chow, Irene Hau-Siu. (2002). Organizational Socialization and Career Success of Asian Managers. *The International Journal of Human Resource Management,* 13(4), 720–737.

Cusumano, M.A. (2004). *The Business of Software.* The Free Press.

Eibun Nihon Daijiten (1996). *Keys to the Japanese Heart and Soul.* Koudansya International. (in Japanese)

House, R. J., Hanges, P. J., Javidan, M., Dorfman, P. W., & Gupta, V. (2004). *Culture, Leadership, and Organizations: The GLOBE Study of 62 Societies,* SAGE Publications.

International Business Machines Corporation (IBM) (1964). *System/360 Announcement.* Retrieved on July 31, 2018 from https://www-03.ibm.co m/ibm/history/exhibits/mainframe/mainframe_PR360.html.

Information Processing Society of Japan (IPSJ) (2009). Information Processing Technology Heritage. Retrieved on July 31, 2018 from http://museum.ipsj .or.jp/computer/main/0006.html (in Japanese).

Information Processing Society of Japan (IPSJ) (2010). *Nihon no konpyuta Shi (The history of Japanese Computers.* Ohmsya (in Japanese).

Japan Productivity Center (1968). America No MIS — Houbei MIS Shisetsudan Houkokusyo — Perican Sya (in Japanese).

Kappelman, L., Johnson, V., McLean, E., & Maurer, C. (2018). The 2017 SIM IT Issues and Trends Study. *MISQ Exec,* 17(1), 53–88.

METI (Minister of Economy, Trade and Industry). Survey of Selected Service Industries, 2011 (in Japanese). Retrieved on July 31, 2018 from http://www.meti.go.jp/english/statistics/tyo/tokusabizi/index.html (in Japanese).

Nakane, C. (1970). *Japanese Society.* University of California Press (in Japanese).

Palvia, P., Jacks, T., Ghosh, J., Licker, P., Romm-Livermore, C., Serenko, A., & Turan, A. H. (2017). The World IT Project: History, trials, tribulations, lessons, and recommendations. *Communications of the Association for Information Systems,* 41(18), 389–413.

Palvia, P., Ghosh, J., Jacks, T., Serenko, A., & Turan, A. (2018). Trekking the globe with the World IT Project. *Journal of Information Technology Case and Application Research,* 20(1), 3–8.

Sasaki, H. (2012). Differences in efficiency between B2Bs and B2Cs in the Japanese IT services industry. In *Proceedings of the 14th IEEE International Conference on Commerce and Enterprise Computing.*

Takeuchi, H., & Nonaka, I. (1986). New Product Development Game. *Harvard Business Review,* 64(2), 137–146.

Chapter 18

Information Technology Issues in Jordan

Aykut Hamit Turan[*,‡], Naciye Güliz Uğur[*,§], and Prashant Palvia[†,¶]

*Sakarya University, Sakarya, Turkey
† The University of North Carolina at Greensboro,
Greensboro, NC, USA
‡ ahturan@sakarya.edu.tr
§ ngugur@sakarya.edu.tr
¶ pcpalvia@uncg.edu

Summary

In this chapter, we report organizational, technological and individual information technology (IT) issues of Jordanian IT workers. Based on our survey, security and privacy, IT reliability and efficiency, business productivity and cost reduction, revenue-generating IT innovations, and alignment between IT and business are the top five crucial organizational IT issues in Jordan. Networks and communications, business intelligence and analytics, business process management systems, collaborative and workflow tools, and big data systems are the top technology issues. Jordanian IT employees seem to be satisfied with their jobs, and their workload is manageable. They exhibit satisfactory levels of job security and plan to remain in the current job and the IT profession.

18.1 Introduction

The focus of the World Information Technology (IT) Project is understanding the major information system (IS) issues in the world in the context of each country's or region's unique cultural, economic, political, religious, and societal environments. In this respect, Jordan offered us a unique opportunity in terms of its religious, cultural and political systems. Jordan is one of the unique nations in the Middle East. The country presents both eastern as well as western characteristics. It is a small Arab country with a constitutional monarchy.

As in other countries, the Jordan team utilized the Standard English instrument from the World IT Project to elicit the important organizational, technological, and individual issues of IT employees. This chapter provides a brief background of the country, its IT development, and then follows with the results from the survey.

18.2 Country Background and History

Jordan, officially the Hashemite Kingdom of Jordan, is located in the Arab part of the Asian continent. It is considered one of the most important historical areas, in which many civilizations have flourished. Jordan was subjected to Roman domination and emerged as a prominent and important community in the region, and later came under the Islamic rule. Jordan gained its independence from the British mandate in 1946.

Jordan is a small and almost landlocked country with a population of about 10 million. Sunni Islam is the dominant religion, practiced by about 95% of the country. The country is run by a constitutional monarchy. The constitution, adopted in 1952 and amended since then provides broad executive and legislative powers to the king. The capital city is Amman, which is located in the central region of the Kingdom and is considered the political and economic center of the country.

Jordan has a relatively well-developed industrial sector, which includes mining, manufacturing, construction, and power. The top export destinations of Jordan are the US, Saudi Arabia, India, Iraq, and the United Arab Emirates. The top import origins are China, Saudi Arabia, the United States, the United Arab Emirates, and Germany. Its tourism sector is vast and is considered an essential part of the economy. It has around 100,000 archaeological and tourist sites. The sector is an abundant source of employment and economic growth. The majority of tourists are from European and Arab countries. In the past few decades, Jordan has become a major medical tourism attraction; it is the region's top medical tourism destination and fifth in the world overall (as rated by the World Bank).

Jordan has a history of accepting refugees from many neighboring countries that have been at war. As a result, over 3 million Palestinian and Syrian refugees are present in Jordan. Lately, the recent surge of refugees from Syria is beginning to put a substantial burden on its economy and its capacity to accept them.

18.3 Information Technology in Jordan

The IT sector in Jordan has witnessed remarkable development over the past few years in conjunction with the global development in the use of information technologies. It can be summed up into four stages: the boom of global technology, the Reach initiative, entrepreneurial support and freedom, and smart technologies.

During the first phase in the 1990s, a group of IT companies emerged, followed by the boom of global technology and the so-called "dot-com" trend. IT also pervaded government agencies and major institutions in telecommunications, banking, education, health, and others. Many of these companies contributed to the development and advancement of the Reach Initiative in 1999/2000 (Omet and Al-Zubi, 2004). This royal initiative included a 5-year action plan to develop the IT industry in Jordan, enhance its competitiveness in regional and global markets and create strong partnerships between the public and private sectors. Since then, several Reach initiatives have been launched. The latest 2025 royal initiative aims to make Jordan a thriving regional ICT hub for the Middle East and North Africa.

The third phase began to emerge in 2008 when the government decided to reduce the minimum initial investment capital to start a business from US$30,000 to US$1,000 and then to US$1 in 2009. This decision has contributed significantly to reducing the barriers to establishing entrepreneurs and companies. The pace of establishment of companies has increased locally as the technology sector does not always require a high initial investment. In the fourth phase, beginning from about 2012/2013 to the present, smart technologies are increasingly being utilized, e.g., the development of smart mobile devices, the Internet and electronic commerce applications, and social networking projects. Many start-up companies and groups of entrepreneurs are looking at opportunities at the regional and global level, and several initiatives have recently emerged.

Jordan has a highly developed telecommunications infrastructure. The telecommunications sector in Jordan is growing at a very rapid pace, and its infrastructure is continuously being upgraded and expanded. The sector remains the most competitive in the Middle East. The Internet penetration rate reached 87% at the end of 2017, and mobile penetration exceeded 168% in the same period. The completion of a fiber optic network linking all regions of the Kingdom is expected by 2022 (Gasaymeh, 2018).

18.4   Methodology

The World IT Project survey (see Palvia *et al.*, 2017; Palvia *et al.*, 2018) was administered to IT employees in organizations in various industries in Jordan, e.g., manufacturing, education, IT, financial institutions and government. Several means were used to contact these employees, e.g., emails, direct contacts and through senior executives. In total, 253 responses were obtained.

In our sample, Jordanian IT employees are mostly in their early adulthood (84% are between 21 and 39 years of age). The second largest group is between 40 and 49 years (12%). The most common job roles were programming, maintenance, and system administrator. A vast majority (95%) has Jordanian citizenship. The sample is largely composed of males (72%). Thus our sample is reflective of the general IT profession. The sample is largely made up of non-management employees. More descriptive statistics are provided in Table 18.1.

18.5   Organizational IT Issues

The survey participants were asked to rate 18 organizational IT-related issues in terms of their importance level based on a 5-point Likert scale, with 1 being the most important and 5 being of no importance. The average rating for each issue was computed. The 18 issues are ranked from 1 to 18 based on their average ratings and are shown in Table 18.2.

Table 18.1: Descriptive Statistics

Characteristics	N	%	Characteristics	N	%
Education:			Years of work experience:		
High school or less	2	0.8	0–4 years	90	35.6
Associate degree	22	8.7	5–9 years	75	29.6
Bachelor's degree	174	68.8	10–19 years	66	26.1
Master's degree	42	16.6	20–29 years	21	8.3
Ph.D.	13	5.1	30+ years	1	0.4
Years of IT experience:			Organizational location:		
0–4 years	98	38.7	IT department employee	167	66.0
5–9 years	88	34.8	IT worker in non-IT department	34	13.4
10–19 years	52	20.6	Contract employee	43	17.0
20–29 years	15	5.9	Consultant	2	0.8
30+ years	—	—	Vendor employee	7	2.8
Work as:			Work position:		
Mostly full time	228	90.1	Not part of management	130	51.4
Mostly part time	20	7.9	In lower management	37	14.6
Mostly over time	5	2.0	In middle management	66	26.1
Been laid off from IT job:			In senior management	20	7.9
Yes	31	12.3			
No	221	87.7			

Table 18.2: Organizational IT Issues in Jordan

Organizational IT Issues	Rank	Mean Rating*	Std. Deviation
Security and privacy	1	1.68	0.906
IT reliability and efficiency	2	1.73	0.841
Business productivity and cost reduction	3	1.78	0.966
Revenue-generating IT innovations	4	1.86	0.850
Alignment between IT and business	5	1.88	0.837
Project management	6	1.90	0.956
Business agility and speed to market	7	1.91	0.877
IT strategic planning	8	1.93	0.908
Continuity planning and disaster recovery	9	1.99	0.959
IT cost reduction	10	2.00	0.984
Attracting and retaining IT professionals	11	2.02	0.959
Knowledge management	12	2.07	1.040
IT service management (e.g., ITIL)	13	2.08	0.958
Business process reengineering	14	2.15	0.915
Enterprise architecture	15	2.25	1.044
Outsourcing	16	2.49	1.086
Globalization	17	2.52	1.064
Bring your own computing device (BYOD)	18	2.63	1.197

*Rating scale ranges from 1 to 5: 1 as most important and 5 as no importance.

According to our sample of IT employees and managers, security and privacy is the most important organizational IT issue. IT reliability and efficiency came up second, followed by business productivity and cost reduction and revenue-generating IT innovations ranking in third and fourth places. The next two issues were alignment between IT and business and project management. The least important organizational issues listed were BYOD, globalization, outsourcing, enterprise architecture, business process reengineering and IT service management.

Security and privacy of information are very important in Jordan since the country is mostly a patriarchal society and administered as a kingdom. It can be argued that in patriarchal and closed societies, information owners would not want to make information openly available to outside parties without a stake in governing. Besides, privacy and security have become very important in the global economy as well. Users have become more aware and concerned with security and privacy issues, especially in the developing economies, given the culture and the lax business practices. Data privacy breaches of citizens in Jordan between 2010 and 2013 have raised the question whether the Jordanian government has the right policies to protect individuals' privacy and security. As a result, the government has enacted more and detailed IT privacy and security laws and regulations.

IT reliability and efficiency are essential for any organization's success and are reflected by our sample. This is an operational issue that is well recognized by the Jordanian IT worker. The Jordanian IT worker is also cognizant of the business role of IT, i.e., to improve business productivity and reduce costs. IT can do so by automating tasks and reengineering business processes. On a more strategic note, the capabilities of IT can be harnessed to bring about innovation and develop revenue-generating models and applications. Jordan and its IT employees recognize this value proposition; in fact, the major premise of the Royal Initiative 2025 is to make Jordanian companies more innovative and creative and enable them to offer cutting-edge products and services.

Among the least important organizational issue was Bring Your Own Device or BYOD. BYOD has emerged as a buzzword to reduce costs of managing IT but has the risk of creating security problems and nightmares to manage diverse equipment. Yet it has not become a major issue, perhaps because it has not reached a critical proportion in the Jordanian industry or because organizations have not actually experienced any major problems. Globalization is also among the bottom issues. We conjecture that globalization has now become a fact and a way of life, and as such IT professionals do not perceive it as a new trend that needs to be managed. Globalization has likely been already integrated into organizational goals and policies and as such does not require any special effort. Outsourcing also came up among the issues at the bottom. In the last decade with IT initiatives, Jordan is trying to be a destination of choice for outsourcing — an outsourcing oasis for the Middle East. Hence, while outsourcing of its own IT activities and processes may not be deemed important by Jordanian companies, outsourcing may be important to IT vendors in Jordan who wish to sign outsourcing contracts with client companies, mostly outside of Jordan.

18.6 Technology and Infrastructure Issues

The survey respondents were also asked to rate several technology issues in terms of their importance based on a 5-point Likert Scale, with 1 being of most importance and 5 being of no importance. Sixteen technology related issues were rated on this scale and are shown in Table 18.3.

The top five most important issues rated by the respondents were networks/telecommunications, business intelligence/analytics, business process management systems, collaboration, and workflow tools and big data systems. The three least important issues are virtualization, service oriented architecture (SOA) and social networking/media.

Table 18.3: Technology and Infrastructure Issues in Jordan

IT-Related Issues	Rank	Mean Rating*	Std. Deviation
Networks/telecommunications	1	1.85	0.807
Business intelligence/analytics	2	1.87	0.898
Business process management systems	3	2.07	0.943
Collaborative and workflow tools	4	2.07	0.914
Big data systems	5	2.08	0.966
Mobile apps development	6	2.09	0.988
Data mining	7	2.15	0.942
Customer relationship management (CRM)	8	2.16	1.094
Cloud computing	9	2.18	1.006
Enterprise resource planning (ERP) systems	10	2.18	0.959
Mobile and wireless applications	11	2.19	1.043
Enterprise application integration	12	2.25	1.023
Software as a service	13	2.30	1.003
Social networking/media	14	2.33	1.116
Service-oriented architecture (SOA)	15	2.37	1.029
Virtualization (desktop or server)	16	2.44	1.024

*Rating scale ranges from 1 to 5: 1 as most important and 5 as no importance.

Networks and telecommunications were rated as the most important issue. According to the National ICT strategy for Jordan, telecommunication, networks, and IT are inextricably intertwined. For a healthy and robust IT industry in Jordan, modern, responsive and competitive telecommunications and networks are a must. The economic benefits of information technologies would not be able to be delivered to the whole country and its businesses without fast, reliable and high-capacity networks.[1]

As data of all variety grow exponentially, business and society are becoming increasingly data-driven. This trend is seen worldwide, especially in the advanced nations, but was also observed in developing countries. While Jordan has launched several initiatives, it is classified as a developing country by the World Bank. According to some researchers, the majority of ICT projects in developing countries would fail (Petrini and Pozzebon, 2003). The main reason for that would be that many IT design projects are implemented with Northern model principles with Southern realities (Heeks, 2002). Business Intelligence and Analytics, rated high in Jordan, have the potential to help firms gather operational data to form reality-based decisions and develop strategies based on their own contexts.

The third high rated technology in Jordan was business process management systems. Business processes are central to improving firm efficiency and effectiveness. They also can serve as a means to innovation in an organization. As is now known and proven, information technologies can play a crucial role in enacting, managing and innovating business processes. The fourth high rated technology issue was collaborative and workflow tools. Collaborative organizations, as well as collaboration between Jordanian public and private IT firms, create and maintain a fruitful competitive environment while attracting further investment into the ICT sector, according to National ICT Strategy Document for Jordan. From a firm's perspective, collaboration and cooperation among employees and customers as well as trade partners can reduce cost and enhance the value and effectiveness of the organization.

Among the least important technology issues were virtualization, service-oriented architecture, and social media/networking. Virtualization may have been rated low because of the decreasing costs of IT hardware, unwillingness to cover the initial costs of virtualization and a lack of

---

[1] http://moict.gov.jo/uploads/Policies-and-Strategies-Directorate/Strategies/Jordan-National-Information-and-Communications-Technology-Strategy-2013-2017.pdf.

immediate need or urgency. The surprising finding was that the Jordanian respondents did not find social media very important. For many companies in the US and the western world, social media has revolutionized the way they get in touch and do business with their customers. One possible explanation is that social media is relatively new in Jordan and users may not have gotten comfortable with doing business using social media, and firms, in turn, may not have established social media presence. We expect that this result would change over time.

18.7   Individual IT Employee Issues

Individual IT employee issues were investigated under seven broad themes. The detailed analysis of the items under each theme is provided in Table 18.4. Note that each item is rated using a scale that ranges from

Table 18.4:   Individual IT Employee Issues in Jordan

Individual Issues	Mean Rating*	Std. Deviation
Job satisfaction		
In general, I like working here.	1.85	0.910
All in all, I am satisfied with my current job.	2.31	1.042
In general, I don't like my current job.	3.82	1.164
Work pressure		
I feel that the number of requests, problems or complaints that I deal with at work is more than expected.	3.02	1.137
I feel that the amount of work I do interferes with how well it is done.	3.01	1.175
I feel busy or rushed at work.	2.79	1.134
I feel pressured at work.	2.70	1.164
Work–life balance		
There is a blurring of boundaries between my job and my home life.	3.17	1.246
My work-related responsibilities create conflicts with my home responsibilities.	3.34	1.314
I do not get everything done at home because I find myself completing job-related work.	3.26	1.271
Workload and burnout		
I feel drained from activities at work.	3.08	1.247
I feel tired from my work activities.	2.96	1.188
Working all day is a strain for me.	2.66	1.264
I feel burned out from my work activities.	3.06	1.249

(*Continued*)

A. H. Turan, N. G. Uğur, & P. Palvia

Table 18.4: (*Continued*)

Individual Issues	Mean Rating*	Std. Deviation
Sense of accomplishment		
I feel I'm making an effective contribution to what this organization does.	2.05	0.874
In my opinion, I do a good job.	1.83	0.861
I have accomplished many worthwhile things in this job.	1.86	0.859
At my work, I feel confident that I am effective at getting things done.	1.77	0.855
Threats to one's job		
I am worried that future technology advancements may pose a threat to my job.	3.20	1.264
I believe that other people may be able to perform my work activities.	2.74	1.017
I am concerned that my job may be eliminated soon.	3.65	1.265
I am concerned that my job may be outsourced soon.	3.58	1.211
Career plans		
I will be with this organization 1 year from now.	2.80	1.291
I will take steps during the next year to secure a job at a different organization.	3.02	1.247
I will be with this organization 5 years from now.	2.74	1.260
I will be working in the IT field 1 year from now.	2.79	1.361
I will take steps during the next year to secure a job outside the IT field.	3.39	1.324
I will be working in the IT field 5 years from now.	2.25	1.215

*Rating scale ranges from 1 to 5: 1 as strongly agree and 5 as strongly disagree.

1 to 5, where 1 is strongly agree and 5 is strongly disagree. Thus a lower score (below 3) shows higher agreement.

The results for the individual issues should be viewed in the context of the IT industry in Jordan. The IT industry is vibrant and well supported by the kingdom. Jordan is an emerging knowledge economy, and banking and finance organizations lead the industry with state-of-the-art applications using cutting-edge technologies and tools.[2] The IT industry has transformed since the 2001 telecom liberalization and continues to attract highly qualified and young talent.

In terms of job satisfaction, the respondents seemed to be well satisfied with their jobs. The IT sector in Jordan pays quite well and young IT professionals usually make good money compared to their peers in other

---

[2]https://en.wikipedia.org/wiki/Economy_of_Jordan.

industries. In addition, many young people in the IT industry are single and do not seem to feel the pressure of their married peers in terms of spending on living expenses.

Looking at work pressure, workload, and burnout together, there seems to be a mixed picture. While the IT employees perceive higher workload and work pressure, they seem to handle it well without being overly drained out or experiencing work burnout. By the same token, there does not seem to be much of a work–home conflict. Note that most participants are usually young and single professionals; thus they may not have many demands at home. This may not be true for older and more experienced employees.

The IT employees in our sample have very high self-efficacy and sense of accomplishment; the average scores are between 1 and 2. Moreover, there is a considerable agreement in this belief as the standard deviations are much lower than the other items. These employees are young, well-educated and aware that there is a high demand for their IT skills in the job market. They also know that they are important contributors to the organization and their skills are valuable to the organization.

Consistent with the above observations, Jordanian IT workers feel very secure in their jobs. While they are realistic in acknowledging that their jobs can be done by others and they are replaceable, they do not see any threats of the job being eliminated, outsourced or replaced by technology advancements. An obvious explanation is a high demand for IT professionals and the supply not being commensurate with the demand.

Looking at the career and turnover data, Jordanian IT employees do not anticipate changing their job or the profession in the short term or even in the long term. As we observe from the previous analysis, the Jordanian IT staff seems to be quite happy and satisfied with their profession, working conditions, job security, and compensation. We expect this rosy outlook to continue given Jordan's emphasis on using IT to boost its economy. For example, the latest royal initiative called Reach 2025 is seeking to digitize Jordan's economy and create some 150,000 new digital jobs by 2025.

18.8 Conclusion

The implementation of the national Jordanian ICT strategy between 2007 and 2011 resulted in increased revenues and created 15,000 direct jobs and more than 80,000 indirect jobs. Over the years, Jordan has made efforts in various subsectors of the ICT sector, including telecommunications, e-education, e-health, the Internet, mobile and gaming content (UNCTAD,

2010). The newest initiative Reach 2025 is building on these achievements and creating an enhanced digital economy for the future.

Against this backdrop, this chapter identified organizational, technological, and individual issues of IT employees in Jordan. We found that the top five organizational issues in Jordan are: security and privacy, IT reliability and efficiency, business productivity and cost reduction, revenue-generating IT innovations, and alignment between IT and business. The top technology issues are networks and communications, business intelligence and analytics, business process management systems, collaborative and workflow tools, and big data systems. The IT employees seem to be very satisfied with their jobs and have a high sense of accomplishment. Their workload is heavy but manageable and they are likely to remain in the current job and the IT profession for a long time. Indeed, IT has much potential in Jordan and IT workers have a bright future.

References

Abuhmaid, A. (2011). ICT training courses for teacher professional development in Jordan. *Turkish Online Journal of Educational Technology-TOJET*, 10(4), 195–210.

Al-Jaghoub, S., & Westrup, C. (2003). Jordan and ICT-led development: towards a competition state?. *Information Technology & People*, 16(1), 93–110.

Gasaymeh, A. (2018). A study of undergraduate students' use of information and communication technology (ICT) and the factors affecting their use: A developing country perspective. *EURASIA Journal of Mathematics, Science and Technology Education*, 14(5), 1731–1746.

Heeks, R. (2002), Information systems and developing countries: Failure. *Success and Local Information Society*, 18(2), 101–112.

Jordantimes (2017). ICT sector contributes 12% of GDP. Retrieved on November 27, 2018, from http://www.jordantimes.com/news/local/ict-sector-contributes-12-gdp.

Omet, G., & Al-Zubi, K. (2004). Financial Reforms and Credit Supply: An Application to the Jordanian Banking System (1982–2002).

Palvia, P., Jacks, T., Ghosh, J., Licker, P., Romm-Livermore, C., Serenko, A., & Turan, A. H. (2017). The World IT Project: History, trials, tribulations, lessons, and recommendations. *Communications of the Association for Information Systems*, 41(18), 389–413.

Palvia, P., Ghosh, J., Jacks, T., Serenko, A., & Turan, A. (2018). Trekking the globe with the World IT Project. *Journal of Information Technology Case and Application Research*, 20(1), 3–8.

Petrini, M. & Pozzebon, M. (2004). *What role is "Business Intelligence" playing in developing countries? A picture of Brazilian companies*. HEC Montréal. Retrieved from http://expertise.hec.ca/gresi/wp-content/uploads/2013/02/Cahier0416.pdf.

Rajab, L. D., & Baqain, Z. H. (2005). Use of information and communication technology among dental students at the University of Jordan. *Journal of Dental Education*, 69(3), 387–398.

Samani, A. & Tahir, M. T. (2012). *A proposed training program in light of the professional training needs of teachers from the faculties of education in Sudanese universities* (Doctoral dissertation).

Tyler, J. (2018). Evolution of Enabling Technologies That Paved the Way for Modern ICT.

UNCTAD (2010). *Jordan Country Report*. The Inter-Sessional Panel of the United Nations Commission on Science and Technology For Development, 15–17 December 2010.

Chapter 19

Information Technology Issues in Lithuania

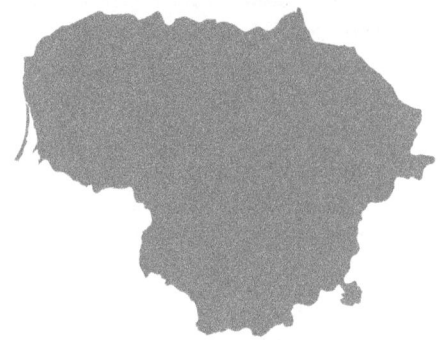

Eglė Vaičiukynaitė[*,‡], Rimantas Gatautis[*], Elena Vitkauskaitė[*,§], and Tim Jacks[†,¶]

*Kaunas University of Technology, Kaunas, Lithuania
†Southern Illinois University Edwardsville, Edwardsville, IL, USA
‡egle.vaiciukynaite@ktu.lt
§elena.vitkauskaite@ktu.lt
¶tjacks@siue.edu

Summary

In this chapter, we provide a better understanding of issues related to organizational, technological, and individual levels as perceived by Lithuanian information technology (IT) employees. This research applied multiple data-collection methods including emails, paper-based surveys, and web-administered surveys. A sample size of 300 IT employees was collected. The findings indicate that the most critical organizational issue in Lithuania involves IT reliability and efficiency while the least important issue is bring your own computing device (BYOD). Regarding technology-related issues, the most significant issue is software as a service. Importantly, Lithuanian IT employees are satisfied with their current job position and seek to maintain it in the near future even though IT professionals have a feeling of pressure at their workplace and face some uncertainty about their future in the IT industry. These findings might act as a catalyst for further analysis and enhance understanding of IT issues in Lithuania.

19.1 Introduction

Previous studies have focused mainly on IT-related issues in organizations mainly in the US and other developed countries. However, the role of information and communication technologies (ICTs) is growing in many countries in Europe. Furthermore, the development trends indicate that the ICT sector represents 4.8% of the European Economy (European Commission, 2018). The ICT sector plays an essential role in Lithuania as well. More specifically, the ICT industry is one of the most prominent sectors in Lithuania (PWC, 2018). However, studies on IT-related issues in Lithuanian organizations are limited. Therefore, this chapter seeks to present the perceptions of Lithuanian IT employees regarding organizational, technological- and individual-related issues.

19.2 Country Background and History

The Republic of Lithuania is situated in the Baltic region along the Baltic sea (on the West side) and bordered by Latvia (on the Northern side), Belarus (on the East and South sides), Poland (on the South side), and Kaliningrad area (on the Southwest side). Lithuania has less than 3 million people (OECD, 2018), while Vilnius is the capital and the largest city. Lithuania was the first Baltic state to declare itself independent in 1990, a year before the end of the Soviet Union. Since that time, Lithuania has integrated into the international community and joined the World Trade Organization in 2001, the European Union in 2005, and the Eurozone in 2015 (OECD, 2018). According to Trading Economics (2018), the gross domestic product (GDP) in Lithuania exceeded US$47 billion in 2017, and it entails 0.08% of the world's economy. Lithuania has many UNESCO world heritage sites (True Lithuania, 2018). Some examples include the historic center in Vilnius, the Curonian Spit, the Hill of Crosses in addition to historical and cultural traditions like Baltic songs and dance celebrations.

19.3   Information Technology in Lithuania

The information communication technology sector, which employs more than 31,500 IT professionals, is one of the most promising industries in Lithuania (PWC, 2018). Thus, the infrastructure of information technologies is well developed. For instance, Lithuania has 4G mobile communications infrastructure including the mobile 4G Internet, the fastest public Wifi, and the largest fiber-optic Internet network penetration rate in Europe (PWC, 2018). Moreover, Lithuania has already been selected by large international IT companies such as Uber, Wix.com, Unity, Adform, Revolut, Bentley, Devbridge Group, Virtustream and many others (Invest Lithuania, 2018).

Following Statista (2018a) findings, Lithuanian smartphone usage increased from 44% in 2014 to 56% in 2016. In a similar vein, the statistics indicate that the share of households in Lithuania that had access to the Internet increased as well for the period from 2007 to 2017 (Statista, 2018b). In 2017, 75% of Lithuanian households had Internet access (Statista, 2018b).

19.4   Methodology

The World IT Project (Palvia *et al.*, 2017; Palvia *et al.*, 2018) instrument included nine categories of constructs: national culture, organizational culture, IT occupational culture, personality (i.e., Big Five), organizational structure, organizational strategy, organizational IT issues, individual issues, technology issues and encompasses 157 close-ended questions (Palvia *et al.*, 2017). Seven questions were open and optional (e.g., country, additional comments about the survey). The instrument was translated into the Lithuanian language, and then back-translation was done to maintain appropriate meaning for the local culture. Moreover, two researchers compared it with the original World IT Project instrument to minimize semantic discrepancies.

All respondents were required to be in the IT profession. This research applied multiple data-collection methods including emails, paper-based surveys, and web-administered surveys. The survey link was posted on an online group for IT specialists titled "Programmers" and sent through email to IT specialists. Finally, the paper-based survey was administered in a face-to-face setting in major Lithuanian cities, particularly in Vilnius, Kaunas, and Klaipeda. This resulted in a higher response rate and better quality of data. More specifically, it was observed that

the completion of the survey required more than 25 minutes. Therefore, some online surveys were eliminated ($n = 15$) because they were
completed in less than 25 minutes. In total, a sample size of 300 IT
employees was collected. The analysis was performed by using IBM SPSS
Statistics v24.

In this sample, Lithuanian IT employees are mostly in early ages
between 21 and 29 years (51.3%) and the average age between 30 and
39 (22%) years (see Table 19.1). The second largest group is between 40
and 49 years (11%), followed by the 50 to 59 years group (8.7%). The most
involved IT roles are programming (39.7%), maintenance (17.3%), system
administrator (7%), and analysis and design (6.7%).

The least common IT roles are security (0.3%) and financial (0.3%),
followed by integration (1%), email/messaging systems (1.3%), consulting
(1.3%) and application support (1.3%). Nearly all respondents were of
Lithuanian nationality, and only two respondents were of a different

Table 19.1:   Descriptive Statistics

Characteristics	N	%	Characteristics	N	%
Age			Education		
18–20	18	6.0	High school or less	30	10.0
21–29	154	51.3	Associate degree	28	9.3
30–39	66	22.0	Bachelor's degree	134	44.7
40–49	33	11.0	Master's degree	100	33.3
50–59	26	8.7	Ph.D.	8	2.7
60+	3	1.0	Years of work experience		
Years of IT experience			0–4 years	99	33.0
0–4 years	136	45.3	5–9 years	87	29.0
5–9 years	65	21.7	10–19 years	60	20.0
10–19 years	54	18.0	20–29 years	35	11.7
20–29 years	30	10.0	30+ years	19	6.3
30+ years	15	5.0	Employee reporting relationship*		
Work as			IT department employee	198	66.0
Mostly full time	220	73.3	IT worker in a non-IT	56	18.7
Mostly part time	42	14.0	department		
Mostly over time	38	12.7	Contract employee	32	10.7
Part of management			Consultant	10	3.3
Not part of management	190	63.3	Vendor employee	2	0.7
In lower management	31	10.3	Been laid off from an IT job		
In middle management	40	13.3	Yes	71	23.7
In senior management	39	13.0	No	229	76.3

*Missing 2 values.

nationality, namely Latvian and Turkish. Importantly, the sample consisted of 79.7% of males.

19.5 Organizational Issues

The IT organizational issues included 18 single-item questions, and respondents rated them on a scale from 1 to 5 where 1 is considered the most important issue while 5 is the least important issue. The average rating value for all issues was calculated. Therefore, 18 organizational issues are ranked from 1 to 18 (see Table 19.2).

Results indicated that IT reliability and efficiency was the most important IT-related organizational issue (1.86). The second most important item was attracting and retaining IT professionals (2.01). It is critical to note that the following three items had very close means: security and privacy (2.02), revenue-generating IT innovations (2.06), and IT strategic planning (2.10). The same tendency was indicated with some other items (see Table 19.2). According to the respondents, the least important organizational issue was bring your own computing device (BYOD) (3.55).

Table 19.2: Organizational IT Issues

IT-Related Issues	Rank	Mean Rating	Std. Deviation
IT reliability and efficiency	1	1.86	0.814
Attracting and retaining IT professionals	2	2.01	0.911
Security and privacy	3	2.02	0.883
Revenue-generating IT innovations	4	2.06	0.974
IT strategic planning	5	2.10	0.892
IT service management (e.g., ITIL)	6	2.13	0.816
Project management	7	2.18	0.917
Business agility and speed to market	8	2.27	0.958
Continuity planning and disaster recovery	9	2.28	0.875
Knowledge management	10	2.30	0.913
Alignment between IT and business	11	2.37	0.896
IT cost reduction	12	2.39	0.960
Business productivity and cost reduction	13	2.42	0.955
Business process reengineering	14	2.61	0.950
Enterprise architecture	15	2.62	0.955
Globalization	15	2.62	0.990
Outsourcing	17	2.89	1.040
BYOD	18	3.55	1.119

*Rating scale ranges from 1 to 5: 1 as most important and 5 as no importance.

IT reliability and efficiency was ranked as the first in Lithuania and is consistent with the results of a study by Luftman *et al.* (2012). According to Luftman *et al.* (2012), the increasing complexities of IT systems contribute to the importance of IT reliability and efficiency. Another reason behind this can be the fact that Europe is increasing the use of various frameworks, guidelines, and standards such as ISAE3402 and SAS70 (Luftman *et al.*, 2012). However, contrary to previous studies (Luftman *et al.*, 2012), attracting and retaining IT professionals and security and privacy were considered as more critical issues.

The higher rating of security and privacy issues was predicted due to several reasons. First, IT companies have to ensure their clients' data privacy, especially for sensitive information which can be useful to competitors. Second, the more the data created and available for other parties, the greater the risk of losing them. Importantly, this issue can be aligned with other IT companies across different countries and in many regions. As a result, IT companies are seeking to implement data security solutions and/or implement more data loss prevention programs.

Regarding attracting and retaining IT professionals, IT specialists are critical for companies, and their percentage increased from 15% in 2014 to 18.1% in 2016 of the total number of a company's employees (Statistics Lithuania, 2018). The majority of young IT professionals are well-educated and speak English (Eurostat, 2018; Go Vilnius, 2018). Moreover, an IT job can be easily relocated, and they can earn a much bigger salary in other countries than in the local market. Therefore, the number of IT job opportunities for young and prospective IT specialists is increasing in Europe as market demand grows. Hence, companies have to offer competitive salaries for IT staff and ensure their retention.

Consistent with previous studies, enterprise architecture and globalization issues were ranked as less important (Luftman *et al.*, 2012). Interestingly, Lithuanian IT specialists rated outsourcing as less important. This can be explained by the fact that some IT services are localized and cannot easily be outsourced to other countries. Moreover, Lithuania over time has built its reputation as an outsourcing destination with cheaper labor.

19.6 Technology Issues

The technology issues included 16 single-items which respondents rated on a scale from 1 to 5 where one is considered the most important while five is the least important. The average rating value for every 16 single-items was calculated. Therefore, 16 technology issues were ranked from 1 to 16 (see Table 19.3).

Table 19.3:   Technology Issues

	Rank	Mean Rating	Std. Deviation
Software as a service	1	1.054	0.998
Networks/telecommunications	2	2.19	1.024
Collaborative and workflow tools	3	2.22	0.909
Virtualization (desktop or server)	4	2.35	2.27
Data mining	5	2.39	1.030
Big data systems	6	2.45	1.125
Service-oriented architecture (SOA)	7	2.48	1.050
Business intelligence/analytics	8	2.57	1.037
Cloud computing	8	2.57	1.072
Customer relationship management (CRM) systems	8	2.57	1.087
Business process management systems	11	2.60	1.002
Enterprise application integration	12	2.62	1.052
Enterprise resource planning (ERP) systems	13	2.66	1.096
Mobile and wireless applications	14	2.72	1.066
Mobile apps development	15	2.77	1.116
Social networking/media	16	2.93	1.140

*Rating scale ranges from 1 to 5: 1 as most important and 5 as no importance.

Results revealed that software as a service was the most important (1.054) among the technology-related issues (see Table 19.3). The top three most important issues rated by Lithuanian IT specialists were networks/telecommunications, collaborative and workflow tools, and virtualization (desktop or server). In sum, all these results present the most relevant trends in this area (Lambert, 2018).

The least important technology issues were mobile and wireless applications (2.72), mobile apps development (2.77), and social networking/media (2.93). These results might reflect a mature market or even considerable competition among global IT companies. More importantly, several global companies can be leading in specific geographies that other companies find it impossible to enter their market.

19.7   Individual IT Employee Issues

The individual IT employee issues involved 7 aspects including job satisfaction, perceived work overload, work–life balance, work exhaustion/strain, professional self-efficacy, job insecurity, and turnover intention. In total, there were 28 different items because each item category has sub-items (see Table 19.4). Lithuanian IT specialists rated each statement on a scale from 1 to 5 where one is considered strongly agree while five is strongly disagree.

Results reveal that respondents were satisfied with their current job (2.04). Despite this, respondents have indicated the moderate influence of feeling pressured at their job (3.29). Importantly, the standard deviations of items for perceived work overload are quite high and can signal there are some differences among organizations. In other words, some organizations may have a different organizational culture or lower-paced industry where

Table 19.4:   Individual IT Employee Issues

Items	Mean	Std. Deviation
Job satisfaction		
In general, I like working here.	1.88	0.880
All in all, I am satisfied with my current job.	2.04	0.940
In general, I don't like my current job/reversed.	4.06	1.022
Perceived work overload		
I feel that the number of requests, problems or complaints that I deal with at work is more than expected.	3.45	1.061
I feel that the amount of work I do interferes with how well it is done.	3.43	1.006
I feel busy or rushed at work.	3.03	1.053
I feel pressured at work.	3.29	1.077
Work–life balance		
There is a blurring of boundaries between my job and my home life.	3.55	1.145
My work-related responsibilities create conflicts with my home responsibilities.	3.91	1.070
I do not get everything done at home because I find myself completing job-related work.	3.51	1.178
Work exhaustion/strain		
I feel drained from activities at work.	3.16	1.017
I feel tired from my work activities.	3.20	1.019
Working all day is a strain for me.	3.67	1.036
I feel burned out from my work activities.	3.54	1.052
Professional self-efficacy		
I feel I'm making an effective contribution to what this organization does.	2.14	0.886
In my opinion, I do a good job.	2.10	0.793
I have accomplished many worthwhile things in this job.	2.10	0.845
At my work, I feel confident that I am effective at getting things done.	2.17	0.841

(*Continued*)

Table 19.4: (*Continued*)

Items	Mean	Std. Deviation
Job insecurity		
I am worried that future technology advancements may pose a threat to my job.	3.75	1.057
I believe that other people may be able to perform my work activities.	2.45	1.002
I am concerned that my job may be eliminated soon.	3.86	1.063
I am concerned that my job may be outsourced soon.	3.78	1.039
Turnover intention		
I will be with this organization 1 year from now.	2.37	1.001
I will take steps during the next year to secure a job at a different organization.	3.38	1.117
I will be with this organization 5 years from now.	2.87	1.066
I will be working in the IT field 1 year from now.	1.76	0.955
I will take steps during the next year to secure a job outside the IT field.	3.48	1.184
I will be working in the IT field 5 years from now.	1.86	0.969

Rating scale ranges from 1 to 5: 1 as strongly agree and 5 as strongly disagree.

employees do not feel pressure at their job. However, it can also depend on the age of employees. For instance, a mature professional may be able to manage his/her feelings of pressure and stress better. In general, IT employees seem to be satisfied with their job conditions and can handle some difficulties.

Regarding IT specialists' ratings of work–life balance, results show that the average score (value) of four items was close to 3.3 (see detailed values in Table 19.4), but the standard deviations were quite high. It also can be assumed that there were some differences between various age groups. Additionally, the respondents were mainly young (21–29) (Table 19.1) and might not face work–life imbalance as much as older IT specialists. In a similar vein, results revealed that the average score of four items of work exhaustion/strain group is 3.39, but the values of standard deviations were also quite high.

Concerning professional self-efficacy, results indicated that IT employees were satisfied with their accomplishments at work. Respondents indicated that they do feel confident about their work (2.17). Importantly, the standard deviation of all items was low.

Despite positive current evaluations of job satisfaction and their self-efficacy, respondents did feel insecure in their work. These feelings can be explained by the fact that the number of IT specialists is high (Eurostat, 2018). Moreover, every year 1,500 new IT specialists graduate in Lithuania (Go Vilnius, 2018). Furthermore, the rapid development of ICTs can also affect their current work position and, thus, can be replaced by an even cheaper workforce from other countries. As a result, IT professionals have to stay on top of new technologies and have the appropriate skills that it takes to commit to being lifelong learners.

The last but not the least indicator is the turnover intention. The results indicated that IT professionals seek to ensure their current position. The standard deviation was high which again signals that there are some differences among respondents' responses. As we have discussed earlier, the salaries are very competitive in the IT industry but also there is strong competition among IT specialists for jobs, and respondents are likely to seek to secure their current position.

19.8   Conclusion

This chapter provides a better understanding of the attitudes towards IT organizational, technological and individual issues by IT workers in Lithuania. The chapter indicates the most and the least critical issues within the diverse IT-related issues. The most critical organizational issue in Lithuania involves IT reliability and efficiency while the least important issue is BYOD. Regarding technology-related issues, the most critical issue is software as a service and the least is the social networking/media issue. Among individual IT employee issues, results indicate that respondents are satisfied with their current job position and seek to maintain it in the near future. Thus, results show that respondents do have a feeling of pressure at their workplace and do face some uncertainty about their future in this industry. These results can act as a catalyst for further analysis, and foster a better understanding of IT issues in the context of Lithuania.

Acknowledgments

Late Professor Dr. Rimantas Gatautis led Kaunas University of Technology researcher team into World IT Project and through the participation in it. We deeply appreciate his inspiration and guidance.

References

European Commission (2018). ICT Research & Innovation. Retrieved from: https://ec.europa.eu/programmes/horizon2020/en/area/ict-research-innovation

Eurostat (2018). ICT specialists in employment. Retrieved from: http://ec.eu ropa.eu/eurostat/statistics-explained/index.php/ICT_specialists_in_emplo yment

Go Vilnius (2018). Retrieved from: http://www.govilnius.lt/business/key-busine ss-sectors-vilnius/ict/

Invest Lithuania (2018). Retrieved from: https://investlithuania.com/key-sector s/technology/

Lambert, S. (2018). 2018 SaaS Industry Market Report: Key Global Trends & Growth Forecast. Retrieved from: https://financesonline.com/2018-saas-in dustry-market-report-key-global-trends-growth-forecasts/

Luftman, J., Zadeh, H. S., Derksen, B., Santana, M., Rigoni, E. H., & Huang, Z. D. (2012). Key information technology and management issues 2011–2012: An international study. *Journal of Information Technology, 27*(3), 198–212.

OECD (2018, July). OECD Economic Surveys: Lithuania 2018. Retrieved from: https://lrv.lt/uploads/main/documents/files/OECD%20Economic%20Sur veys%20-%20Lithuania%202018.pdf

Palvia, P., Jacks, T., Gosh, J., Licker, P., Romm-Livermore, C., Serenko, A., & Turan, A. H. (2017). The World IT Project: History, trials, tribulations, lessons, and recommendations. *Information Systems, 41*, 389–413.

Palvia, P., Ghosh, J., Jacks, T., Serenko, A., & Turan, A. (2018). Trekking the globe with the World IT Project. *Journal of Information Technology Case and Application Research, 20*(1), 3–8.

PWC (2018). Business guide Lithuania 2018. Retrieved from: https://www.pwc. com/lt/lt/assets/publications/PwC_Business_Guide_Lithuania_2018.pdf

Statista (2018a). Connected device usage rate in Lithuania 2014 to 2017, by device. Retrieved from: https://www.statista.com/statistics/347067/conn ected-device-usage-lithuania/

Statista (2018b). Share of households with internet access in Lithuania from 2007 to 2017. Retrieved from: https://www.statista.com/statistics/377738/hous ehold-internet-access-in-lithuania/

Statistics Lithuania (2018). Įmonės, kuriose dirba IT specialistai. Retrieved from: https://osp.stat.gov.lt/statistiniu-rodikliu-analize?indicator=S4R087#/

Trading Economics (2018). Lithuania GDP. Retrieved from: https://tradingecon omics.com/lithuania/gdp

True Lithuania (2018). UNESCO sites in Lithuania. Retrieved from: http://www. truelithuania.com/unesco-sites-in-lithuania-4438

Chapter 20

Information Technology Issues in Republic of Macedonia

Nikola Levkov[*,§], Mijalche Santa[*,¶], Tim Jacks[†,‖],
and Aykut Hamit Turan[‡,**]

[*]Ss Cyril and Methodius University, North Macedonia,
Republic of Macedonia
[†]Southern Illinois University Edwardsville,
Edwardsville, IL, USA
[‡]Sakarya University, Sakarya, Turkey
[§]nikolal@eccf.ukim.edu.mk
[¶]mijalce@eccf.ukim.edu.mk
[‖]tjacks@siue.edu
[**]aykut.turan@gmail.com

Summary

The information technology (IT) industry is recognized as a key sector in the Republic of Macedonia. It demonstrates continuous growth in terms of number of IT companies and employees. We present the key results from our World IT Project of Macedonian IT workers regarding organizational, technological and individual IT issues in this chapter. The most important organizational issues are knowledge management, lack of IT staff, and very high IT employee turnover rate in the Macedonian sample. Business intelligence/analytics and software as a service are also identified among the most important technology and infrastructure issues. Finally, regarding the individual IT employee issues, it should be noted that most of the IT professionals in Macedonia in general are quite satisfied with and they like their current job. Thus, employment in IT industry in Macedonia continues to be very attractive for young people.

20.1   Introduction

Small developing countries have their own characteristics that enable or limit their successful engagement in development and implementation of information technologies. There are progressive examples of small developed countries that have achieved a high level of sophistication in implementing information technologies. However, there is a need to identify how a small developing country can achieve this progress. The first step is to identify the human potential of the country. To achieve this, there is a need for primary research on employee-related information technology (IT) issues. Republic of Macedonia is an example of a small developing country that is making attempts to improve its digitalization level. Thus, primary research on employee-related IT issues could have a valuable contribution. In this chapter, we will present the results of a research that investigated the perceptions of Macedonian IT employees with regard to organizational IT, technological, and individual issues in the scope of the World IT Project.

20.2   Country Background and History

The Republic of Macedonia is a small land-locked country in the middle of the Balkan peninsula. Throughout its history, the present-day territory of Macedonia has been a crossroads for both traders and conquerors moving between the European continent and Asia Minor. It gained its independence in 1991 and, in 1993, the Republic of Macedonia joined the United Nations. In 2003, Macedonia became a member of the WTO, and in 2005, it became a candidate country for accession to the European Union (EU). However, Greece blocked the initiation of the accession negotiations with the EU, as well as Republic of Macedonia's admission to NATO. Despite this, access to EU remains the main driver for political and economic reforms in the country.

Regarding the economy, Macedonia is classified as an upper-middle-income country with a small, open market economy that has made great strides in reforming its economy over the past decade (World Bank, 2018). The economy is characterized by macroeconomic stability and low inflation. The key industries in the Republic of Macedonia are manufacturing, trade and agriculture. European countries are the major trade partners of Macedonia, with Germany accounting for more than 45% of the total export. Export growth is driven by foreign direct investments and export-oriented companies (Santa and Kekenovski, 2013).

20.3   Information Technology in Republic of Macedonia

Information communications technology (ICT) in Macedonia is acknowledged by the government and society as an important driver for Macedonia's development. Through the years, we see constant improvement in Macedonia's network readiness. The country reached a rank of 46 out of 139 countries in 2016 (Baller *et al.*, 2016).

According to the data of the State Statistical Office of the Republic of Macedonia (DZS, 2017, 2017), as of January 2017, 67.1% of the households and 91.2% of the enterprises with 10 or more employees had broadband connections to the Internet (via fixed or mobile broadband connection). During 2016, of the persons ever using the Internet, 24.8% ordered/bought goods or services over the Internet in the last 12 months, and the majority of them (64.1%) bought clothes or sports goods. While 10.3% of enterprises with 10 or more employees had e-commerce, i.e., buying or selling goods or services over computer networks (via websites or EDI-type systems), 6.0% of enterprises did online selling, and 5.8% of enterprises did online purchasing. Also, 54.2% of the enterprises used social media (e.g. Facebook, LinkedIn, Twitter, Present.ly, YouTube, Flickr, Picassa, Wiki-tools, etc.), i.e., had a user profile, an account or a user license for using certain social media.

Macedonia has a high level of E-Government Development Index. According to UNPAN (Nationen, 2016), by 2016 Macedonia had made significant progress in engaging citizens through ICTs in policy, decision-making, and service design and delivery in order to make it participatory, inclusive, and deliberative. Thus, it has high E-Participation Index (EPI) of 0.6102, but it is still not considered to be in Stage 3 (Nationen, 2016). Thus, there is still room for improvement.

Regarding the ICT industry, the total ICT market value in Macedonia had an estimated USD 500 million in 2016. Hardware was the largest share (55%). ICT services were the second largest portion (30%), and software comprised 15% of the ICT market in the country (InvestMacedonia, 2016). To support the development of the software industry, in 2010 the "Export Promotion Strategy for the Macedonian Software and IT Services Industry" was developed. The overall goal of the strategy is to establish Macedonia as a well-recognized brand for specialized, high quality outsourcing services and software products within Europe. Furthermore, the ICT industry is recognized as a key industry for foreign direct investments. As a result, there is a constant increase of IT companies providing services to international companies. As of 2017, there were more than 1,650 ICT companies in Macedonia. This is a tenfold increase from 10 years ago. There are approximately more than 14,690 employees in the ICT industry. In the next section, we will present the findings of the survey we administered among these employees.

20.4   Methodology

Collecting primary data for the World IT Project (Palvia *et al.*, 2018) through online surveys was a challenging task for the two country investigators. We modified and used original instrument by Palvia *et al.* (2017). At the very beginning of the data collection process, we encountered some of the common difficulties related to collection of survey data. Some of the challenges we faced were related to the creation of the mailing list with potential respondents and being able to obtain a high rate of response. Also, companies usually are not very open for this kind of collaboration, especially in developing countries, mainly because of a low level of trust within the business sector and between business and the state government. Therefore, we asked for help from some professional IT associations to provide us with e-mail contacts of their members. In most companies, we were using our personal contacts with management executives to motivate their IT staff to fill out the online survey. In total, 304 responses were collected between March and July 2016, and after cleaning up the data a total of 294 usable observations were included in the study.

Macedonian is considered the national language of the country, yet all national minorities who are above 25% of the population in municipalities where they live can use their mother tongue as the official language. Before implementing the survey we had serious discussions regarding the choice of the language in which to conduct the online survey. At the end, we decided to administer the survey instrument in its original English language version, because today most of the Macedonian IT workers are very fluent in English. Also, English is taught in Macedonia from elementary school and we wanted to avoid the potential problems with the translation such as semantic noise or lack of adequate synonyms for IT terminology in the Macedonian language. The data collection process for IT workers in Macedonia was done through an online survey platform using SurveyMonkey. All questions from the survey instrument were imported into the SurveyMonkey platform and pre-testing was done with three of our colleagues from our university. Based on their suggestions, we made minor corrections. We removed mainly typing errors in the online survey.

To increase response rate, we have used IT professionals and associations mailing list in addition to our own personal contacts (mainly our LinkedIn contacts) which we have established through our collaboration with the IT industry professionals (such as: IT managers, developers, network administrators, etc.). Occasionally we were interacting with our contact persons by phone or email to remind them to fill out the online survey, or to remind their colleagues to fill out the online survey. We did not provide any gifts or incentives to the respondents to complete the survey, but we explained clearly the objective and importance of the project through our email and phone communication with the respondents.

The two Macedonian investigators worked as a team and divided the data collection endeavor almost equally between them. They were constantly communicating personally, through phone and email and they coordinated their effort to succeed in the data collection process. At the end, all results from the filled surveys were automatically coded within the SurveyMonkey platform and were directly exported into an Excel spreadsheet. We then made a visual inspection in order to validate the accuracy of the coded variables. Table 20.1 shows the descriptive statistics of the IT professionals from Macedonia.

The data presented above in Table 20.1 provide several interesting insights regarding demographic characteristics of IT professionals from Macedonia included in the study. First, a significant number of IT professionals included in the study as survey respondents are with a bachelor's (65%) or master's (27.2%) degree. This indicates that most of the IT professionals working in the IT industry in Macedonia have a formal

Table 20.1:   Descriptive Statistics

Characteristics	N	%	Characteristics	N	%
Education:			Years of work experience:		
High school or less	11	3.7	0–4 years	120	40.8
Associate degree	11	3.7	5–9 years	97	33.0
Bachelor's degree	191	65.0	10–19 years	58	19.7
Master's degree	80	27.2	20–29 Years	18	6.1
Ph.D.	1	0.4	30+ years	1	0.4
Years of IT experience:			Organizational location:		
0–4 years	148	50.3	IT department employee	177	60.2
5–9 years	96	32.7	IT worker in non-IT department	24	8.1
10–19 years	38	12.9	Contract employee	49	16.7
20–29 years	11	3.7	Consultant	35	11.9
30+ years	1	0.4	Vendor employee	9	3.1
Work as:			Work position:		
Mostly full time	244	83.0	Not part of management	148	50.3
Mostly part time	14	4.8	In lower management	38	13.0
Mostly over time	36	12.2	In middle management	68	23.1
Been laid off from IT job:			In senior management	40	13.6
Yes	37	12.6			
No	257	87.4			

academic education. Second, the majority of the respondents have modest IT experience (83%) with 0–9 years working in the IT field and at the same time majority work as full-time employees (83%) and also have never been laid off from IT job (87.4%). The distribution of their total working experience is not significantly different when it is compared to the distribution of their IT experience. Third, a great number (60%) of the IT professionals who took part in the survey work as IT department employees, or as a contract (16.7%) or consultant (11.9%) employee. Fourth, half of the respondents (50.3%) included in the study are not part of the management, while only 23.1% are part of the middle management. Only a small number of the respondents (13.6%) are part of senior management.

20.5   Organizational IT Issues

The data in Table 20.2 show the organizational IT issues for IT professionals from Macedonia in order of importance. Security and Privacy is the only item in the top five which matches an item in the top five of the 2017 SIM IT Key Issues and Trends study (Kappelman *et al.*, 2018). Furthermore,

Table 20.2:   Organizational IT Issues in Macedonia

Organizational IT Issues	Rank	Mean Rating*	Std. Deviation
Knowledge management	1	1.79	0.71
Project management	2	1.84	0.72
IT reliability and efficiency	3	1.85	0.71
Security and privacy	4	1.86	0.79
Attracting and retaining IT professionals	5	1.93	0.79
IT strategic planning	6	1.98	0.78
Alignment between IT and business	7	2.04	0.71
Business agility and speed to market	8	2.10	0.75
Revenue-generating IT innovations	9	2.12	0.69
Continuity planning and disaster recovery	10	2.12	0.80
IT service management (e.g., ITIL)	11	2.17	0.78
Business process reengineering	12	2.17	0.78
Business productivity and cost reduction	13	2.25	0.73
Enterprise architecture	14	2.30	0.73
Globalization	15	2.43	0.83
IT cost reduction	16	2.54	0.88
Outsourcing	17	2.75	0.98
Bring your own computing device (BYOD)	18	3.35	1.04

*Rating scale ranges from 1 to 5: 1 as most important and 5 as no importance.

only five items in the top 10 match items in the top 10 of the 2017 SIM issues list.

The highest-ranked item is knowledge management which does not appear in the 2017 SIM issues list. For Macedonian IT organizations, this is a key issue because most of the IT departments of large companies are confronted with a lack of IT staff and a very high rate of IT employee turnover. The lack of IT staff often puts pressure on IT professionals to do multiple tasks (for example: system design, development, and implementation) and means not having enough time to well document the system functional requirements and IS changes. This reality followed with very high IT employee turnover rate is creating a lot of problems for companies in Macedonia. When the knowledge for a certain IT system or software development project resides only within the IT professional, knowledge is lost when he or she leaves the organization. The IT industry in Macedonia has the highest average salary and the quest for highly skilled IT professionals is very competitive among companies. Nevertheless, there is a high mobility of IT professionals coming from abroad, since there is huge demand for IT professionals on the global market. Also, the use of intranet systems as corporate yellow pages provide very useful information to IT professionals regarding various IT projects and people expertise in certain IT domains. The use of collaborative tools for knowledge sharing and communication is very important for Macedonian software companies working often on outsourcing projects. Hence, building knowledge-based systems is ranked as the number one priority for Macedonian companies as an attempt to protect themselves from IT employee turnover, through preservation of tacit knowledge residing in the heads of IT professionals.

The second highest-ranked issue is project management. Project management skills are very crucial for the IT industry and especially for the software industry in Macedonia. A large number of software companies in Macedonia work with clients from abroad (mostly the US and EU) who have outsourced software development projects to Macedonian companies. The success of those companies is highly dependent on the quality of their project management skills, because they need to effectively manage their available manpower and timelines to deliver the required functionality of the software product within the time and budget constraints. Also, many companies in Macedonia have low levels of IT maturity and lack of experience in successful IT project management. Often, smaller software development companies combine multiple roles in one person (for example, project manager and developer) and IT professionals often do not have enough time to devote solely to project management. Thus, the need for better IT project management skills is highly demanded and important for the Macedonian IT and software industry. The project management and knowledge management issues are highly related with the need of Macedonian IT companies to effectively respond to changing client requirements.

IT reliability and efficiency was ranked as the third most important issue by the IT professionals in Macedonia. The IT sectors from larger companies — especially from telecommunications, insurance, and banking industries — are primarily concerned with providing reliable and efficient services to their clients. This is a challenging task for IT professionals in Macedonia confronted with relatively low IT budgets for development and implementation of new IT systems. Often, problems with having a good business case and ROI become major issues in financing new IT projects. Therefore, the software industry is working mainly for foreign clients on outsourced projects and domestic companies cannot afford to budget serious IT projects from local IT companies. The relatively small size of the companies and their small profits limit their potential to compete for IT services on the Macedonian market with foreign companies with substantially larger budgets.

Security and privacy is ranked as the fourth important issue, which is also ranked as the first in the 2017 SIM issues list. This issue is a global challenge, faced by IT companies across the world, and is not specific to any particular country. The Facebook-Cambridge Analytica data scandal surely adds fuel to the fire globally and Macedonian IT companies are not isolated from global trends.

The fifth ranked issue is attracting and retaining IT professionals. Although this issue has global roots as a result of high demand for IT professionals on the global market, it has also inherent local roots. Macedonia, as a post-communist country, has had a long transition period followed by political crises, internal arms conflicts, high rate of unemployment, and poverty. All these factors have a strong influence on the decision of young and skillful IT professionals to leave the country, work and live abroad in more developed Western countries. This created high mobility in the local IT market and a high rate of employee turnover in the IT profession. Therefore, it is one of the most challenging issues for IT companies in Macedonia. Many IT companies in Macedonia do not offer a lot of other work benefits and incentives, professional training, and good working climate. Most of the foreign investments in the IT sector in Macedonia have brought different working practices and better working environments for IT professionals. Therefore, a lot of IT professionals from Macedonia who are going abroad are not only motivated by large salaries, but also seeking better working environments and incentives.

Some of the reasons behind the significant differences between the SIM list of issues and issues identified in the Macedonian sample result from high mobility in the IT employee market. This creates challenges for Macedonian companies to keep the best IT professionals for prolonged periods within their companies. A large number of IT professionals work in the software industry primarily working on outsourced projects. The main costs for those companies are the salaries of IT professionals and struggling to reduce the costs and at the same time to improve the quality of delivered products, which is a major challenge for them. In summary, the Macedonian IT industry must invest greater effort in developing their own innovative software products in order to have products with higher added value and higher profits. Using IT professionals only as a "brute force" for coding the functional requirements in the form of outsourced software projects will not create fast and substantial growth for the IT industry and economic development. IT companies will still fight for IT professionals in the near future to keep them within their own companies and preventing them from emigrating abroad for higher salaries and a better standard of living.

20.6   Technology and Infrastructure Issues

Table 20.3 shows the technology and infrastructure issues from Macedonia ranked in order of importance. Two of the top five responses (Business intelligence/Analytics and Software as a Service) from Macedonia match those in the top five of the 2017 SIM IT Key Issues and Trends study (Kappelman *et al.*, 2017). Five of the top 10 in Macedonia list match the top 10 in the USA list. Overall, we can confirm similarity to a certain extent in the technology and infrastructure issues between Macedonia and the 2017 SIM report.

The top issue in both lists is business intelligence and analytics. AI as a base for business intelligence has become a globally widespread disruptive technology and will be more widely available in the near future. For Macedonian telecommunication, banking, and retail industries, it is becoming a very important priority and a lot of IT investments within these industries are focused on developing business intelligence capacities. Also, smaller IT service companies provide business analytics services for external clients (for example, business analytics support for improving customer relationship

Table 20.3:   Technology and Infrastructure Issues in Macedonia

IT Related Issues	Rank	Mean Rating*	Std. Deviation
Business intelligence/analytics	1	2.06	0.85
Mobile and wireless applications	2	2.16	0.95
Software as a service	3	2.16	0.87
Collaborative and workflow tools	4	2.19	0.86
Customer relationship management (CRM)	5	2.20	0.99
Enterprise application integration	6	2.22	0.84
Business process management systems	7	2.29	0.89
Service-oriented architecture (SOA)	8	2.30	0.95
Enterprise resource planning (ERP) systems	9	2.34	0.94
Networks/telecommunications	10	2.38	1.05
Big data systems	11	2.39	1.00
Mobile apps development	12	2.42	1.03
Cloud computing	13	2.43	0.94
Virtualization (desktop or server)	14	2.45	0.93
Social networking/media	15	2.47	0.93
Data mining	16	2.49	1.00

*Rating scale ranges from 1 to 5: 1 as most important and 5 as no importance.

management). Providing services in business intelligence requires highly skilled IT professionals and necessary IT infrastructure. Educational institutions in Macedonia are under continuous reforms, but still struggling to find the right path in developing highly skillful IT experts in the field of business intelligence and analytics. Overall, the assumption behind the high rank of this issue is mainly a result of the high need for business intelligence and analytics skills and services. The capacity of the IT industry in Macedonia for providing high quality of this type of service is still at a modest level. Also, ranking big data systems as the eleventh issue and data mining as the last, the 16th issue, shows that the Macedonian IT industry uses business intelligence and analytics on a more modest level, in the form of reporting. Although some larger companies from telecom and banking industry have their own data warehouse centers, the small size of Macedonian companies does not allow implementation of big data concepts. Hence, the importance of business intelligence and analytics for Macedonian companies is the direct result of global trends.

The second highest-rated technology and infrastructure issue in Macedonia was mobile and wireless applications. The high rank of mobile applications is part of the global trend, yet there is growth in providing this type of application on the Macedonian market. Mobile banking recently became a reality in the banking industry in Macedonia. Although mobile apps development was ranked as the 12th issue, there is a growing number of software companies in Macedonia working on development of mobile applications for global markets as their own innovative products or working for foreign clients on outsourced projects. The fourth generation (4G) of broadband cellular network technology in Macedonia provided the required infrastructure for deploying more mobile applications to the market. Recently, it was announced that preparation activities were taking place for introducing 5G technology. Following global trends, mobile phones in Macedonia have surely become the most important devices. This gave incentive to the IT industry to offer a wide range of mobile applications on the market, but it is still important in the future to improve back-end services in order to provide suitable IT infrastructure to successfully deploy more mobile applications.

Software as a service (Saas) is at the third highest position, while collaborative and workflow tools appear in the fourth place. SaaS issues appeared as a very important challenge for business clients as well as for IT service vendors. Often business clients do not have budgets to finance buying of software licenses and require IT vendors to offer a business model in the form of SaaS. On the other hand, Macedonian IT vendors do not have good ROI to offer IT services in the form of SaaS because the Macedonian market is too shallow for this type of business model. Collaborative and workflow tools are very important, especially for the software industry in Macedonia working mainly over the Internet for foreign clients. Also, some larger companies get support for their IT systems from abroad and the collaboration through this type of system is crucial for successful development and maintenance of their IT systems.

Customer relationship management (CRM) systems are ranked as the fifth issue. A large number of software companies in Macedonia work on development and maintenance of CRM systems. In developing countries such as Macedonia, in certain cases using CRM systems from well-known vendors is too expensive for local companies. Often global companies allow their branches in Macedonia to use CRM systems developed as local solutions from local vendors. In other cases, Macedonian software companies provide CRM systems to one or few branches of a global company which operates on many developing markets where the use of well-known brands of CRM systems is not cost effective, taking into account the small size of the market on which the branch operates.

It is interesting to mention that virtualization, social networking and data mining were ranked as the three least important issues. Virtualization (desktop or server) is ranked as one of the least important issues because Macedonian IT professionals do not consider it as a great challenge. Most of the companies in banking and telecommunication industry successfully use virtualization (more server virtualization) to save on IT infrastructure costs. The use of social media for marketing in the corporate world in Macedonia is still on a modest level, rarely having professional teams working on social media marketing. Nevertheless, first signs of developing applications for social networking platforms (such as Facebook) can be witnessed on Macedonian space. Data mining was ranked as the least important technology issue from Macedonian IT professionals. The small size of Macedonian companies limits the opportunities for implementing big data concepts and data mining. Two important factors have a strong influence for such low importance of data mining: (1) modest size of the data and Macedonian companies and (2) lack of expertise to implement such business analytics tools.

Macedonian IT professionals rank certain technology and infrastructure issues as more important compared to those in 2017 SIM IT Key Issues and Trends study, mainly as a result of the intertwined influence of global trends and local conditions in the IT industry.

20.7 Individual IT Employee Issues

Table 20.4 shows the summary of the responses regarding individual issues of IT professionals. The lower average score below 3.00 means higher agreement with each statement in terms of satisfaction. The analysis of responses regarding the job satisfaction of IT professionals in Macedonia shows that most of them in general are quite satisfied with and they like their current job. The IT industry in Macedonia provides the highest average salary

Table 20.4: Individual IT Employee Issues in Macedonia

Individual Issues	Mean Rating*	Std. Deviation
Job satisfaction		
In general, I like working here.	1.99	0.87
All in all, I am satisfied with my current job.	2.14	0.93
In general, I don't like my current job.	3.89	1.05
Work pressure		
I feel that the number of requests, problems or complaints that I deal with at work is more than expected.	3.14	1.03
I feel that the amount of work I do interferes with how well it is done.	2.83	1.03
I feel busy or rushed at work.	2.95	1.08
I feel pressured at work.	3.26	1.09
Work–life balance		
There is a blurring of boundaries between my job and my home life.	3.29	1.07
My work-related responsibilities create conflicts with my home responsibilities.	3.63	1.08
I do not get everything done at home because I find myself completing job-related work.	3.39	1.16
Workload and burnout		
I feel drained from activities at work.	3.27	1.05
I feel tired from my work activities.	3.23	1.10
Working all day is a strain for me.	3.16	1.09
I feel burned out from my work activities.	3.47	1.01
Sense of accomplishment		
I feel I'm making an effective contribution to what this organization does.	2.05	0.87
In my opinion, I do a good job.	1.86	0.71
I have accomplished many worthwhile things in this job.	1.97	0.80
At my work, I feel confident that I am effective at getting things done.	1.90	0.72

(*Continued*)

Table 20.4: (*Continued*)

Individual Issues	Mean Rating*	Std. Deviation
Threats to one's job		
I am worried that future technology advancements may pose a threat to my job.	3.60	1.09
I believe that other people may be able to perform my work activities.	2.62	0.98
I am concerned that my job may be eliminated soon.	4.00	0.99
I am concerned that my job may be outsourced soon.	3.94	0.96
Career plans		
I will be with this organization 1 year from now.	2.54	1.10
I will take steps during the next year to secure a job at a different organization.	3.15	1.10
I will be with this organization 5 years from now.	3.07	1.03
I will be working in the IT field 1 year from now.	1.89	0.94
I will take steps during the next year to secure a job outside the IT field.	3.71	1.11
I will be working in the IT field 5 years from now.	1.98	0.91

*Rating scale ranges from 1 to 5: 1 as strongly agree and 5 as strongly disagree.

compared to other industries, therefore, there is a great interest from students to study computer science and work as an IT professional. Also, there is a trend of retraining and changing professional backgrounds in the labor market towards the IT profession. Working as an IT professional in Macedonia is quite attractive and a well-paid profession for Macedonian labor market conditions.

The answers related to work pressure indicate that most IT professionals in Macedonia feel modest work pressure and do not feel very busy at work. There were similar surprising results regarding the scores for work–life balance and burnout. We were expecting to witness more problems in work–life balance and work overload among IT professionals in Macedonia. Apart from our initial expectations, the scores showed that most IT professionals do not feel burned out or drained from work and have the feeling that they can easily keep a balance between their work and private life. Most IT professionals have a high sense of accomplishment that they do a good job and effectively contribute towards the achievement of organizational goals.

IT professionals in Macedonia do not feel threatened regarding their job position. The lack of IT professionals in the market gives them a strong feeling of security. Also, the trend of emigration of IT professionals contributes towards shrinking the pool of available IT professionals in the labor market. Macedonia, as a developing country, is highly politically unstable and is not attractive for IT professionals from other countries, hence, Macedonian IT employees do not feel the threat from foreign peers. Although the salaries for IT professionals in Macedonia are very high in terms of Macedonian standards, they are quite low compared to the salaries of IT experts in developed countries such as the US or EU countries. All these local conditions give strong security to IT professionals to not feel that their job positions are threatened.

Most IT professionals did not show a strong interest to leave the organization or IT profession. The results on these issues are in line with their relatively high job satisfaction, job security and good work–life balance. It is easy to notice that most IT professionals did not consider leaving the IT profession and working in some other field in the future. They show a bit stronger interest in leaving the organization, but not the IT profession. Employment in the IT sector in Macedonia remains highly attractive and, therefore, most IT professionals do not consider changing their field of work.

Employment in the IT industry in Macedonia continues to be very attractive for young people. The Faculty of Computer Science and Engineering, at University Ss. Cyril and Methodius in Skopje (the largest university in Macedonia), has the largest number of student enrollments for 2018. The opportunities to work as a freelancer from home and providing self-employment income from Google ads, Facebook, and social media marketing becomes very attractive for future IT professionals. Many projects for software development, system integration and maintenance are outsourced to Macedonian software companies, which provide regular streams of income for the software industry. The IT maturity in larger companies in Macedonia is on a modest level and IT sectors within those companies do not have a key role in company development. Although global trends impact Macedonian business as well, the old style of management inherited from the communist system still looks on IT benefits with certain caution, relying more on personnel and human management skills. Some industries like telecommunication, finance and banking sector, pharmacy, retail industry, and service sector rely more on IT systems and business applications for success.

In recent years, there have been a larger number of IT startup companies, trying to offer their own creative and innovative product or service on the market. But again, IT service companies and software development companies working on outsourced projects are taking the major stake in the IT industry in Macedonia.

20.8   Conclusion

The top organizational concerns in Macedonia are knowledge management, project management, IT reliability and efficiency, security and privacy and attracting and retaining IT professionals. Top technology concerns include business intelligence and analytics, mobile and wireless applications, software as a service, collaborative and workflow tools and customer relationship management (CRM) systems. IT professionals from Macedonia are generally satisfied with their jobs and salaries because IT industry in Macedonia holds the highest average salary compared to other industries. The IT employee market is very dynamic, characterized by high mobility and high turnover rate mainly led by the process of emigration. Although most IT professionals are satisfied with their salaries taking in consideration the living conditions in Macedonia, they are starting to look for a better working climate, which is often offered by foreign companies in the IT sector.

References

Baller, S., Dutta, S., & Lanvin, B. (2016). The Global Information Technology Report 2016: Innovating in the Digital Economy. Retrieved from http://www.deslibris.ca/ID/10090686.

DZS. (2017). Државен завод за статистика - соопштение: Користење на информати чко-комуникациски технологии во домаќинствата и кај поединци, 2017. Retrieved September 5, 2018, from http://www.stat.gov.mk/PrikaziSoopstenie.aspx?rbrtxt=77.

InvestMacedonia. (2016). Information and Communications Technology. Retrieved September 5, 2018, from http://www.investinmacedonia.com/investment-opportunities/information-and-communications-technology.

Kappelman, L., Nguyen, Q., McLean, E., Maurer, C., Johnson, V., Snyder, M., & Torres, R. (2017). The 2016 SIM IT issues and trends study. *MIS Quarterly Executive*, 16(1).

Nationen, V. (Ed.). (2016). E-government in support of sustainable development. New York.

Palvia, P., Jacks, T., Ghosh, J., Licker, P., Romm-Livermore, C., Serenko, A., & Turan, A. H. (2017). The World IT Project: History, trials, tribulations, lessons, and recommendations. *Communications of the Association for Information Systems*, 41(18), 389–413.

Palvia, P., Ghosh, J., Jacks, T., Serenko, A., & Turan, A. (2018). Trekking the globe with the World IT Project. *Journal of Information Technology Case and Application Research*, 20(1), 3–8.

Santa, M., & Kekenovski, L. (2013). Hidden Champions of the Republic of Macedonia. In P. McKiernan & D. Purg (Eds.), *Hidden Champions in CEE and Turkey.* Berlin, Heidelberg: Springer, pp. 245–260. Retrieved from http://link.springer.com/chapter/10.1007/978-3-642-40504-4_16.

DZS (2017). Државен завод за статистика - соопштение: Користење на информати чко-комуникациски технологии во деловните субјекти, 2017. Retrieved September 5, 2018, from http://www.stat.gov.mk/PrikaziSoopstenie.aspx?rbrt xt=76.

Chapter 21

Information Technology Issues
in Malaysia

Ainin Sulaiman[*,‡], Noor Ismawati Jaafar[*,§],
and Prashant Palvia[†,¶]

[*] University of Malaya, Kuala Lumpur, Malaysia
[†] The University of North Carolina at Greensboro,
Greensboro, NC, USA
[‡] ainins@um.edu.my
[§] isma_jaafar@um.edu.my
[¶] pcpalvia@uncg.edu

Summary

This chapter discusses three main categories of information technology (IT) issues, namely, organizational, technological and individual. The empirical data for this study was obtained via a survey questionnaire of 300 IT professionals in various firms in the Multimedia Super Corridor area. The top three most important organizational issues are IT reliability and efficiency, security and privacy, and IT strategic planning. In terms of technology issues, mobile and wireless applications, networks/telecommunications, and business intelligence/analytics were perceived to be the most important. Generally, the IT professionals are satisfied with their job, feel a sense of accomplishment towards their job and are able to balance their work–life responsibilities.

21.1   Introduction

A review of the literature pertaining to information technology (IT) research showed that there are an abundance of studies that focused on developed countries such as the USA, Australia, Europe, Japan and Korea as well as some developing countries such as China, India, and the Middle East. Although there are studies pertaining to IT that have been conducted in Malaysia, most of them are on IT adoption or usage of a particular innovation from the users' perspective (either the organizations or individuals). Besides, there are not many studies that focus on the IT professionals. In addition, it was also observed that only a few examined organizational, technological and individual issues related to IT. Hence, the aim of this chapter is to highlight the organizational, technological, and individual issues related to IT among Malaysian IT professionals.

21.2   Country Background

Malaysia, formerly known as Malaya, is a small nation in the South East Asia region. Its neighboring countries include Singapore, Thailand, Indonesia and the Philippines. Malaya gained its independence from the British in 1957. It is a federal constitutional elective monarchy, the founding member of the Association of Southeast Asian Nations (ASEAN) and the Organization of Islamic Cooperation (OIC), a member of the Asia-Pacific Economic Cooperation, the Commonwealth of Nations, and the Non-Aligned Movement.

Malaysia consists of a federation of 13 states and three federal territories. The capital city of Malaysia is Kuala Lumpur, situated in the Federal Territory of Kuala Lumpur while the federal administrative center is Putrajaya located in the Federal Territory of Putrajaya. According to the Malaysian Department of Statistics, the country's population was 31.2 million in 2016. However, it is estimated that Malaysia had 32.4 million in 2018 as compared to 32.0 million in 2017 with an annual population growth rate of 1.1% (Department of Statistics Malaysia, 2018). The male population in 2018 outnumbered the females with 16.7 million (male) and 15.7 million (female). The sex ratio in 2018 remained at 107 males per 100 females. The percentage of working population aged 15–64 years old (working age) increased from 69.6% in 2017 to 69.7% in 2018. Malaysia has a diverse population consisting of three main ethnicities: Malays, Chinese and Indians. The composition of Bumiputera (Malays) ethnic recorded an increase of 0.3 percentage points in 2018 as compared to 2017, which accounts for 69.1% of total citizens. However, the composition of Chinese and Indian ethnicities dropped by 0.2 and 0.1 percentage points, respectively.

The Malaysian Constitution established Islam as the country's religion; however, however the population is free to practice any other religions. The official and national language is Malaysian, a standardized form of the Malay language, however, English is spoken widely, particularly among the urban population.

Malaysia's economy has traditionally been fueled by its natural resources, i.e., agriculture, forestry and minerals (Raman and Yap, 1996); however, it has expanded to other sectors such as science, tourism, and commerce. Malaysia is currently the world's largest center of Islamic Finance. Malaysia Gross Domestic Product (GDP) was worth US$314.5 billion in 2017, and the GDP annual growth rate increased by 4.8% from 2000 until 2018. The World Bank and the International Monetary Fund (IMF) recognize Malaysia as an emerging economy and a developing country.

21.3 Information Technology in Malaysia

As a developing country, Malaysia aspires to be a developed nation by 2025. One of the sectors that has been recognized to help achieve this is the IT industry, as the industry has continued to contribute substantially to the country's GDP. This is also evident as the roles and expectations of the IT industry have always been incorporated in Malaysia Development Plans and Policies (MDPP), particularly from 1996 onwards. The MDPPs are 5-year development plans and policies, the country is in its Eleventh Malaysia Plan (1996–2020). In the 11th Malaysia Plan, ICT is recognized as a facilitator to several strategies such as inclusiveness towards an equitable society, improving well-being for all, accelerating human capital development for an advanced nation as well as transforming public service for productivity. ICT has also received a significant mention in the nation's yearly financial budget. For example, in the 2018 budget, the then Prime Minister announced an allocation of RM270 million for the Police Force to upgrade their ICT equipment and communication systems. In addition, tax payers and SME owners are given a capital allowance for the purchase of ICT equipment which is claimable for a period of 4 years, the beginning year of assessment being 2018–2020.

According to Raman and Yap (1996), the origin of IT policies and plans in Malaysia can be traced back to the early 1980s. Since then many initiatives have been taken by the government to promote the ICT industry. Perhaps, among the most conspicuous initiatives, is the establishment of the Multimedia Super Corridor (MSC) in 1996. The MSC was envisioned to be a catalyst for ICT development in the country. It was supposed to create a new breed of entrepreneurs, attract global IT players to use the MSC as a test bed for new ideas, products and services. The Malaysian Digital Economy Corporation (MDEC) is a one-stop hub for all applications and services related to MSC Malaysia. There are now 3,241 active companies (as of February 2018) with MSC status (MDEC, 2018). It was reported in the MSC Malaysia Annual Report 2016 that 167,044 jobs were created and RM47.1 billion was generated. As a high-profile ICT project, the MSC's aim is to build a world-leading technology environment that facilitates Malaysia's evolution into a knowledge-based society, by attracting and nurturing leading-edge and world-class companies. The state certainly hopes the MSC will be a catalyst that lifts Malaysia further up the economic value chain away from the limits of low-wage and low-value added manufacturing of IT hardware. But, the aspirations for the MSC go even further than these economic goals (Muhamad, 2015).

Under the MSC initiatives, e-government was introduced. The e-government initiative was implemented and monitored by Malaysia Administrative Modernization and Management Planning Unit (MAMPU), a central agency in Malaysia Government. The development of e-government capabilities is an important undertaking because it is not only rapidly changing the way that governments supply information, deliver services, and deal with the public, but they are also becoming an integral part of government strategies (Zhang *et al.*, 2014). Among the e-government applications are e-BR1M, e-Kasih, e-Rezeki, e-Zakat, e-Jakim, e-Usahawan and Myeg (https://www.myeg.com.my). Besides the e-government initiatives, the IT industry in Malaysia like in other countries is now open to many new areas of applications such as Big Data, the internet of things (IoT) and cloud computing. Malaysia, in 2017, began to focus on the Fourth Industrial Revolution (4IR), Fintech and Block Chain. According to IDC Malaysia, the best prospects for the Malaysian IT industry are Big Data in the Cloud, Enterprise Mobility and Device Deployment, IoT, Cognitive Cyber security, Datacenter Vision and Fintech (IDC Malaysia, 2018).

According to a report by the Department of Commerce (2018), structural changes are taking place within the information communication technology (ICT) industry in Malaysia, as smaller ICT devices are replacing traditional computers and mobile technology is becoming more sophisticated (allowing for real-time interactive multimedia content to take place). This is also substantiated with the increasing number of social media applications, internet of things, cloud computing and big data analytics. Workers in organizations have been affected by the changes of technology assimilation in their work place. Thus, this study highlights the main IT issues that are faced by IT professionals in Malaysia. It is hoped that the findings would be able to shed light on some less known, yet important, issues that could be improved to enhance the skills and potential performance of IT professionals.

21.4 Methodology

Empirical data for the study was collected via a questionnaire which was developed for the World IT Project (Palvia *et al.*, 2017; Palvia *et al.*, 2018). As the targeted respondents were IT professionals, it was expected that they would be fluent in English. Nevertheless, the questionnaire was translated to Malay (Malaysia's official language) using the backward translation method, as in some cases the targeted respondents were not very fluent in English. One of the authors translated the questionnaire from English to Malay, and a language lecturer translated it back to English. Subsequently, it was compared with the original to check for inconsistencies, and the differences were resolved.

The questionnaire was distributed to IT professionals in several firms around the Multimedia Super Corridor area as it is a hub for innovative producers and users of multimedia technology. The researchers first identified the organizations based on a list of firms in MSC Malaysia Directory (https://www.mdec.my/directory/msc-malaysia-direc tory). They then established a contact in the identified firms and asked them to identify between 2 and 10 IT professionals (depending on the size of the firm) to fill the questionnaire. In some firms, the researchers distributed the questionnaire face to face while in others the contact person distributed it. To ensure consistency, the researchers went through the questionnaire with the contact persons and wrote down their telephone numbers in case they had any queries. A total of 350 questionnaires were distributed over a period of six months. No gifts or incentives were provided for completing the questionnaire. The respondents were assured that the data would be used at an aggregate level for academic purposes alone, and their identities would not be disclosed in any form whatsoever.

At the end of the survey, 300 IT professionals participated. 38.7% and 41.3% were between the ages of 21–29 and 30–39 years, respectively, and the

Table 21.1:   Descriptive Statistics

Characteristics	N	%	Characteristics	N	%
Years of work experience:			Years of IT experience:		
0–4 years	81	27.0	0–4 years	105	35.0
5–9 years	104	34.7	5–9 years	108	36.0
10–19 years	87	29.0	10–19 years	71	23.7
20–29 years	25	8.3	20–29 years	14	4.7
30+ years	3	1.0	30+ years	2	0.7
Level of education:			Organizational location:		
High school or less	33	11.0	IT dept. employee	135	45.0
Associate degree	99	33.0	IT worker in non-IT department	99	33.0
Bachelor's degree	159	53.0	Contract employee	33	11.0
Master's degree	9	3.0	Consultant	14	4.7
Employment status:			Vendor employee	19	6.3
Mostly full time	277	92.3	Work position:		
Mostly part time	17	5.7	Not part of management	96	32.0
Mostly over time	6	2.0	In lower management	76	25.3
Been laid off:			In middle management	112	37.3
Yes	78	26.0	In senior management	16	5.3
No	222	74.0			

remaining were 40 and above. The detailed descriptive statistics are shown in Table 21.1. It can be seen from Table 2.11 that a majority of them have worked between 5 and 19 years (63.7%) and a very small percentage have worked for more than 30 years. The percentage for years of IT experience however differs as the majority have less than 10 years IT experience (71%). The distribution in terms of the level of education illustrates that 86% of the sample have either a Bachelor's or Associate degree. Only 3% have a Master's degree. About 45% are IT department employees, and most of them are employed on a full time basis.

21.5   Organizational IT Issues

The IT professionals were asked to rate the importance of each of the 18 organizational IT issues on a 5-point scale. Based on the mean score value, the 18 items were ranked (Table 21.2). From Table 21.2, it can be seen that the top three items were IT reliability and efficiency, security and privacy and IT strategic planning, while the bottom three were Bring your own device, Outsourcing and Enterprise architecture.

Table 21.2: Organizational IT Issues in Malaysia

Organizational IT Issues	Rank	Mean*	Std. Deviation
IT reliability and efficiency	1	2.08	0.82
Security and privacy	2	2.09	0.85
IT strategic planning	3	2.12	0.84
Knowledge management	4	2.13	0.82
Business productivity and cost reduction	5	2.14	0.81
IT service management (e.g., ITIL)	6	2.15	0.81
Alignment between IT and business	7	2.16	0.80
Business agility and speed to market	8	2.16	0.81
Project management	9	2.18	0.84
Continuity planning and disaster recovery	10	2.19	0.84
Revenue-generating IT innovations	11	2.23	0.81
Globalization	12	2.28	0.84
Attracting and retaining IT professionals	13	2.32	0.86
Business process reengineering	14	2.36	0.80
IT cost reduction	15	2.37	0.85
Enterprise architecture	16	2.40	0.83
Outsourcing	17	2.58	0.90
Bring your own computing device (BYOD)	18	2.63	0.96

*Rating scale ranges from 1 to 5: 1 as most important and 5 as no importance.

Comparisons were made with Kappelman *et al.* (2018)'s study, and it was observed that in their study security and privacy was the top issue, while in this study it is the second top issue. In terms of top 10 issues, three issues (besides security and privacy) in this study were also rated among the top 10 in their study, namely business productivity and cost reduction (fifth in 2017 and sixth in 2016); alignment between IT and business (second in 2017 and first in 2016); and business agility and speed to market (ninth in 2017 and fifth in 2016). The differences between Kappelman *et al.* (2018)'s and this study may be due to the fact that the IT industry in Malaysia is still lagging behind and has not matured compared to the US.

IT reliability and efficiency is perceived to be the most important issue in Malaysia. The IT professionals believe that IT applications and innovations are developed to increase the performance and productivity of an organization; hence their reliability and efficiency are crucial. Generally, all organizations require that their IT applications be safe and secure from internal and external threats, and that their data is protected. Hence, it is not surprising that security and privacy is the second top issue of importance. BYOD was perceived by the IT professionals as the least important issue, as all of them have their own mobile devices and laptops, hence, using them for work-related task is not an issue at all.

In summary, the IT professionals place more emphasis on organizational issues that are directly related to IT innovations/application, i.e., their reliability, efficacy as well as security and privacy. This may be because the largest number of IT professionals in our sample were from the IT department; hence, the emphasis on the reliability, efficiency and security of IT infrastructure and applications.

21.6 Technology and Infrastructure Issues

The IT professionals were also asked to rate the technology issues that they perceive to be important to them, on a 5-point scale. Table 21.3 summarizes the findings. It was found that the top three technology issues are related to Mobile and wireless applications, networks and telecommunications, and business intelligence/analytics; while the bottom three were collaborative and workflow tools, ERP systems, and cloud computing.

The IT professionals indicated that mobile and wireless applications are important as with the advent of the Internet, more and more applications are using mobile and wireless devices. In addition, in order to use the mobile and wireless devices efficiently and effectively, the underlying networks and telecommunication services need to be reliable. With the increased usage of social media applications and internet of things, organizations are exposed to a huge amount of data (many times, big data) every day; hence, the IT

Table 21.3:   Technology and Infrastructure Issues in Malaysia

Technology Issues	Rank	Mean*	Std. Deviation
Mobile and wireless applications	1	2.13	0.83
Networks/telecommunications	2	2.15	0.86
Business intelligence/analytics	3	2.17	0.83
Software as a service	5	2.27	0.85
Enterprise application integration	6	2.29	0.83
Virtualization (desktop or server)	7	2.30	0.82
Mobile apps development	8	2.32	0.82
Business process management systems	9	2.34	0.82
Big data systems	10	2.34	0.84
Data mining	11	2.34	0.89
Social networking/media	12	2.35	0.90
Service-oriented architecture (SOA)	13	2.38	0.85
Collaborative and workflow tools	14	2.40	0.82
Enterprise resource planning (ERP) systems	15	2.40	0.83
Cloud computing	16	2.45	0.84

*Rating scale ranges from 1 to 5: 1 as most important and 5 as no importance.

professionals believe that business intelligence/analytics is very crucial as it allows them to conduct important analyses, such as trends and sentiment analysis, in an efficient and effective manner. Cloud computing was rated the least important technology as cloud computing adoption in Malaysia is still rather low.

Comparisons were made between this study and Kappelman *et al.* (2018)'s study. Business intelligence/analytics was rated high in both studies; in the Kappelman study, it received the topmost rank. However, this study found cloud computing and ERP to be the least important, whereas in Kappelman *et al.* (2018), cloud computing and ERP were ranked third and fifth, respectively. This is a significant departure in the two countries.

In summary, it can be said that the Malaysian IT professionals are responding to the current trends in IT, where the introduction of the social media, big data and data analytics have brought about profound changes in the business landscape. In today's landscape, data are in abundance and are growing, therefore, they realize that business intelligence and analytics are very important in modern times. In our sample, 45% of the IT professionals were part of middle and senior management; thus, they must have the need to conduct deep analysis to help them in their decision-making. Mobile and wireless applications were rated the highest technology issue, as the usage of mobile and wireless device is on the rise in Malaysia, and so also the many business applications that can run on them.

21.7 Individual IT Employee Issues

In addition to rating the organizational and technology issues, the IT professionals were also required to rate 28 items related to their individual concerns. The items were grouped into seven categories: job satisfaction, work pressure, work–life balance, workload and burnout, sense of accomplishment, threats to job, and career plans. Table 21.4 shows the results.

Table 21.4: Individual IT Employee Issues in Malaysia

Items	Mean*	Std. Deviation
Job satisfaction		
In general, I like working here.	2.11	0.88
All in all, I am satisfied with my current job.	2.29	0.89
In general, I don't like my current job.	3.49	1.12

(*Continued*)

Table 21.4: (*Continued*)

Items	Mean*	Std. Deviation
Work pressure		
I feel that the number of requests, problems or complaints that I deal with at work is more than expected.	2.65	0.92
I feel that the amount of work I do interferes with how well it is done.	2.58	0.88
I feel busy or rushed at work.	2.74	0.94
I feel pressured at work.	2.89	0.92
Work–life balance		
There is a blurring of boundaries between my job and my home life.	2.99	1.00
My work-related responsibilities create conflicts with my home responsibilities.	3.23	1.03
I do not get everything done at home because I find myself completing job-related work.	3.15	1.07
Workload and burnout		
I feel drained from activities at work.	2.98	0.97
I feel tired from my work activities.	2.95	0.99
Working all day is a strain for me.	2.91	0.95
I feel burned out from my work activities.	2.89	0.91
Sense of accomplishment		
I feel I'm making an effective contribution to what this organization does.	2.34	0.79
In my opinion, I do a good job.	2.20	0.78
I have accomplished many worthwhile things in this job.	2.29	0.77
At my work, I feel confident that I am effective at getting things done.	2.21	0.79
Threats to one's Job		
I am worried that future technology advancements may pose a threat to my job.	2.96	1.06
I believe that other people may be able to perform my work activities.	2.44	0.89
I am concerned that my job may be eliminated soon.	3.05	1.03
I am concerned that my job may be outsourced soon.	2.92	1.04
Career plans		
I will be with this organization 1 year from now.	2.60	0.95
I will take steps during the next year to secure a job at a different organization.	2.93	1.03
I will be with this organization 5 years from now.	2.71	0.95
I will be working in the IT field 1 year from now.	2.76	1.01
I will take steps during the next year to secure a job outside the IT field.	3.06	1.00
I will be working in the IT field 5 years from now.	2.80	1.03

*Ratings scale ranges from 1 to 5: 1 as strongly agree and 5 as strongly disagree.

Our scale is a 5-point scale; therefore, an average below the mid value of 3.0 suggests agreement with the stated item. Overall, the IT employees generally paint a positive picture of their jobs and the IT profession.

The Malaysian IT employees are happy and satisfied with their jobs and feel like they are making worthwhile contributions to their employers. Their level of confidence in carrying out their work is high. In spite of these positives, they seem to be overburdened by work and report some level of fatigue and burnout. On the other hand, they seem to be comfortable balancing the demands of work and home life and do not report much conflict between the two.

The scores for job security and threats to the job are in the mid-range, indicating that the IT employees are not overly secure in their positions and acknowledge the possibility of their jobs being replaced by someone else, eliminated or even outsourced. In spite of all these factors, the employees show loyalty to the current jobs and the organizations. Most of them plan to stay with the current job both short term and long term. Most of them also plan to stay in the IT field for a long time.

As mentioned earlier, most of the respondents in our survey were between 21 and 39 years of age, and many of them are single. Thus, even though they may have to work hard in the office, they do not have problems balancing their work with home life. In addition, people in this age-range generally find work they are passionate about; as such, they have no problems carrying out their job responsibilities in spite of the pressures of work. In terms of career planning, they are satisfied with their current jobs and plan to stay with the current organization and the IT profession for a long time. In Malaysia, for full time employees, the fear of unemployment is minimal, unless of course if the firm ceases to exist.

21.8 Conclusion

This chapter highlighted the organizational, technological and individual issues from the perspective of over 300 IT professionals in Malaysia. The three most important organizational issues are: IT reliability and efficiency, security and privacy, and IT strategic planning. Mobile and wireless applications, networks/telecommunications, and business intelligence/analytics are the most important technology issues. The IT professionals are generally satisfied with their jobs, feel a sense of accomplishment towards their jobs and are able to balance their work–life responsibilities. Thus, in general, the IT employees reflect issues associated with a growing and thriving field.

Acknowledgments

The authors would like to thank Universiti Malaya for providing us research funding to conduct this study. Grant No: UM.C/625/1/HIR/ASH/033.

References

Department of Commerce, USA (2018). Malaysia - Information & Communications Technology, retrieved from https://www.export.gov/article?id=Malaysia-Information-Communications-Technology. Accessed on 15th September 2018.

Department of Statistics Malaysia (2018). Selected Demographic Indicators Malaysia, 2018. Available at https://www.dosm.gov.my/v1/index.php. Accessed on 24th September 2018.

IDC Malaysia, retrieved from https://www.idc.com/asean. Accessed on 15th September 2018.

Kappelman L., Johnson, V., McLean, E., & Maurer, C. (2018). The 2017 SIM IT Issues and Trends Study, *MISQ Executive*, 17(1), 53–88.

Malaysian Digital Economy Corporation retrieved from https://www.mdec.my/msc-malaysiaMSC Accessed on 15th September 2018.

Muhamad, R. (2015). The Development Of Ict And Its Political Impact In Malaysia. *Journal of Borneo Social Transformation Studies (JOBSTS)*, 1(1), 83–98.

Multimedia Super Corridor, MSC Malaysia Directory. Retrived from https://www.mdec.my/directory/msc-malaysia-directory. Accessed on 15th September 2018.

Palvia, P., Jacks, T., Ghosh, J., Licker, P., Romm-Livermore, C., Serenko, A., & Turan, A. H. (2017). The World IT Project: History, trials, tribulations, lessons, and recommendations. *Communications of the Association for Information Systems*, 41(18), 389–413.

Palvia, P., Ghosh, J., Jacks, T., Serenko, A., & Turan, A. (2018). Trekking the globe with the World IT Project. *Journal of Information Technology Case and Application Research*, 20(1), 3–8.

Raman, K. S. & Yap, C. S. (1996). From a resource rich country to an information rich society: An evaluation of information technology policies in Malaysia, *Information Technology Development*, 7, 109–131.

Zhang, H., Xu, X., & Xiao, J. (2014). Diffusion of e-government: A literature review and directions for future directions. *Government Information Quarterly*, 31(4), 631–636.

Chapter 22

Information Technology Issues in Mexico

Gustavo Parés[*,‡], Ricardo Luis Parés Arce[*],
Vicente Cubells Nonell[*], Eder Espinos Diaz[*],
Luis Humberto Rojas Pineda[*], and Prashant Palvia[†,§]

[*]*Nearshore Delivery Solutions,*
Mexico City, Mexico
[†]*University of North Carolina at Greensboro,*
Greensboro, NC, USA
[‡]*gustavo@nearshoremx.com*
[§]*pcpalvia@uncg.edu*

Summary

This chapter investigates the top organizational, technological, and individual issues of information technology (IT) workers in Mexico, a country just south of the US border. It utilized the standard World IT Project survey instrument, which was carefully translated into Spanish by the local research team. Our findings indicate that the top five organizational issues are: IT reliability and efficiency, security and privacy, alignment between IT and business, project management, and IT strategic planning. The top five technology issues are: customer relationship management (CRM) systems, networks and telecommunications, business intelligence/analytics, enterprise application integration, and software as a service. As per individual issues, IT employees seem to be extremely satisfied with their jobs with a high sense of accomplishment, feel secure in their jobs and plan to be in the IT industry for a long time. One concern relates to gender equality and opportunities presented to women in the IT industry. Women and men are not treated equally in the workplace; while men are hired directly and hold full-time positions, women are hired as external consultants or on an outsourcing basis from a third party.

22.1 Introduction

Mexico, officially the United Mexican States, is located in the southern part of North America and is bordered in the north by the United States. The US and Mexico have a shared history going back to the Texas Revolution (1835–1836) and the Mexican–American War (1846–1848). Mexico's economy is growing and it is quickly becoming an emerging market. The Mexican information technology (IT) industry is large and expanding, although it is still behind major players like India and the United States. Its proximity to the US as well as being in the same time zone makes it an attractive IT outsourcing destination.

Given the importance of Mexico in the world economy, the World IT Project included Mexico in its survey. This chapter provides an overview of the country's background and history, followed by a description of its IT industry, and then presents the results of the survey. These results include organizational IT issues, technology issues as well as individual employee issues.

22.2 Country Background and History

Located in North America, Mexico is home to 123 million people. Its capital is Mexico City, one of the most populated cities and metropolitan areas in the world. Interestingly, Mexico has the world's largest population of Spanish speakers. The Mexican society exhibits both extreme wealth and extreme poverty, and a small middle class. Yet, Mexico holds an important status as a regional power in the Latin American region and a relevant actor in North America, especially because of its 2,000 miles border with the US and the free trade agreement linking the country with the US and Canada.

An important aspect of Mexico is its rich ancient culture, with civilizations that can be traced back to 1,400 BC. That heritage is still present, not only in the social composition of the country, but in its cuisine, historical landmarks, and traditions such as the "Day of the Dead."

Historically, Mexico's economy has focused mainly on oil extraction; nevertheless, in recent years the country has moved into other industries, especially automotive and electronics. This turn has had a strong influence in the composition of graduates and the specialized talent required to serve these industries. By the time the World IT Project was launched in Mexico, 113,000 engineers were graduating annually from Mexican universities (Human Capital Report, 2015), 11% of them in IT-related programs (Rascón, 2013). Opportunities for graduates are distributed geographically in several areas of the country. The central part, known as Bajío, has an automotive focus, while the northeast has a strong manufacturing and industrial makeup. The main IT hubs are located in Mexico City and the surrounding states, and in Guadalajara, locally referred as the Mexican Silicon Valley.

According to official studies, during 2004, there were approximately 1,660 companies that were dedicated to IT-related services. For the year 2016, the number reached to more than 4,000 (IT and software services in Mexico, 2019). This has been an impressive growth period and bodes well for the IT industry.

22.3   Methodology

The data collection effort for the World IT Project (Palvia *et al.*, 2017; Palvia *et al.*, 2018) endeavor in Mexico faced several challenges that started with the translation of the questionnaire and continued with engagement with the participants. First, to address the language issue, the local team translated the document from English to Spanish by bilingual native speakers and official translators. The translators pointed out that some of the terms, especially those related to feelings, would be better understood by the Mexicans if their translation corresponded to the Mexican Spanish variations, rather than the most common Spanish from Spain. Parallel to this process, the team started contacting large companies known to have a large number of workers in their IT departments, as well as IT companies themselves.

In spite of our best efforts to contact such companies, most of the IT-related respondents came from an alternative channel. The team partnered with one of the largest IT oriented media companies in Mexico, which distributed the questionnaire through emails and posted it on their website. This proved to be an effective way to draw attention to the research project, as we got over 550 responses. Almost 98% of the respondents came from this channel. The media company focuses mainly on niche technologies and specialized publications, which drew the kind of audience that is interested in sharing their own experiences and perspectives. With this approach, we were able to have over 300 responses in less than a month.

The data collection started in January 2015 and lasted until May of the same year. Looking back at that process, it would have been interesting to share the questionnaire not only in Spanish, but in English as well. This is because of the high number of foreign IT professionals that work for Mexican companies and foreign companies with delivery centers in Mexico. However, on the positive side, our sample is made up of predominantly Mexicans, which is the primary interest of the World IT Project.

Table 22.1 shows descriptive statistics of the respondents in the survey. In total, there were 343 usable responses. As can be seen, most employees

Table 22.1: Descriptive Statistics

Characteristics	N	%	Characteristics	N	%
Education:			Years of work experience:		
High school or less	11	3.2	0–4 years	62	18.6
Unfinished BSc. degree	35	10.2	5–9 years	60	18.0
Bachelor's degree	202	58.9	10–19 years	127	38.2
Master's degree	89	26.0	20–29 years	68	20.2
Ph.D.	6	1.8	30+ years	16	4.8
Years of IT experience:			Organizational location:		
0–4 years	73	21.4	IT department	180	53.3
5–9 years	73	21.4	IT worker in non-IT department	33	9.8
10–19 years	130	38.0	Contract employee	42	12.4
20–29 years	50	14.2	Consultant	72	21.3
30+ years	16	4.7	Vendor employee	11	3.2
Work as:			Work position:		
Mostly full time	304	88.6	Not part of management	143	41.9
Mostly part time	26	7.6	In lower management	57	16.7
Mostly over time	13	3.8	In middle management	67	19.7
Been laid off from IT job:			In senior management	74	21.7
Yes	81	23.6			
No	262	76.4			

have much experience in IT, have college degrees, work in IT departments, and work full time. Interestingly, about one-fourth of them were previously laid off from IT jobs.

22.4   Organization IT Issues

Table 22.2 reflects the organizational priorities in the Mexican information industry. It is worth noting that these data were collected in a survey which included both employees as well as managers at different levels, thus giving us a glimpse of the general views of the Mexican IT industry.

Comparisons were made with the study of Kappelman *et al.* (2018). In their study, security and privacy was the top issue. This issue is important in Mexico as well, and it ranked second in our study. It was exceeded only by reliability and efficiency of information systems. This can be explained due to the general trend in the country of providing diverse IT services both in the internal Mexican market and the global market. The next three: alignment between IT and business, project management and IT strategic planning, are all related to the growing importance of strategic IT planning in companies. Moreover, there was high agreement on their importance, as

Table 22.2:   Organizational IT Issues in Mexico

Organizational IT Issues	Rank	Mean Rating*	Std. Deviation
IT reliability and efficiency	1	1.51	1.03
Security and privacy	2	1.55	0.75
Alignment between IT and business	3	1.64	0.74
Project management	4	1.67	0.84
IT strategic planning	5	1.69	0.84
Revenue-generating IT innovations	6	1.70	0.78
IT service management (e.g., ITIL)	7	1.73	1.07
Attracting and retaining IT professionals	8	1.74	0.84
Continuity planning and disaster recovery	9	1.74	0.88
Business productivity and cost reduction	10	1.74	0.73
Knowledge management	11	1.75	1.02
Business agility and speed to market	12	1.80	0.90
Business process reengineering	13	1.92	0.79
Enterprise architecture	14	1.92	0.79
IT cost reduction	15	1.97	0.79
Globalization	16	2.05	0.86
Bring your own computing device (BYOD)	17	2.17	0.82
Outsourcing	18	2.62	1.15

*Rating scale ranges from 1 to 5: 1 as most important and 5 as no importance.

exhibited by the low standard deviations. Continuity planning and disaster recovery appears in ninth place, not as high as we expected but higher than many other countries. This is explained by the devastating national disasters caused by earthquakes and climate events. The situation is bringing about new and successful investments to protect against such events.

Outsourcing appears last in our list, given that most of the companies participating in the survey provide IT services and do not themselves require subcontracting. Globalization was ranked low as well and is indicative of the fact that a low percentage of the personnel focus on exports in the Mexican IT industry and are not directly affected by global trends. IT cost reduction was not much of an issue either, as Mexico enjoys a significant advantage in personnel costs as compared to advanced nations.

22.5 Technology and Infrastructure Issues

Table 22.3 shows the ranks of the sixteen technology issues that were evaluated by the respondents.

The high ranking of customer relationship management systems indicates the central importance for Mexican companies to use IT services and tools in order to know their market, understand clients, define strategies to

Table 22.3: Technology and Infrastructure Issues in Mexico

IT-Related Issues	Rank	Mean Rating*	Std. Deviation
Customer relationship management (CRM) systems	1	1.68	0.90
Networks/telecommunications	2	1.75	1.04
Business intelligence/analytics	3	1.77	0.99
Enterprise application integration	4	1.83	1.08
Software as a service	5	1.84	0.93
Collaborative and workflow tools	6	1.86	1.02
Business process management systems	7	1.93	1.02
Mobile and wireless applications	8	1.95	0.94
Enterprise resource planning (ERP) systems	9	2.02	0.87
Data mining	10	2.05	0.86
Service-oriented architecture (SOA)	11	2.06	0.96
Mobile apps development	12	2.09	1.05
Virtualization (desktop or server)	13	2.11	0.94
Cloud computing	14	2.16	0.80
Big data systems	15	2.24	0.87
Social networking/media	16	2.42	0.95

*Rating scale ranges from 1 to 5: 1 as most important and 5 as no importance.

increase customer loyalty, develop appropriate market segmentation, and improve business relationships with customers, service users, colleagues, partners and suppliers.

Networks and telecommunications were rated the second most important issue, which reflects the fact that there is still a need for expansion of Internet services in companies and businesses of all sizes and there are whole areas of the country that are yet to be properly included in the new era of remote services and integration.

Business and analytics were rated third, giving a strong indication that the areas of artificial intelligence and all its related applications like data science are becoming more important in the minds of people related to the IT industry. Mexico is just starting to have companies in these areas but there is a growing interest, sometimes motivated more by the influence of other countries, especially the US, than by a knowledge of the applications and real necessities. Customers are still learning about the potential applications of business intelligence, data analytics and cognitive computing, but they seem to be more influenced by global trends and the accompanying hype. Nevertheless, many companies, especially the large ones, are adopting or investigating data-driven models and are learning about data governance.

Enterprise application integration was rated fourth. Many companies have had a somewhat chaotic or at least suboptimal development in their IT areas, as they adapt to the new era of computing and Internet applications. Application development seemed to have been a reaction to immediate problems rather than following a well-defined plan. As a result, different systems and applications have proliferated that now need integration. We anticipate that this will remain an important issue for the IT industry for years to come.

Software as a service came out as fifth in our rankings. This is becoming increasingly an important trend in the Latin American market, along with infrastructure as a service (IaaS) having a rapid rate of growth of double digits in the last few years.

Big data systems ranked second from the bottom. A possible explanation for this is that even though companies are becoming more aware of the key importance of data, there is still an acute lack of processes to consolidate and populate large datasets. This kind of massive dataset may not exist or even if it exists, it may not be easily accessible. Furthermore, there is the requirement of high computing power to process big data which many

companies may not currently have or may not want to acquire due to high
costs.

22.6 Individual IT Employee Issues

Table 22.4 shows the results of IT employees' assessment of individual
issues. IT employees are extremely satisfied with their jobs and a have a
strong sense of accomplishment. On top of it, they do not feel extreme work

Table 22.4:   Individual IT-Related Issues

Individual Issues	Mean Rating*	Std. Deviation
Job satisfaction		
In general, I like working here.	1.79	0.86
All in all, I am satisfied with my current job.	1.95	0.96
In general, I don't like my current job.	4.18	0.95
Work pressure		
I feel that the number of requests, problems or complaints that I deal with at work is more than expected.	3.49	1.12
I feel that the amount of work I do interferes with how well it is done.	3.28	1.21
I feel busy or rushed at work.	3.30	1.14
I feel pressured at work.	3.27	1.16
Work–life balance		
There is a blurring of boundaries between my job and my home life.	3.76	1.19
My work-related responsibilities create conflicts with my home responsibilities.	3.68	1.17
I do not get everything done at home because I find myself completing job-related work.	3.55	1.27
Workload and burnout		
I feel drained from activities at work.	3.32	1.20
I feel tired from my work activities.	3.26	1.20
Working all day is a strain for me.	3.08	1.24
I feel burned out from my work activities.	3.66	1.15
Sense of accomplishment		
I feel I'm making an effective contribution to what this organization does.	1.77	0.83
In my opinion, I do a good job.	1.59	0.70
I have accomplished many worthwhile things in this job.	1.85	0.90
At my work, I feel confident that I am effective at getting things done.	1.73	0.77

(*Continued*)

Table 22.4: (*Continued*)

Individual Issues	Mean Rating*	Std. Deviation
Threats to one's Job		
I am worried that future technology advancements may pose a threat to my job.	3.72	1.28
I believe that other people may be able to perform my work activities.	2.25	0.98
I am concerned that my job may be eliminated soon.	3.77	1.17
I am concerned that my job may be outsourced soon.	3.53	1.32
Career plans		
I will be with this organization 1 year from now.	2.35	1.25
I will take steps during the next year to secure a job at a different organization.	2.88	1.36
I will be with this organization 5 years from now.	3.03	1.36
I will be working in the IT field 1 year from now.	1.34	0.92
I will take steps during the next year to secure a job outside the IT field.	3.55	1.38
I will be working in the IT field 5 years from now.	1.62	0.95

*Rating scale ranges from 1 to 5: 1 as strongly agree, to 5 as strongly disagree.

pressure or burnout and there do not seem to be conflicts with their home life. Although they know that others can do their jobs, they feel secure in their jobs and do not see any imminent threats. Given all the positives, they feel committed to their jobs and the IT profession, although they are willing to change jobs in the longer run.

We can thus surmise that, like in India, IT employees in Mexico are among the privileged members of society. While this is generally true, a review of the different sectors of the Mexican industry shows that the situation of the worker, whether employee, manager or entrepreneur, faces great challenges that affect their safety, working conditions and compensation. For example, among the range of small and large companies located in Mexico, inequality is very strong. In 2005, a profit of approximately US$2 billion was invoiced among the 150 highest-income companies. About 80% of this profit was among only 30 of them. In the other 1500 companies, the profits were minimal. (Hualde, 2007, 2013)

Thus some degree of dissatisfaction exists in a situation of structural inequality. This is reflected in the standard deviation of the various issues. By the same token, IT employees in some companies, especially the smaller ones, may be overworked and may not feel as secure. Education, universities, and public policies may have a role in order to bridge the inequality. As pointed out by Mochi (2004), "it is necessary to considerably expand the quantity and quality... of the relationship between University and business." Fortunately the industry now has more developed public policies, such as PROSOFT, but the optimal conditions for the workers are yet to be reached.

Upon drilling down our data further, we also noted gender inequality in the IT profession in Mexico. In our sample alone, the male participation dwarfed female participation by 82% to 18%. Although there is a tendency for women to gain ground within the IT industry, the female employment is still low. The cycle starts with a much lower number of women enrolling in technical education. For example, in Computer Science and Engineering in 2016, the percentages of women graduating from the largest higher education institution were only 14.3% and 25%, respectively (Perfil de los alumnos egresados del nivel licenciatura de la UNAM). Furthermore, women and men are not treated equally in the workplace; while men are hired directly and hold full-time positions, women are hired as external consultants or on an outsourcing basis from a third party.

22.7　Conclusion

The Mexican IT industry has been expanding rapidly in the last few years, driven by the necessities of an economy still in the process of adapting to the era of Internet, data and analytics. The strong relationship with the US and Canada will continue to permeate new markets and practices. Offshore outsourcing will grow, together with data analytics. At the same time, the country will continue with the renewal of outdated IT services and the process of enterprise applications integration. Given this optimistic outlook, the IT industry is thriving and its employees envision a bright future.

References

Hualde, A. & Gomis, R. (2007) PYME de *software* en la frontera norte de México: desarrollo empresarial y construcción institucional de un *cluster*. *Revista Problemas del Desarrollo*, 38(150). Retrieved from http://www. revistas.unam.mx/index.php/pde/article/view/7676.

Hualde, A. (2013). Jordy Micheli Thirión, Telemetrópolis. Explorando la ciudad y su producción inmaterial, México: UAM-Azcapotzalco/Gedisa. Espacialidades, 3(2), 254–259. Retrieved from http://espacialidades.cua.uam. mx/ojs/index.php/espacialidades/article/view/79/75.

Human Capital Report (2015) World Economic Forum.

IT and software services in Mexico. ProMexico. Retrieved on January 22, 2019 from http://mim.promexico.gob.mx/swb/mim/Servicios_de_TI_y_software/_lang/en.

Kappelman, L., Johnson, V., McLean, E., & Maurer, C. (2018). The 2017 SIM IT Issues and Trends Study. *MISQ Exec*, 17(1), 53–88.

Mochi, P. (2004). La industria del software en México. *Problemas del Desarrollo, Revista Latinoamericana de Economía*, 35(137). Retrieved from http://www.revistas.unam.mx/index.php/pde/article/view/7533/7022.

Palvia, P., Jacks, T., Ghosh, J., Licker, P., Romm-Livermore, C., Serenko, A., & Turan, A. H. (2017). The World IT Project: History, trials, tribulations, lessons, and recommendations. *Communications of the Association for Information Systems*, 41(18), 389–413.

Palvia, P., Ghosh, J., Jacks, T., Serenko, A., & Turan, A. (2018). Trekking the globe with the World IT Project. *Journal of Information Technology Case and Application Research*, 20(1), 3–8.

Perfil de los alumnos egresados del nivel licenciatura de la UNAM (Octubre 2015–Septiembre 2016). Reporte Global UNAM. Retrieved from http://www.planeacion.unam.mx/Publicaciones/pdf/perfiles/egresados/p_eg2015-2016.pdf.

Rascón, O. *et al.* (2013). Estado del arte y prospectiva de la ingeniería en México y en mundo. Academia de Ingeniería de México. Retrieved from http://www.ai.org.mx/sites/default/files/16.documento-de-conclusiones-y-recomendaciones-del-estudio-eapimm.pdf.

Tello Leal, E. (2007). Las tecnologías de la información y comunicaciones (TIC) y la brecha digital: su impacto en la sociedad de México. *Revista de Universidad y Sociedad de Conocimiento, Universitat Oberta de Catalunya*, 4(2). Retrieved from https://www.raco.cat/index.php/Rusc/article/viewFile/78534/102611.

Chapter 23

Information Technology Issues
in New Zealand

Jocelyn Cranefield*,**, Mary Ellen Gordon*,††, Zlatko Kovačić†,‡‡,
Gillian Oliver‡,§§, Alexander Serenko§,¶,¶¶,
and Aykut Hamit Turan‖,‖‖

*Victoria University of Wellington, Wellington, New Zealand
†Independent Researcher, New Zealand
‡Monash University, Clayton VIC, Australia
§University of Toronto, Toronto, Canada
¶University of Ontario Institute of Technology, Oshawa, Canada
‖Sakarya University, Sakarya, Turkey
**Jocelyn.cranefield@vuw.ac.nz
††Maryellen.gordon@vuw.ac.nz
‡‡Zlatko@mystatisticalconsultant.com
§§gillian.oliver@monash.edu.au
¶¶a.serenko@utoronto.ca
‖‖ahturan@sakarya.edu

Summary

Our results suggest that New Zealand's information technology (IT) workforce is aging and evolving to become more gender-balanced (28% of survey responses were from women) and more diverse (45% of survey respondents were born in another country). It is generally a happy workforce: on average, survey respondents reported that they were satisfied with their jobs and that they felt a sense of accomplishment, without expressing excessive concern about work pressure, workload, work–life balance, or losing their jobs. Respondents of this survey were concentrated in financial services, the public sector, and educational organizations. Those in financial services tended to be particularly focused on outward-looking organizational issues and mobile app development, and those working for public sector organizations tended to be more particularly focused on inward-looking organizational issues and enterprise-level technologies. The only issues where those working for educational institutions attributed greater importance than other respondents were bring your own computing device (BYOD) and globalization.

23.1   Introduction

New Zealand has a multi-faceted information technology (IT) workforce. It includes many people who have moved to the country relatively recently as well as many New Zealanders who have been working in IT for decades. The IT workforce includes many people working in large, government organizations, as well as others working in entrepreneurial start-ups or doing the technical work required to create some of the world's best-known films, such as the Lord of the Rings Trilogy. The country is relatively small, and many organizations employing IT workers in the cities are near one another, so there are ample opportunities for professionals to interact with one another. That combination of diversity and proximity has allowed information technology to evolve in New Zealand from a support function to a major and growing part of the economy.

23.2   Country Background and History

New Zealand is an island country of 4.8 million that is geographically distant from most of its main trading partners, which include Australia, the European Union, USA, China, South Korea, Japan and Canada.

New Zealand has a mixed economy based on free-market principles (following massive economic reforms in the 1980s and 1990s), and it is very reliant on global trade. Hence, the country has worked hard to gain a reputation for being a good place to do business. In 2017, it was ranked first in the world for ease of doing business, tied for first place for being non-corrupt, fourth for transparency, and thirteenth for competitiveness (MBIE, 2017). Since the 1970s, there has been a shift away from a traditional agricultural and manufacturing economy towards a services-oriented economy (Statistics NZ, 2018c). As of 2016, the services sector made up about two-thirds of GDP (The Treasury, 2016).

One in seven of New Zealanders identify as indigenous Māori, 74% have European ancestry, 12% are of Asian descent, and 7% identify as Pacific Islanders. Over 86% of the population live in urban areas, and over 53% live in the four largest cities. New Zealand has a beautiful natural environment, which makes it attractive to tourists, and the quality of life makes it attractive to immigrants. According to the 2013 census, 25% of the population were born overseas. Recent years have seen high immigration, with incoming net migration reaching 73,000 in the year to June 2017 (The Guardian, 2017). IT workers are well-represented among those gaining visas through a skilled migrant category (MBIE, 2018).

Compared with many other countries, New Zealand has a high percentage of small and medium-sized enterprises. In 2017, only 1% of enterprises had 50 or more employees, however, those organizations employed around 45% of the country's workforce (Statistics NZ, 2017).

In this study, we had enough survey respondents working in educational, financial, and government/public organizations to compare those types of organizations to one another. To help to interpret the results, we briefly introduce these three sector contexts:

Public tertiary education institutions employed over 36,000 people in 2013 (Ministry of Education, 2018). The sector is well-regarded, but it faces rising costs while centralized rules and requirements make it hard to innovate and adapt (NZ Productivity Commission, 2017).

Banking contributes $7.2 billion to the economy, and banks employ over 25,000 people. New Zealand ranked first in financial market development and third for soundness in the World Economic Forum's Global Competitive Report (NZ Bankers Association, 2018).

NZ's *government/public sector* employs around 348,000 people across 2,900 agencies; 14% of the country's workforce (State Services Commission, 2018). The sector is undergoing a major transformation aimed to deliver citizen-centric services across all of government.

23.3   Information Technology in New Zealand

The technology sector is a major and growing area of the New Zealand economy. It had exports of $960 million in 2016 and employed an estimated 29,700 workers (NZ Immigration, 2018). Some of New Zealand's best-known IT exports come from the cloud accounting company, Xero (founded in Wellington), and digital and special effect work produced by Weta Digital, Park Road, and other Wellington-based companies related to film director Peter Jackson and his collaborators.

New Zealand also has an active IT-related start-up ecosystem. Organizations that participated in the country's largest incubator/accelerator have a collective valuation of over NZ$55 million on annual revenue of over NZ$25 million and employ over 250 staff (CreativeHQ, 2018).

The country's public sector has embraced digital government, with many services delivered online. It has also created an integrated data infrastructure, which enables joint analysis of data from many public-sector organizations while protecting the privacy of individuals and organizations (Statistics New Zealand, 2018d). New Zealand is a leader in open government data, ranking in the top five countries for data readiness and impact (Open Data Barometer, 2018).

New Zealand's traditional exports (e.g., food and beverages, wool, tourism) are also increasingly enabled by IT (MBIE, 2013). Overall, in the country's wider economy, nearly 75,000 people work in IT-related roles (NZ Immigration, 2018), representing just under 4% of the total New Zealand workforce (Statistics New Zealand, 2018e). Increasing use of ICT and digitization across sectors is generating growth in roles such as software engineering and development, project management, business analysis and analytics (NZ Immigration, 2018). The areas of computer system design and interactive gaming are rapidly growing subsectors of the IT economy, while at the level of occupations, growth in IT jobs has been driven by roles such as "Software and Applications Programmers" and "ICT Business and Systems Analysts".

The growth in demand for IT is matched by a significant skills shortage: in 2014, more firms in the IT sector reported having vacancies, and that these vacancies were harder to fill, than in any other sector in the economy (NZ Immigration, 2018). New Zealand had an innovative approach to addressing that issue. In 2017, the capital city of Wellington offered 100 free trips to the city for applicants for 265 available technology-related jobs and was able to select those people from 48,000 applicants from around the world, including people in senior roles with well-known technology companies (Stuff, 2017).

23.4 Methodology

We gathered survey data between September and December 2016. In keeping with the global World IT Project's (Palvia *et al.*, 2018; Palvia *et al.*, 2017) requirements, our survey population was drawn from a cross-section of New Zealand's larger organizations (those with a minimum of 10 IT employees). Our sampling strategy was designed to optimize coverage of New Zealand's main industries and geographic regions.

We identified suitably large IT employers using a list of New Zealand's largest IT employers (The CIO 100 www.cio.co.nz) mapped onto target industry areas (identified using government GDP statistics). We selected around 46 target organizations to approach, based in major cities in the North and South Islands of New Zealand. For the targeted companies, we made personal contact (generally by phone) with senior IT Executives (the CIO, CDO, Group Manager IT, or IT Manager), outlining the scope and goals of the World IT Project and invited them to participate by sharing the survey with their IT staff and championing its importance. The CIO-as-survey-intermediary approach was chosen as a way of mitigating the burden of a 160-item questionnaire on individual workers.

We gave each organization a survey deadline of approximately one week, so as to create a sense of urgency to participate, and sent a link to a media article about the study. We sent a follow-up email a few days later, and again a day before the deadline, asking the CIOs themselves to follow up with their staff. The level of engagement with the survey from within each organization was therefore dependent on CIO advocacy as well as individual time and interest.

For the analysis that follows, we checked for relevant individual- or organization-based demographic differences in the information presented in the tables. Patterns of statistically significant differences are discussed

Table 23.1: Descriptive Statistics

Characteristics	N	%	Characteristics	N	%
Education:			Years of work experience:		
High school or less	95	18.4	0–4 years	20	3.9
Associate degree	90	17.4	5–9 years	51	9.9
Bachelor's degree	254	49.2	10–19 years	128	24.8
Master's degree	72	14.0	20–29 years	157	30.4
Ph.D.	5	1.0	30+ years	160	31.0
Years of IT experience:			Organizational location:		
0–4 years	52	10.1	IT department employee	441	85.5
5–9 years	75	14.5	IT worker in non-IT department	7	1.4
10–19 years	183	35.5	Contract employee	42	8.1
20–29 years	130	25.2	Consultant	18	3.5
30+ years	76	14.7	Vendor employee	8	1.6
Work as:			Work position:		
Mostly full time	477	92.4	Not part of management	319	61.8
Mostly part time	23	4.5	In lower management	72	14.0
Mostly over time	16	3.1	In middle management	68	13.2
Been laid off from IT job:			In senior management	57	11.0
Yes	121	23.4			
No	395	76.6			

in the text. Those that were not statistically significant were not part of a larger pattern (e.g., only one statistically significant difference among a set of related variables), or were self-evident (e.g., contractors, consultants, and employees of vendors agree less strongly than regular employees of an organization that they will be with the organization in 1 or 5 years) are not discussed.

As shown in Table 23.1, we ended up with 516 usable responses. Most respondents had at least a bachelor's degree or more (64%), 20 or more years of overall work experience (61%), and 10 or more years of IT experience (75%). Most worked full-time (92%), as IT department employees (86%), are not part of management (62%), and had never been laid off from an IT job (77%).

Overall, 28% of the responses came from women. Compared to the male respondents, they had significantly less IT experience, with 29% of them having 20 or more years of IT experience, compared to 44% of the men. There were no other statistically significant gender differences in any of the variables shown in Table 23.1 (based on chi-square tests, $p < 0.05$).

As mentioned in the preceding sections, New Zealand has a fairly large migrant population, which is reflected in the IT workforce. As a consequence, only 55% of respondents were born in New Zealand with the other 45% coming from 45 different countries. There were 15 or more respondents who were born in the UK (61), India (36), South Africa (18) and the Philippines (16), so those countries of origin are broken out and compared in some of the subsequent analysis.

Compared to respondents who were born in New Zealand, those from India and the Philippines had more education and fewer years of work experience. About 97% of respondents who were born in India had at least a Bachelor's degree, as did 93% of those born in the Philippines. That compares to 55% of respondents from New Zealand. In contrast, only 14% of respondents from India and 25% of those from the Philippines had at least 20 years of work experience, compared to 66% of respondents who were born in New Zealand. Similar patterns held for IT work experience. These differences reflect an age difference. Three quarters (75%) of respondents who were born in India and nearly two-thirds (63%) of those who were born in the Philippines were younger than 40, compared to less than a third (31%) of New Zealand born respondents. Compared to respondents born in other countries, those born in the UK were disproportionately likely to have been laid off from IT jobs (39%) and those from the Philippines were disproportionately unlikely to have been (6%). There were no other statistically significant differences in the variables in Table 23.1 (based on chi-square tests, $p < 0.05$).

23.5 Organizational IT Issues

As shown in Table 23.2, the organizational IT issues rated as most important by New Zealand respondents were, on average, IT reliability and efficiency, security and privacy, and alignment between IT and business.

Perceptions regarding the importance of organizational IT issues varied based on the type of organization a respondent worked for. The overall patterns were for those working in financial organizations to focus on more outward-looking issues and those in public-sector organizations to focus on more inward-looking issues. Specifically, respondents who work for financial institutions rated the following organizational issues as having greater average importance than those who work for other types of organizations (based on ANOVA F statistics, $p < 0.05$):

- Security and privacy (mean = 1.52);
- Alignment between IT and business (mean = 1.58);

Table 23.2:   Organizational IT Issues in New Zealand

Organizational IT Issues	Rank	Mean Rating*	Std. Deviation
IT reliability and efficiency	1	1.60	0.64
Security and privacy	2	1.71	0.78
Alignment between IT and business	3	1.73	0.70
IT strategic planning	4	1.92	0.77
Attracting and retaining IT professionals	5	1.94	0.74
Continuity planning and disaster recovery	6	2.00	0.84
Knowledge management	7	2.04	0.75
Project management	8	2.21	0.80
Business agility and speed to market	9	2.23	0.86
Enterprise architecture	10	2.30	0.88
Business productivity and cost reduction	11	2.35	0.76
Business process reengineering	12	2.40	0.82
IT service management (e.g., ITIL)	13	2.50	0.93
IT cost reduction	14	2.60	0.86
Revenue-generating IT innovations	15	2.66	1.06
Globalization	16	3.21	0.99
Bring your own computing device (BYOD)	17	3.30	1.13
Outsourcing	18	3.60	0.99

*Rating scale ranges from 1 to 5: 1 as most important and 5 as no importance.

- Business agility and speed to market (mean = 1.82);
- Revenue-generating IT innovations (mean = 2.29).

Those working in public-sector organizations rated the following as significantly more important, on average (based on ANOVA F statistics, $p < 0.05$):

- Project management (mean = 2.13);
- Business productivity and cost reduction (mean = 2.19);
- Business process reengineering (mean = 2.26);
- IT service management (mean = 2.34);
- IT cost reduction (mean = 2.47).

On average, the only organizational IT issues those working in education organizations rated as significantly more important than those in other sectors (based on ANOVA F statistics, $p < 0.05$) were BYOD (mean = 2.94) and globalization (mean = 3.15).

The average importance attributed to different IT issues varied based on whether or not respondents were managers, and if so, at what level (based on ANOVA F statistics, $p < 0.05$). The higher in management a respondent was, the more importance (on average) he or she attributed to alignment between IT and business, business agility and speed to market, and BYOD. A similar pattern held up through middle management for IT strategic planning, attracting and retaining IT professionals, enterprise architecture, business productivity/cost reduction, and outsourcing.

This overall pattern of results suggests a possible need to provide more information about the importance of different organizational issues to individual contributors and lower level managers.

Individual perceptions of the importance of organizational issues varied by country of origin. Compared to those born in other countries, respondents born in India attributed greater average importance (based on ANOVA F statistics, $p < 0.05$) to: business productivity and cost reduction, business agility and speed to market, revenue-generating IT activities, IT cost reduction, globalization, and outsourcing. Those born in the Philippines attributed the greatest importance to IT service management and those born in India or the Philippines rated project management as more important, on average, than did those from other countries.

That pattern of results suggests that people from countries that have become recipients of significant IT outsourcing may be particularly attuned to profitability-related organizational issues.

Comparing these New Zealand results with the results of the 2017 SIM IT Issues and Trends study (Kappelman *et al.*, 2017), the top three issues in both studies included security and privacy and alignment between IT and business. Business agility also ranked as the ninth most important organizational issue in both studies.

The SIM study focused on the perspective of top IT executives, whereas the New Zealand study is based on opinions from a range of IT professionals, so we also compared results using only senior IT executives in New Zealand, but the only difference between that and the overall comparison was that, on average, the New Zealand senior executives rated business agility as even more important than either the SIM sample or the full New Zealand sample.

Beyond that, the ten top-rated issues identified by New Zealand respondents to the World IT study included more protective concerns (e.g., IT reliability and efficiency, continuity planning and disaster recovery, and attracting and retaining IT professionals); whereas (95% US-based) respondents to the SIM IT Issues and Trends study included more pro-active concerns (innovation and digital transformation) and more concerns related to cost reduction. These differences may reflect differences in the types of organizations respondents were drawn from. One quarter of New Zealand

Table 23.3:   Technology and Infrastructure Issues in New Zealand

IT Related Issues	Rank	Mean Rating*	Std. Deviation
Networks/telecommunications	1	2.25	0.96
Enterprise application integration	2	2.26	0.88
Mobile and wireless applications	3	2.29	0.93
Collaborative and workflow tools	4	2.30	0.79
Business intelligence/analytics	5	2.34	0.92
Virtualization (desktop or server)	6	2.46	1.00
Software as a service	7	2.46	0.94
Cloud computing	8	2.55	0.96
Business process management systems	9	2.56	0.94
Customer relationship management (CRM)	10	2.64	1.00
Service-oriented architecture (SOA)	11	2.66	0.94
Big data systems	12	2.69	1.03
Mobile apps development	13	2.74	1.04
Data mining	14	2.77	1.05
Enterprise resource planning (ERP)	15	2.84	1.01
Social networking/media	16	3.17	1.08

*Rating scale ranges from 1 to 5: 1 as most important and 5 as no importance.

respondents worked in government organizations, compared to only 6% of participating organizations in the SIM study, and as noted above we found that, at least in New Zealand, those who work for such organizations tend to have more inward-looking concerns.

23.6   Technology and Infrastructure Issues

As shown in Table 23.3, the three technology issues rated as most important were, on average, networks/telecommunications, enterprise application integration, and mobile and wireless applications.

Compared to those who work for other types of organizations, respondents who work for government institutions rated the following technical issues as having greater average importance (based on ANOVA F statistics, $p < 0.05$):

- Enterprise application integration (mean = 2.07);
- Business intelligence and analytics (mean = 2.17);
- Business process management systems (mean = 2.34);
- Enterprise resource planning systems (mean = 2.63).

Respondents working for financial institutions attributed greater average importance (based on ANOVA F statistics, $p < 0.05$) to mobile app development (mean $= 2.42$) and service-oriented architecture (mean $= 2.42$).

The more senior the respondents were in their organization, the greater average importance they attributed to cloud computing. Among managers, the more senior the respondents were, the greater importance (on average) they attributed to business intelligence/analytics, data mining, social networking/media, software as a service, and service-oriented architecture (based on ANOVA F statistics, $p < 0.05$).

As with organizational issues, average importance placed on technology issues varied (based on ANOVA F statistics, $p < 0.05$) based on where respondents were born. Those born in India attributed greater average importance to cloud computing, ERP systems, mobile app development, networks/telecommunications, social, media, virtualization, software as a service, and service-oriented architecture. Those born in the Philippines attributed greater average importance to CRM systems and those born in South Africa attributed greater importance to data mining.

Some of these differences appear to be confounded with age since some of the technologies are relatively new and, as discussed previously, respondents born in India and the Philippines were disproportionately young compared to other respondents; however, the association between country of origin and the importance placed on the technology-related issues described is stronger than for age (i.e., there are only six statistically significant differences in mean technology importance ratings for age compared to 11 for country of origin).

23.7 Individual IT Employee Issues

As shown in Table 23.4, on average, respondents agreed that they liked working for their organization and were satisfied with their current jobs.

Female respondents recorded greater average job satisfaction than male respondents on all three measures shown in Table 23.4 (based on ANOVA F statistics, $p < 0.05$). A similar pattern held for senior managers, who also reported greater job satisfaction on all three measures compared to those lower in their organizational structures.

On average, respondents felt only moderate work pressure; however, compared to those from other countries, respondents who were born in South Africa more strongly agreed that they felt busy/rushed or pressured at work and those who were born in India disagreed more strongly

Table 23.4: Individual IT Employee Issues in New Zealand

Individual Issues	Mean Rating*	Std. Deviation
Job satisfaction		
In general, I like working here.	1.83	0.73
All in all, I am satisfied with my current job.	2.05	0.82
In general, I don't like my current job.	4.08	0.89
Work pressure		
I feel that the number of requests, problems or complaints that I deal with at work is more than expected.	3.33	1.07
I feel that the amount of work I do interferes with how well it is done.	3.05	1.12
I feel busy or rushed at work.	2.96	1.07
I feel pressured at work.	3.04	1.04
Work–life balance		
There is a blurring of boundaries between my job and my home life.	3.50	1.20
My work-related responsibilities create conflicts with my home responsibilities.	3.73	1.09
I do not get everything done at home because I find myself completing job-related work.	3.84	1.07
Workload and burnout		
I feel drained from activities at work.	3.12	1.14
I feel tired from my work activities.	3.03	1.15
Working all day is a strain for me.	3.56	1.05
I feel burned out from my work activities.	3.60	1.09
Sense of accomplishment		
I feel I'm making an effective contribution to what this organization does.	1.87	0.65
In my opinion, I do a good job.	1.76	0.59
I have accomplished many worthwhile things in this job.	1.84	0.65
At my work, I feel confident that I am effective at getting things done.	1.92	0.67
Threats to one's job		
I am worried that future technology advancements may pose a threat to my job.	3.62	1.06
I believe that other people may be able to perform my work activities.	2.53	1.02
I am concerned that my job may be eliminated soon.	3.81	1.02
I am concerned that my job may be outsourced soon.	3.93	0.97
Career plans		
I will be with this organization 1 year from now.	2.32	1.06
I will take steps during the next year to secure a job at a different organization.	3.53	1.04

(Continued)

Table 23.4:   (*Continued*)

Individual Issues	Mean Rating*	Std. Deviation
I will be with this organization 5 years from now.	3.10	1.04
I will be working in the IT field 1 year from now.	1.79	0.87
I will take steps during the next year to secure a job outside the IT field.	4.09	0.91
I will be working in the IT field 5 years from now.	2.18	1.03

*Rating scale ranges from 1 to 5: 1 strongly agree and 5 strongly disagree.

with those last two work pressure statements (based on ANOVA F statistics, $p < 0.05$).

Perhaps counter-intuitively, female respondents disagreed more strongly than male respondents with the statements that "There is a blurring of boundaries between my job and my home life." and "My work-related responsibilities create conflicts with my home responsibilities."

Compared to respondents who were born in other countries, those who were born in India or the Philippines disagreed more strongly (on average, based on an ANOVA F statistic, $p < 0.05$) that "I feel drained from activities at work" and "I feel tired from my work activities."

There were also country of origin differences in the sense of accomplishment respondents felt. Those who were born in the Philippines (followed by those born in South Africa) agreed most strongly with the last three statements about accomplishment, while those born in New Zealand and the UK agreed least strongly.

Country of origin was also related to perceived threats. Respondents who were born in the UK (followed by New Zealand) were least concerned about their jobs being eliminated or outsourced.

Compared to other respondents, those working in public-sector organizations agreed less strongly that they would be with their current organization in 1 year and 5 years (based on ANOVA F statistics, $p < 0.05$).

23.8   Conclusion

According to survey respondents, the top organizational IT issues in New Zealand were IT reliability and efficiency, security and privacy, and alignment between IT and business, although the second and third of these issues were perceived as more important by those working in financial services organizations than other types of organizations. The top technology issues identified by survey respondents were networks/telecommunications, enterprise application integration, and mobile and wireless applications, though enterprise application integration was a particular concern for those working in public-sector organizations.

New Zealand's IT workforce is already being heavily influenced by migrants, who made up 45% of survey respondents. Those from India, in particular, appear to be bringing with them a greater sense of urgency, perhaps developed through exposure to India's much larger, more competitive, IT industry. They are also disproportionately young and well educated: suggesting that their influence will only grow as more of them are promoted into managerial positions.

For now, New Zealand IT workers appear to be fairly well satisfied with their roles, without feeling overly pressured or fearful about losing their jobs.

References

Creative HQ (2018). https://creativehq.co.nz/startups/. Accessed July 2018.

Kappelman, L., Nguyen, Q., McLean, E., Maurer, C., Johnson, V., Snyder, M., & Torres, R. (2017). The 2016 SIM IT Issues and Trends Study. *MIS Quarterly Executive*, 16 (1).

Ministry of Education (2018). Resources statistics page, Education Counts website. https://www.educationcounts.govt.nz/statistics/tertiary_education/r esources. Accessed July 2018.

New Zealand Bankers Association (2018). Banking industry facts and figures. http://www.nzba.org.nz/consumer-information/nzba-assistance/ban king-industry-facts-figures/. Accessed July 2018.

New Zealand Immigration (2018). Information technology. NZ Immigration website, NZ government. https://www.newzealandnow.govt.nz/work-in-nz/nz -jobs-industries/information-technology-jobs. Accessed July 2018.

New Zealand Productivity Commission (2017). New models of tertiary education. https://www.productivity.govt.nz/sites/default/files/New %20models%20of%20tertiary%20education%20FINAL.pdf   Accessed July 2018.

NZ State Services Commission (2018). Workforce. http://www.ssc.govt.nz/public -service-workforce-data/hrc-workforce. Accessed July 2018.

NZ Government (2013). ICT strategy and action plan to 2017. Available at https://www.ict.govt.nz/assets/Uploads/Government-ICT-Strategy-a nd-Action-Plan-to-2017.pdf.

New Zealand Treasury (2016). New Zealand economic and financial overview 2016. https://treasury.govt.nz/sites/default/files/2010-04/nzefo -16.pdf. Accessed July 2018.

NZ State Services Commission (2018). Workforce. SSC website. Available at http: //www.ssc.govt.nz/public-service-workforce-data/hrc-workforce

Open Data Barometer (2018). Key findings. https://opendatabarometer.org/2nd Edition/summary/. Accessed July 2018.

Palvia, P., Jacks, T., Ghosh, J., Licker, P., Romm-Livermore, C., Serenko, A., & Turan, A. H. (2017). The World IT Project: History, trials, tribulations, lessons, and recommendations. *Communications of the Association for Information Systems*, 41(18), 389–413.

Palvia, P., Ghosh, J., Jacks, T., Serenko, A., & Turan, A. (2018). Trekking the globe with the World IT Project. *Journal of Information Technology Case and Application Research*, 20(1), 3–8.

Statistics New Zealand (2017). New Zealand business demography statistics: At February 2017–tables: https://www.stats.govt.nz/information-releases /new-zealand-business-demography-statistics-at-february-2017. Accessed July 2018.

Statistics New Zealand (2018a). https://www.stats.govt.nz/topics/population. Accessed July 2018.

Statistics New Zealand (2018b). http://archive.stats.govt.nz/browse_for_stats/ec onomic_indicators/NationalAccounts/Contribution-to-gdp.aspx. Accessed July 2018.

Statistics New Zealand (2018c). http://archive.stats.govt.nz/browse_for_stats/ec onomic_indicators/NationalAccounts/Contribution-to-gdp.aspx. Accessed July 2018.

Statistics New Zealand (2018d). https://www.stats.govt.nz/integrated-data/int egrated-data-infrastructure/. Accessed July 2018.

Statistics New Zealand (2018e). Quarterly Employment Survey supplementary tables for June 2018 quarter.

Stuff (2017). https://www.stuff.co.nz/business/91188888/more-than-48000-f rom-around-the-world-apply-for-a-looksee-at-wellington. Accessed July 2018.

Chapter 24

Information Technology Issues in Nigeria

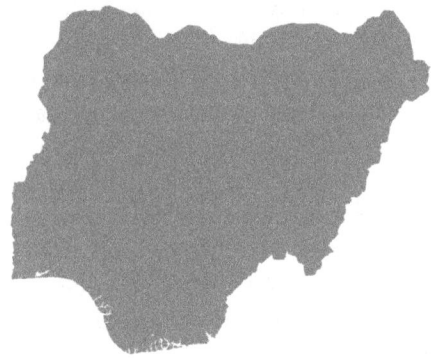

Olayinka David-West[*,‡], Eugene Ohu[*,§], and Prashant Palvia[†,¶]

[*]Lagos Business School, Pan-Atlantic University, Ajah - Lagos, Nigeria
[†]University of North Carolina at Greensboro, Greensboro, NC, USA
[‡]ydavid-west@lbs.edu.ng
[§]eohu@lbs.edu.ng
[¶]pcpalvia@uncg.edu

Summary

Based on our implementation of the World Information Technology (IT) Project survey, the three top organizational issues from the perspective of IT employees in Nigeria are: IT reliability and efficiency, business process reengineering, and business agility and speed to market. The top three technology issues are: cloud computing, mobile and wireless applications, and business intelligence/analytics. Nigerian IT employees are relatively satisfied with their jobs, and feel highly accomplished in terms of what they contribute to their organizations. Despite a perception of increased pressure from work, there is a moderate experience of work and family balance, even though some still experience burnout. The threat of job losses or outsourcing is low although job rotations are expected. On the whole, IT workers are relatively mobile and open to new opportunities in and outside of the IT industry.

24.1   Introduction

The use and adoption of information technology (IT) is global, but the issues related to its implementation, management and measurement have contextual variations, which may be political, environmental and economic in nature. In developing economies, deficiencies in infrastructure, talent and finance further compound these IT issues (Avgerou, 2008). The development of indigenous innovations to bridge these gaps has not been well studied. Such is the case with Nigeria, an emerging economy, where access to relevant knowledge is scarce, but which has one of the most dynamic burgeoning IT hubs on the African continent. This study on IT issues in Nigeria forms part of the World IT Issues initiatives that seeks to provide fresh insights into the perceptions of IT issues across organizational, technological and individual dimensions, and to highlight some innovative approaches to mitigate these issues.

24.2   Country Background and History

Nigeria, located on the West Coast of Africa, is Africa's most populous nation, and with the 2014 rebasing of the GDP, it also became Africa's largest economy. Nigeria, a former British Colony (until independence in 1960) and member of the Commonwealth, was created as a result of the amalgamation of the Southern Nigeria and Northern Nigeria Protectorates in 1914 (Kirk-Green, 1967). Today, it operates as a Federal Republic with 36 states and a centrally located Federal Capital Territory (FCT), called Abuja. Geographically, Nigeria comprises six geopolitical zones and numerous ethnic tribes and languages. Although English is the official language, Nigeria is a multi-ethnic country with more than 250 local dialects. The dominant local languages are Hausa, Igbo, Yoruba and an English patois called "Pidgin English", spoken by the majority. Roughly distributed in half, Christianity and Islam are the two dominant religions practiced among other indigenous religions.

With a nominal GDP of about 400 billion USD, Nigeria derives the lion share of her income from crude oil deposits. Other income sources include agriculture (20.9%), industry (20.4%) and services and other activities (58.5%). Nigeria has a predominantly youthful population with over 45% under 18 and almost 70% below 35 years of age (National Bureau of Statistics and Federal Ministry of Youth Development, 2012). Unemployment in Nigeria remains on the rise at 20.98% in 2016 (National Bureau of Statistics, 2018; National Bureau of Statistics, 2017).

The World Bank (2018) income categorization places Nigeria as a lower middle-income economy, challenged with low human development index (HDI) as well as infrastructure deficits, especially power, roads and transportation.

24.3 Information Technology in Nigeria

The beginning of information and communication technologies (ICT) in Nigeria dates back to the monopoly days of original equipment manufacturers, such as IBM, NCR and DEC who either had a local presence or made products available through local dealership/partnership. The microcomputer democratized the hardware industry through the entry of a new breed of suppliers, often buying and reselling through a gray market of ICT products and accessories (Ogunsola and Aboyade, 2005). One effect of this was the emergence in the 1990s of a large computer trading center in the Ikeja area of Lagos State called the "Computer Village", which coincided with the start of local assembly of computers, there and elsewhere in the country. The formation of the Computer and Allied Products Dealers Association of Nigeria (CAPDAN) was therefore a natural aftermath (Chete *et al.*, 2014). The advent of Cloud Computing led to the appearance of local co-location and data centers.

Before the liberalization and privatization of the telecommunications industry in 2003, the Nigeria Telecommunications Company (NITEL) dominated the landscape with analogue and mostly fixed telephony, which were insufficient and inefficient. The auctioning of GSM licenses marked the start of a new era in communications (Ogunsola and Aboyade, 2005; Asongu, 2013). Today, more than 90% of telephone and Internet usage is on GSM powered mobile devices. Broadband speed tests in Nigeria currently show it to be at a sub-optimal level with signs of decline (Speedtest Global Index, 2018; Telegeopraphy, 2018).

The software and applications industry comprises of the following subcategories: OEMs, indigenous software developers, vendors and value-added resellers (VARs). The range of software applications varies from productivity and utility to business systems — functional or enterprise. The ICT industry is managed and governed by the following organizations: National Communications Commission (NCC), National Information Technology Development Agency (NITDA), Computer Professionals Registration Council of Nigeria (CPN), Association of Telecommunications Companies of Nigeria (ATCON), Nigeria Computer Society (NCS), Information Technology Association of Nigeria (ITAN), Institute of Software Practitioners of Nigeria (ISPON), and Nigeria Internet Registration Association (NiRA).

The services sector includes professional services firms and independent consultants providing advisory and applications-related services like change management, software selection and the like to the user community. In support of emerging technology ventures, a variety of business incubation institutions offering services ranging from space and infrastructure to mentoring and development have emerged.

Regarding overall performance of the IT industry, Nigeria ranked 119 in the world on the global Networked Readiness Index published by the World Economic Forum (World Economic Forum, 2016) and has shown varying degrees of progress (or decline) since ratings began in 2009. There is significant heterogeneity in terms of individual components of networked readiness — in particular, a move up in readiness and a decline in impacts. The improvement in readiness is due to approaching full mobile coverage and falling broadband prices.

24.4 Methodology

An online version of the World IT Project instrument (Palvia *et al.*, 2017; Palvia *et al.*, 2018) was posted on Surveymoz.com, and the link was shared electronically with prospective respondents. Participants were professionals who attended various executive programs at a business school in Nigeria. Data collection was arduous for various reasons. One constraint was the lack of sufficient in-house IT employees in many organizations. Some respondents complained that the questionnaire was too long, which contributed to the attrition rate. Data collection began in 2015–2017. Out of

Table 24.1: Descriptive Statistics

Characteristics	N	%	Characteristics	N	%
Education:			Years of work experience:		
High school or less			0–4 years	77	85.6
Associate degree			5–9 years		
Bachelor's degree	51	56.0	10–19 years		
Master's degree	38	41.8	20–29 years	11	12.2
Ph.D.	2	2.2	30+ years	2	2.2
Years of IT experience:			Organizational location:		
0–4 years	81	90.0	IT department employee	68	73.9
5–9 years			IT worker in non-IT	9	9.78
10–19 years			department		
20–29 years			Contract employee	2	2.17
30+ years	9	10.0	Consultant	12	13.0
Work as:			Vendor employee	1	1.09
Mostly full time	87	97.8	Work position:		
Mostly part time	1	1.1	Not part of management	14	15.4
Mostly over time	1	1.1	In lower management	21	23.1
Been laid off from IT job:			In middle management	33	36.3
			In senior management	23	25.3
Yes	4	4.3			
No	89	95.7			

167 responses, 100 were complete and 93 were usable. Telephone calls and emails were made to respondents in order to improve their participation.

The organizations surveyed were mainly demand-oriented organizations adopting IT for business operations and services. The IT organizations (i.e., suppliers of services) account for 20% of the respondents. Respondents were mostly younger employees; approximately two-thirds were under the age of 40. Practically all had at least a bachelor's degree and had full time jobs. Additional details of the respondents are shown in Table 24.1.

24.5 Organizational IT Issues

Table 24.2 shows the mean ratings of organizational issues as well as the ranks. It is noteworthy that all organizational issues are relatively close in their mean ratings. There are two possible explanations: one that the issues are actually relatively close in importance (which is a more credible explanation), or two that the IT employees are not fully informed about organizational concerns and priorities. In any case, the results should be viewed with caution.

Table 24.2:   Organizational IT Issues in Nigeria

Organizational IT Issues	Rank	Mean Rating*	Std. Deviation
IT reliability and efficiency	1	1.95	0.96
Business process reengineering	2	1.95	0.73
Business agility and speed to market	3	2.02	0.88
Security and privacy	4	2.04	1.05
IT strategic planning	5	2.05	0.88
Alignment between IT and business	6	2.05	0.99
Knowledge management	7	2.05	1.00
Enterprise architecture	8	2.07	0.81
Business productivity and cost reduction	9	2.07	0.90
Attracting and retaining IT professionals	10	2.07	1.13
Revenue-generating IT innovations	11	2.09	0.79
Project management	12	2.10	1.03
IT service management (e.g., ITIL)	13	2.14	1.01
Continuity planning and disaster recovery	14	2.21	1.10
IT cost reduction	15	2.25	0.78
Globalization	16	2.31	0.85
Outsourcing	17	2.35	1.11
Bring your own computing device (BYOD)	18	2.39	1.13

*Rating scale ranges from 1 to 5: 1 as most important and 5 as no importance.

Reliability and efficiency of IT systems, ranked first, is a critical success factor for most organizations where IT enables a significant proportion of business operations. The ability to maintain sufficient IT reliability and efficiency is predominantly dependent on resources and capabilities, which includes infrastructure and human capacity that require significant financial resources. While there is variability in importance across industries, healthcare, professional services, manufacturing, financial services and education report the highest values. The importance of the second-ranked Business Process Reengineering (BPR) in Nigerian organizations is related to the high adoption of foreign developed applications whose underlying processes (and paradigms) are often in conflict with local business processes and controls. Also, the incremental cost of software modifications or change management practices required to adopt proprietary, off-the-shelf software applications can be considerable. Business agility and speed to market is the third most important IT issue relating to the rate at which IT business transformation occurs. Adapting to the Nigerian market system can be difficult for foreign IT businesses, a trait not unrelated to the developing nature of the country.

Security and privacy of IT infrastructure and data rank fourth in Nigeria. In this era of cloud computing, this issue is most pertinent where businesses often ask about data security on third-party infrastructure. IT strategic planning and the related, alignment between IT and business, follow as the next two issues. The importance of these issues has arisen from questions relating to the business value of IT. With the significant proportion of IT expenses being denominated in foreign currency, macroeconomic factors like exchange rate volatility play a significant role in IT budgets and further highlight the payoff question. Hence, strategic planning, IT and business alignment and other governance issues are of vital importance.

Knowledge management and the retention of tacit knowledge ranks seventh among the IT issues in Nigeria. This may be because of increased employee turnover and mobility due to the availability of better job opportunities at home and abroad. This issue is closely followed in the 10th place by the difficulty in attracting and retaining IT professionals, a factor that has consequences for organizational IT reliability, efficiency and for knowledge management.

The growth of IT operations and business dependency makes holistic enterprise analysis and planning necessary. The knowledge of Enterprise Architecture (EA) practices that use frameworks like Zachmann Enterprise Architecture is however scarce. Business productivity and cost reduction, seeming myths of IT deployments, are among the top 10 issues identified. The use of IT to enhance work and productivity as well as reduce costs is a key issue for business leaders. Surprisingly, the reduction of IT costs itself was ranked lower; it may be inevitable given the uncontrollable nature of IT costs due to economic and external factors.

Ranked at the bottom of the list are: continuity planning and disaster recovery, IT cost reduction, globalization, outsourcing, and BYOD. Business continuity planning and disaster recovery are one of the IT standards mandated for the banking industry. The gradual digitization of business processes has increased business dependence on IT and hence the importance of continuity of IT operations; yet it was not ranked high by our respondents as it is a cost whose benefit is not realized immediately or may never be realized.

IT has been a veritable globalization catalyst especially in emerging markets like Nigeria, where it has empowered customers and others thus increasing product and service competition. While ranked low, the globalization effects of IT have begun to create opportunities for Nigerian companies, especially banks, to expand their reach to other countries within the continent and beyond. Outsourcing, especially business process outsourcing

(BPO), is a closely related phenomenon to globalization. In Nigeria, BPO and offshoring have had limited impact on the delivery of IT services. While some international organizations have outsourced IT operations, the majority still insource or utilize alternative external contracting mechanisms to supplement IT capacity.

The issue ranked the least important is BYOD. The ubiquity of computing devices, especially mobile devices in emerging markets such as Nigeria, has resulted in a mobile-first environment with the result of an improved management of the BYOD approach. Apparently, it has not caused headaches for management or the employees.

24.6   Technology and Infrastructure Issues

The ranks on technology issues are presented in Table 24.3. Once again, there is little variability in the mean ratings provided by the respondents.

While cloud computing refers to the provision of software, platforms or infrastructure as a service, the differential ranking of "software as a service" (# 13) and this top-ranked issue relates to infrastructure and platforms. The rising cost of infrastructure and hardware provisioning has improved the attractiveness of cloud based services as well as tier 4 data centers/co-location sites. It has also resulted in technological innovations, thus leading

Table 24.3:   Technology and Infrastructure Issues in Nigeria

IT-Related Issues	Rank	Mean Rating*	Std. Deviation
Cloud computing	1	2.11	0.97
Mobile and wireless applications	2	2.11	1.07
Business intelligence/analytics	3	2.12	1.09
Collaborative and workflow tools	4	2.12	1.02
Networks/telecommunications	5	2.17	1.15
Big data systems	6	2.19	1.03
Customer relationship management (CRM)	7	2.26	1.05
Data mining	8	2.26	1.13
Service-oriented architecture (SOA)	9	2.30	1.04
Enterprise application integration	10	2.30	1.15
Mobile apps development	11	2.31	1.19
Enterprise resource planning (ERP) systems	12	2.32	0.96
Software as a service	13	2.33	1.09
Business process management systems	14	2.34	1.18
Virtualization (desktop or server)	15	2.40	1.21
Social networking/media	16	2.49	1.21

*Rating scale ranges from 1 to 5: 1 as most important and 5 as no importance.

to the development of self-service products and services and an increased demand for constant access in Nigeria where grid power is scarce.

Since the liberalization of the telecommunications industry, mobile penetration has surpassed PC penetration, making Nigeria a mobile-first market (Asongu, 2013). The provision of mobile and wireless applications has become a technology priority (ranked second in our survey), requiring suitable enterprise architecture and infrastructure to support applications delivery. Also there is an increasing demand for evidence-based decision-making, thus making business intelligence an important issue for business. This is evident in the position occupied by the umbrella term "business intelligence and analytics" (ranked third) or related technologies addressing specific areas — "big data systems" (ranked sixth) and "data mining" (ranked eighth). There is a growing need for credible data and an attendant need for people with capabilities for information management; for data scientists who can design big data systems as well as write programs capable of extracting business insights from data.

The global nature of workforce teams, mobility and other factors have somewhat altered the nature of work, increasing telecommuting as a work option as against working from a fixed location. Thus, the deployment of collaborative and workflow tools (ranked fourth) to support the work of global teams across multiple time zones is necessary to ensure organizational productivity. These tools include voice and video conferencing (individual and group-based), document sharing and management and collaborative workflow (groupware) systems.

The organizational utility of IT services within organizations as well as their extension beyond the organization and Nigeria's mobile-first context heightens the relevance and importance of high-speed networks/telecommunications — local area, metropolitan, wide and global (Internet). Within industries that operate in urban and rural locations, the increased infrastructure costs can be a significant limiting factor. Even though there is a national broadband policy aiming to increase penetration, nationwide fibre optic cable installation has been stalled by inadequate business cases for these endeavors, and issues of right of way (RoW).

The growth of the middle class in Nigeria has created consumer/retail markets that require effective management strategies. Hence, the management of data using customer relationship management (CRM) systems takes on added importance. CRM has proven beneficial in business-to-consumer (B2C) relationships like e-commerce, financial services, healthcare, hospitality, etc. Depending on the age of the organizations, IT deployments vary in architecture, design and age. The replacement challenges associated with some of these applications and systems have warranted the need for alternative integration strategies, such as service-oriented architecture (SOA) and enterprise application integration (EAI). The relationship between these two issues may be inferred in their rank positions, at ninth and 10th, respectively.

Among the lower-ranked issues are business process management systems, virtualization, and social networking/media. The lower importance given to business process management systems is indicative of some misalignment between business and IT. In Nigeria, the majority of the businesses adopted computing systems in the PC era, thus the need to operate multiple environments through virtualization is minimal and not considered a priority. Even though social networking/media are seemingly low priority technologies, the evolving nature of new media and its applications to marketing and corporate naming and shaming are keeping its use on the rise.

24.7   Individual IT Employee Issues

The perceptions of individuals working in the Nigerian IT industry were examined across seven themes, as reported in Table 24.4. As per the scale, a lower score represents higher agreement with each statement.

In general, the IT workers surveyed are extremely satisfied and happy with their current employers and jobs. However, there seems to be an apparent contradiction. While they are happy and satisfied, they report high levels of work pressure, workload, and moderate levels of work–home conflict and even some burnout. They seem to be coping up well as they indicate a high sense of accomplishment and feel their contributions are of value to the organization. While IT jobs are not unique and are interchangeable, threats arising from technology advancements, job eliminations and outsourcing are not prevalent. Thus, the IT employees do not foresee threats to their job security. Another contradiction is that the IT workers are mobile, open to changing organizations, roles and even relocating to other nations. They even anticipate working outside the IT profession.

These opposing results can be explained by the observation that while Nigeria has a vibrant and burgeoning IT industry and has a high demand for IT workers, the job itself is very demanding and requires heavy workload and long hours. The growth in the number of suppliers through entrepreneurial activity has resulted in Nigeria being one of the leading

Table 24.4: Individual IT Employee Issues in Nigeria

Individual Issues	Mean Rating*	Std. Deviation
Job satisfaction		
In general, I like working here.	1.92	0.72
All in all, I am satisfied with my current job.	1.94	0.84
In general, I don't like my current job.	3.04	1.23
Work pressure		
I feel that the number of requests, problems or complaints that I deal with at work is more than expected.	2.76	1.15
I feel that the amount of work I do interferes with how well it is done.	2.56	0.94
I feel busy or rushed at work.	2.61	1.07
I feel pressured at work.	2.64	1.07
Work–life balance		
There is a blurring of boundaries between my job and my home life.	2.70	1.06
My work-related responsibilities create conflicts with my home responsibilities.	2.80	1.17
I do not get everything done at home because I find myself completing job-related work.	2.79	1.06
Workload and burnout		
I feel drained from activities at work.	3.07	1.13
I feel tired from my work activities.	2.59	1.12
Working all day is a strain for me.	2.85	1.16
I feel burned out from my work activities.	2.34	1.10
Sense of accomplishment		
I feel I'm making an effective contribution to what this organization does.	1.85	0.96
In my opinion, I do a good job.	1.83	0.91
I have accomplished many worthwhile things in this job.	1.84	0.97
At my work, I feel confident that I am effective at getting things done.	2.85	1.42
Job security		
I am worried that future technology advancements may pose a threat to my job.	3.14	1.30
I believe that other people may be able to perform my work activities.	2.90	1.35
I am concerned that my job may be eliminated soon.	3.46	1.32
I am concerned that my job may be outsourced soon.	3.29	1.37
Career planning/progression		
I will be with this organization 1 year from now.	3.17	1.42
I will take steps during the next year to secure a job at a different organization.	2.95	1.40

(*Continued*)

Table 24.4: (*Continued*)

Individual Issues	Mean Rating*	Std. Deviation
I will be with this organization 5 years from now.	3.23	1.39
I will be working in the IT field 1 year from now.	4.25	0.97
I will take steps during the next year to secure a job outside the IT field.	3.58	1.25
I will be working in the IT field 5 years from now.	3.99	0.95

*Rating scale ranges from 1 to 5: 1 as strongly agree, to 5 as strongly disagree.

IT hubs on the continent. Hence, the need for competent and reliable IT personnel is growing and is a critical success factor for enterprises (Asongu and Nwachukwu, 2018, Huang *et al.*, 2003). As the demand worldwide and in the region is also high, skilled and competent workers can also find attractive offers outside the country.

24.8   Conclusion

This chapter provided an understanding of important issues relevant to IT employees in Nigeria across organizational, technological, and individual dimensions. IT reliability and efficiency, business process reengineering and business agility and speed to market are perceived as the most important IT issues across the organizational dimension. The technology issues perceived to be of greatest significance are cloud computing, mobile and wireless applications, and business intelligence and analytics. Nigerian IT workers are highly satisfied with their jobs and feel positive about their value contribution to the organization. Interestingly, while they have a strong sense of job security, they also have an active desire for career progression.

References

Asongu, S. A. (2013). How has mobile phone penetration stimulated financial development in Africa?. *Journal of African Business*, 14(1), 7–18.

Asongu, S. A., & Nwachukwu, J. C. (2018). Openness, ICT and entrepreneurship in sub-Saharan Africa. *Information Technology & People*, 31(1), 278–303.

Avgerou, C. (2008). Information systems in developing countries: A critical research review. *Journal of information Technology*, 23(3), 133–146.

Chete, L. N., Adeoti, J. O., Adeyinka, F. M., & Ogundele, O. (2014). *Industrial development and growth in Nigeria: Lessons and challenges* (No. 2014/019). WIDER Working Paper.

Huang, J., Makoju, E., Newell, S., & Galliers, R. D. (2003). Opportunities to learn from 'failure' with electronic commerce: A case study of electronic banking. *Journal of Information Technology*, 18(1), 17–26.

Kirk-Greene, A. H. M. (1967). The peoples of Nigeria: the cultural background to the crisis. *African Affairs*, 66(262), 3–11.

National Bureau of Statistics & Federal Ministry of Youth Development. (2012). *2012 National Baseline Youth Survey*. Nigeria, p. 146.

National Bureau of Statistics. (2017). *Labor Force Statistics Vol.1: Unemployment and underemployment report (Q1-Q3 2017)*.

National Bureau of Statistics. (2018). *Nigerian Gross Domestic Product Report (Q2 2018)*. Nigeria: NBS, p. 89.

Ogunsola, L. A., & Aboyade, W. A. (2005). Information and communication technology in Nigeria: Revolution or evolution. *Journal of Social Sciences*, 11(1), 7–14.

Palvia, P., Jacks, T., Ghosh, J., Licker, P., Romm-Livermore, C., Serenko, A., & Turan, A. H. (2017). The World IT Project: History, trials, tribulations, lessons, and recommendations. *Communications of the Association for Information Systems*, 41(18), 389–413.

Palvia, P., Ghosh, J., Jacks, T., Serenko, A., & Turan, A. (2018). Trekking the globe with the World IT Project. *Journal of Information Technology Case and Application Research*, 20(1), 3–8.

Speedtest Global Index. (2018). Nigeria's Mobile and Broadband Internet Speeds. Retrieved August 29, 2018, from http://www.speedtest.net/global-index/nigeria

Telegeopraphy. (2018). Submarine Cable Map. Retrieved from https://www.sub marinecablemap.com/ [Accessed August 29, 2018].

The World Bank. (2018). Data- Country Classification. Retrieved from http s://datahelpdesk.worldbank.org/knowledgebase/articles/906519 [Accessed September 10, 2018].

World Economic Forum. (2016a). Economies- Nigeria. Retrieved from http://wef .ch/1SOiFe5 [Accessed August 30, 2018].

Chapter 25

Information Technology Issues in Pakistan

Ijaz A. Qureshi[*,¶], Moazzam Hussain[*,∥], Hassan Raza[†,**],
Shaista Shahid[*,††],
Prashant Palvia[‡,‡‡], and Aykut Hamit Turan[§,§§]

[*]University of Sialkot, Sialkot, Pakistan
[†]University of Malaya, Kuala Lumpur, Malaysia
[‡]University of North Carolina at Greensboro,
Greensboro, NC, USA
[§]Sakarya University, Sakarya, Turkey
[¶]ijaza.qureshi@gmail.com
[∥]moazzamhussains@gmail.com
[**]86hassanraza@gmail.com
[††]Shaista.Shahid@uskt.edu.pk
[‡‡]pcpalvia@uncg.edu
[§§]aykut.turan@gmail.com

Summary

Country Introduction: Pakistan, officially the Islamic Republic of Pakistan, is a country in South Asia. It is the world's sixth-most populous country with a population exceeding 212,742,631 people. In area, it is the 33rd-largest country, spanning 881,913 square kilometres.[1] Pakistan is a vibrant economy in terms of its information technology (IT) industry and services. This chapter focuses on organizational IT issues, technology issues, and individual IT employee issues in Pakistan. Revenue-generating IT innovations, IT strategic planning, and business productivity and cost are the highest-ranked organizational IT issues in Pakistan. Among the top technology issues are business intelligence/analytics, customer relationship management (CRM) systems, and mobile and wireless applications. Bring your own device (BYOD), outsourcing, and globalization are the least important organizational IT issues, while networks/telecommunications, big data systems, and enterprise resource planning (ERP) systems are the least important technology issues. IT industry workers in Pakistan seem to be satisfied but have moderate levels of burnout and workload pressure at their jobs. Most of the IT workforce in Pakistan is motivated to remain in the IT industry in the future.

25.1   Introduction

A developing country's demographics are different from those of developed countries in terms of market characteristics, economics, culture, politics, educational background of the population, and technology adoption. Pakistan being developing economy with almost 96% Muslim population provides unique insights into the adoption and growth of information technology (IT) in such an environment. Ironically, the literature regarding Pakistan's IT-related issues is sparse and difficult to find. Thus, the inclusion of Pakistan among the GITMA World IT Project's 37 countries was a fortunate choice (Palvia *et al.*, 2017). The country investigators from Pakistan used the instrument and procedures provided by the World IT Project's leaders to collect survey data in Pakistan. While the types of data collected in the survey were broad and extensive, in this chapter, we provide the organizational IT issues, infrastructure and technology issues, and individual issues of the IT workforce in Pakistan. Explanation and discussion of the findings are included to provide deeper insights.

---

[1] *Source*: Wikipedia.org.

25.2   Country Background and History

Pakistan, or the Islamic Republic of Pakistan, is located in South Asia. Pakistan was established on August 14, 1947, yet it has a vast history dating back thousands of years. Pakistan was established because of the ongoing struggle of Muslims and their vision to have a homeland for themselves in South Asia. The history of the region, where Pakistan is established, is reflected in the history of the country itself. In the region, economic, political and religious formations have been developed by Muslims with their unique impressions. Shariah was adopted as the law of the state and was enforced.

Urdu is the national language of Pakistan along with English. Pakistan is a developing where half of production of goods and services is dependent on non mechanical sources semi-industrialized economy whose main industries are manufacturing, agriculture, and remittances (PBS, 2018). The total GDP of Pakistan accounts for 0.49% of the world economy. The GDP of Pakistan has continuously increased from US$71.19 billion in 1960 to an all-time high of US$304.95 billion in 2017. Pakistan has a very young population profile, and 60% of its 200 million population are between 15 and 29 years of age. This demographic provides Pakistan with enormous human and knowledge capital potential (MOITT, 2018).

25.3   Information Technology in Pakistan

IT in Pakistan is a rising industry. Every day, new startups are being incubated by one of the many incubators that are in existence in Pakistan. Many people are coming out of the typical mindset of getting a job in a company and working there for the rest of their lives. They are willing to take risks, and more importantly, the ones with money are also willing to invest in the start-ups.

A packaging company in Pakistan started to use computers first back in 1957 and initiated the process of computerization in the country (PJSS, 2009). Later, computers were installed in Karachi in the mid-sixties, and now there are more than 1800 mainframe and minicomputers in both government and the private sector. It is estimated that Pakistan imports almost 500,000 new personal computers per year. IT is used in industries like Education, Agriculture, Defense, and the Entertainment industries (PKKH, 2017). Generally, computer prices are not very high, and the country has a very high growth rate in computer usage and penetration

(PP, 2008). Furthermore, software developers, programmers, BPO specialists, and application developers are demanding that its IT industry be the source of differentiation in the country and in the region.

Pakistan was among the top IT freelance developers in the world in 2015. Its IT exports have increased by 70% since 2015 (MOITT, 2018). Now Pakistan makes about $2.4 billion in IT exports annually (MOITT, 2018). Today, Pakistan's IT Industry revenue has reached $3.0 billion. The cellular network industry makes up a large part of IT industry, presenting a promising and valuable IT landscape (MOITT, 2018). Pakistan's IT industry has won many international awards and recognition at international forums while representing the country very well.

E-government was introduced in Pakistan in 2002. E-government services in the country offer a wide range of networks and mobile computing to transform the government's operations and enhance efficiency to offer better value to their customers. The government facilitates the IT industry through numerous sustainable development and accelerated digitization projects. It also tries to enhance research and innovation via software technology parks, subsidized bandwidth, international marketing, certifications, internships, and training. The government also plans to position the country on the global outsourcing map by participating in global forums and conferences (MOITT, 2018).

25.4 Methodology

The standard instrument of the World IT Project (Palvia *et al.*, 2017; Palvia *et al.*, 2018) survey instrument was used to collect data without translating into the local language. The national language is Urdu, but English has become the common language in schools, colleges, and universities — as the national curricula are in English. The survey data was collected from different metropolitan cities, as most of the firms are located in these cities. Managers and executives of these firms were approached through personal contacts.

Initially, the response rate was not very good but later, it has improved as questions were answered in meetings with senior executives. After every week, the person in charge in each firm was contacted to get the status of completed questionnaires. Participants took approximately 30–45 minutes to complete the survey, although there were some inquiries about the length of the questionnaire.

In all, 301 completed questionnaires were received. Most of the respondents were males (82%). The majority fell under the age category of

Table 25.1:   Descriptive Statistics

Characteristics	N	%	Characteristics	N	%
Education:			Years of work experience:		
High school or less	7	2.3	0–4 years	132	43.9
Associate degree	15	5.0	5–9 years	122	40.5
Bachelor's degree	155	51.5	10–19 years	44	14.6
Master's degree	124	41.2	20–29 years	2	0.7
Ph.D.	0	0	30+ years	1	0.3
Years of IT experience:			Organizational location:		
0–4 years	148	49.2	IT department employee	234	77.7
5–9 years	124	41.2	IT worker in non-IT department	31	10.3
10–19 years	27	9.0	Contract employee	15	5.0
20–29 years	1	0.3	Consultant	20	6.6
30+ years	1	0.3	Vendor employee	1	0.3
Work as:			Work position:		
Mostly full time	265	88.0	Not part of management	176	58.5
Mostly part time	20	6.6	In lower management	41	13.6
Mostly over time	16	5.3	In middle management	74	24.6
Been laid off from IT job:			In senior management	9	3.0
Yes	53	17.6	Missing	1	0.3
No	247	82.1			
Missing	1	0.3			

21–29 years (67%), followed by the age category of 30–39 (30%). Programming was listed as the priority IT role of IT employees (43%). Other prominent IT roles were analysis and design, testing, database administration, and telecommunications. Further details of the descriptive statistics are provided in Table 25.1.

25.5   Organizational IT Issues

Table 25.2 presents the organizational IT issues in Pakistan ranked in order of importance based on a 5-point Likert scale, with 1 being the most important and 5 being of no importance. The items are ranked from 1 to 18 based on their mean scores.

The highest-ranked item is revenue generating IT innovations, which does not appear in the top five of the https://www.researchgate.net/publ ication/323723507_The_2017_SIM_IT_issues_and_trends_study list (Kappelman *et al.*, 2018). This issue is not rated among the top in most countries across the world. One reason may be the need for overseas companies to establish and set firm footprints in the developed country's IT market. New and profitable IT innovations provide a higher chance to penetrate

Table 25.2:   Organizational IT Issues in Pakistan

Organizational IT Issues	Rank	Mean Rating*	Std. Deviation
Revenue-generating IT innovations	1	1.73	0.87
IT strategic planning	2	1.80	0.85
Business productivity and cost reduction	3	1.81	0.71
Project management	4	1.81	0.89
Knowledge management	5	1.88	0.84
Alignment between IT and business	6	1.89	0.80
Business agility and speed to market	7	1.90	0.79
IT reliability and efficiency	8	1.90	0.89
Security and privacy	9	1.90	0.93
Enterprise architecture	10	1.96	0.91
IT cost reduction	11	1.97	0.812
Business process reengineering	12	1.97	0.818
IT service management (e.g., ITIL)	13	2.04	0.99
Continuity planning and disaster recovery	14	2.05	0.92
Attracting and retaining IT professionals	15	2.05	0.93
Globalization	16	2.13	1.02
Outsourcing	17	2.36	1.09
Bring your own computing device (BYOD)	18	2.84	1.32

*Rating scale ranges from 1 to 5: 1 as most important and 5 as of no importance.

the mature IT industry worldwide. IT strategic planning came in second place. Lack of IT planning is a major problem in organizations (PJSS, 2009); together, these two issues represent the increasing and strategic value of IT in organizations.

The next three issues are business productivity and cost reduction, project management, and knowledge management. These issues relate to the ability of the IT organization to deliver services on time, within budget and at a high level of quality. As client requirements change, the IT organization has to be able to respond to the changes rapidly; this creates a requirement for new knowledge to be disseminated effectively.

Alignment between IT and business, business agility and speed to market, IT reliability and efficiency are the next three issues in the rankings. These issues highlight the need for rapid response to current market needs and client requests in the global IT market. Security and privacy, and enterprise architecture are in the middle at number 9 and 10, respectively. In developed countries, security and privacy are typically at the top of the list. In Pakistan, we hope that this issue will eventually get attention, as in the past the Government's uncertain rules, regulations and policies have led to bureaucracy and less attention to these matters.

Regarding the least important issues, BYOD, outsourcing and global-ization came at the bottom. Globalization is now a fact and being part of the global village is a way of life. Thus, it may not be regarded as a seri-ous issue anymore. IT professionals integrate easily now as the technology advances and more online open resources are available. IT cost reduction was also not a major concern as the cost of IT products is continuously decreasing as the technology continues to advance.

25.6 Technology and Infrastructure Issues

Table 25.3 shows the IT and infrastructure issues in Pakistan. The respon-dents were asked to rate several technology issues in terms of their impor-tance based on a 5-point Likert scale, with 1 being of most importance and 5 being of no importance. Sixteen technology related issues were rated on this scale and are as follows.

Business intelligence/analytics is at the top of the list. This area is a global trend and seems to have an increasing demand. Several IT orga-nizations, both large and small, are specializing in this area and are in

Table 25.3: Technology and Infrastructure Issues in Pakistan

IT-Related Issues	Rank	Mean Rating*	Std. Deviation
Business intelligence/analytics	1	1.89	.882
Customer relationship management (CRM) systems	2	1.98	.912
Mobile and wireless applications	3	1.99	.934
Data mining	4	2.02	.935
Software as a service	5	2.04	.876
Service oriented architecture (SOA)	6	2.06	.943
Mobile apps development	7	2.08	.922
Social networking/media	8	2.08	1.010
Enterprise application and integration	9	2.09	.895
Collaborative and workflow tools	10	2.11	.854
Cloud computing	11	2.15	.926
Business process management systems	12	2.15	.937
Virtualization (desktop or server)	13	2.16	1.010
Enterprise resource planning (ERP) systems	14	2.17	.892
Big data systems	15	2.23	.986
Networks/telecommunications	16	2.27	1.065

*Rating scale ranges from 1 to 5: 1 as most important and 5 as of no importance.

collaboration with IT firms in developed countries. The Ministry of Information Technology and Telecommunications, Pakistan released its Pakistan's National IT Policy in 2017 (Ministry of Science and Technology, 2000), and emphasized that necessary measures should be taken to deal with the latest trends and technologies in the field (TechJuice, 2017). A related technology, data mining, came at the fourth rank and plays a vital role in decision-making by engineers and IT professionals. To move from quality control to quality assurance and reduced error occurrence, companies need to exploit their existing knowledge and previous experiences more effectively. Data mining analysis offers many potential benefits in this context (Shahbaz, 2008).

Customer Relationship Management (CRM) systems was the second highest issue in Pakistan. Both operational and analytical CRM are important in Pakistan. The CRM area constitutes a huge portion of the development and maintenance services provided by Pakistani companies. Mobile/wireless applications was the third highest-ranked technology. The importance of this area not only reflects the growth in mobile applications globally but also indicates its growth in the Pakistani market. The cost of mobile and Internet access is dropping steadily due to increased domestic competition among mobile service providers. In most large cities, the IT infrastructure is good and can support demand. However, in smaller towns and in rural areas, IT infrastructure is not as strong or reliable. This is true for both wired and wireless networks. The Pakistani Government has invested in this sector and collaborated with IT companies and mobile operators to facilitate this sector to grow.

Software as a service, service-oriented architecture (SOA), mobile apps development; social networking/media are somewhat in the middle of the list. On average, these issues are of less importance than the top five, but demand from international customers is still present, and many of the Pakistani IT companies are providing services in these sectors as well.

Enterprise application and integration, collaborative and workflow tools, cloud computing, business process management systems, and virtualization are among the less important issues for Pakistan's IT industry and workforce. The least important issues are big data systems, networks/telecommunications, and ERP systems. While networks/telecommunications is a little difficult to explain, the reasons for the others may have to do with the high upfront investment that is required. While ERP systems may be popular in developed nations, they require high investment. Therefore, the emphasis in Pakistan seems to be on the integration of existing applications and new applications in CRM.

25.7 Individual IT Employee Issues

Table 25.4 shows a summary of the responses about individual IT employee issues in Pakistan. These issues were investigated under seven broad themes, namely: job satisfaction, work pressure, work–life balance, workload and burnout, sense of accomplishment, threats to one's job, and career plans. Each item was rated on a 5 point scale, where a lower score below 3 means agreement and above 3 means disagreement.

With regards to job satisfaction, most respondents seem to be satisfied with their jobs. While the standard deviations are relatively low, there are some who are not content at the current job. The work pressure scores are below the mid-point of 3, indicating that the participants seem to perceive high workloads in their jobs. Most new employees are in programming. In programming, depending on the organization, there is a general tendency in Pakistani organizations for managers to give more work and put more pressure on new employees. However, new employees are more than willing to take the opportunity to learn new programming skills and complete the assigned workload.

In terms of work–life balance and work exhaustion factor, the average scores are below the mid-point. This means that most employees experience work–home conflict as well as work burnout. While younger workers may not experience much work–home conflict, older and more experienced IT employees working for high demand industries, such as financial institutions and software companies, may find difficulty in balancing work and life needs. By the same token, IT employees working for private organizations and call centers may have issues managing work–life balance and pressure at work, while employees working in public organizations may not have exhausting workloads.

The respondents' results on sense of accomplishment are positive, and most IT workers feel they are making a contribution to the organization. Nevertheless, job security and threats to their job seem to be concerns for Pakistani IT employees. They see a persistent threat to their jobs due to advancements in technology and the ever-increasing number of IT graduates. However, they still feel that their jobs will not be eliminated or outsourced soon, most likely because Pakistan is an outsourcing destination

Table 25.4:   Individual IT Employee Issues in Pakistan

Individual Issues	Mean Rating*	Std. Deviation
Job satisfaction		
In general, I like working here.	1.95	0.893
All in all, I am satisfied with my current job.	2.09	0.951
In general, I don't like my current job.	3.01	1.371
Work pressure		
I feel that the number of requests, problems or complaints that I deal with at work is more than expected.	2.85	1.217
I feel that the amount of work I do interferes with how well it is done.	2.17	1.020
I feel busy or rushed at work.	2.63	1.228
I feel pressured at work.	3.00	1.188
Work–life balance		
There is a blurring of boundaries between my job and my home life.	2.68	1.123
My work-related responsibilities create conflicts with my home responsibilities.	2.70	1.250
I do not get everything done at home because I find myself completing job-related work.	2.53	1.205
Workload and burnout		
I feel drained from activities at work.	2.74	1.128
I feel tired from my work activities.	2.69	1.131
Working all day is a strain for me.	2.70	1.242
I feel burned out from my work activities.	2.98	1.286
Sense of accomplishment		
I feel I'm making an effective contribution to what this organization does.	2.05	0.919
In my opinion, I do a good job.	1.81	0.849
I have accomplished many worthwhile things in this job.	1.84	0.817
At my work, I feel confident that I am effective at getting things done.	1.89	0.948
Threats to one's job		
I am worried that future technology advancements may pose a threat to my job.	2.96	1.299
I believe that other people may be able to perform my work activities.	2.44	1.074
I am concerned that my job may be eliminated soon.	3.08	1.297
I am concerned that my job may be outsourced soon.	3.18	1.373
Career plans		
I will be with this organization 1 year from now.	2.35	1.021
I will take steps during the next year to secure a job at a different organization.	2.62	1.126

(*Continued*)

Table 25.4: (*Continued*)

Individual Issues	Mean Rating*	Std. Deviation
I will be with this organization 5 years from now.	2.81	1.184
I will be working in the IT field 1 year from now.	2.37	1.17
I will take steps during the next year to secure a job outside the IT field.	2.88	1.244
I will be working in the IT field 5 years from now.	2.35	1.12

*Rating scale ranges from 1 to 5: 1 as strongly agree and 5 as strongly disagree.

for multinational companies MNCs are vendors and not the clients of outsourcing activity.

The results of IT employees' career plans are somewhat mixed. But overall, they suggest that Pakistani IT employees do not anticipate changing their job or the profession either in the short term or the long term. The current boom in mobile communications and the IT services sector provides more financial stability and independence to IT workers, and the profession remains lucrative.

25.8 Conclusion

This chapter provided insights into the organizational IT issues, technology issues and individual concerns of IT employees in Pakistan. The most important IT-related organizational issues in Pakistan are; revenue-generating IT innovations, IT strategic planning, and business productivity and cost reduction. Top technology concerns include business intelligence/ analytics, customer relationship management (CRM) systems, and mobile and wireless applications. At the individual level, the results show that the IT employees in Pakistan despite having high work loads are satisfied with their jobs and yet have high workloads. They also have issues with work–life balance and exhaustion, yet they have a strong sense of accomplishment in life and plan to remain on the job and the profession for a long time. In the end, the IT industry is a very attractive career choice for these employees.

References

Kappelman, L., Johnson, V., McLean, E., & Maurer, C. (2018). The 2017 SIM IT Issues and Trends Study. *MISQ Exec*, 17(1), 53–88.

Ministry of Science and Technology (2000), IT Policy and Action Plan" Retrieved from URL: http://www.sindh.gov.pk/PakistanITPolicya ndActionplan.pdf [Accessed January 22].

MOITT (2018). "Digital Pakistan Policy". Retrieved from URL: http://www .moitt.gov.pk/policiesdetails.aspx [Accessed October 10, 2018].

Palvia, P., Jacks, T., Gosh, J., Licker, P., Romm-Livermore, C., Serenko, A., & Turan, A. H. (2017). The World IT Project: History, trials, tribulations, lessons, and recommendations. *Communications of the Association for Information Systems*, 41(18), 389–413.

Palvia, P., Ghosh, J., Jacks, T., Serenko, A., & Turan, A. (2018). Trekking the globe with the World IT Project. *Journal of Information Technology Case and Application Research*, 20(1), 3–8.

PBS (2018). "Pakistan Bureau of Statistics." Retrieved from URL: https://tradi ngeconomics.com/pakistan/gdp-growth [Accessed 11 October, 2018].

PKKH (2017). Malik, S., "Developments-pakistan-beginning-present." Retrieved from URL: www.pakistankakhudahafiz.com [Accessed 10 October, 2018].

PP (2008). "Computers-and-its-usage-pakistan". Retrieved from URL: www.pr opakistani.pk [Accessed October 10, 2018].

PJSS (2009). Shaukat, M., Information technology in Pakistan: An analysis of problems faced in IT implementation by Pakistan's banking and manufac-turing companies. *Pakistan Journal of Social Sciences*, 29(1), 13–22.

Shahbaz, M. (2008). "Data mining for engineering sector in Pakistan: Issues and implications." In *Proceedings of the World Congress on Engineering and Computer Science*, San Francisco, USA, October 22–24, 2008.

TechJuice (2017). "Challenges faced by IT sector of Pakistan." Retrieved from URL: https://www.techjuice.pk/challenges-faced-by-it-sector-of-pakistan [Accessed October 15 2018].

Chapter 26

Information Technology Issues in Peru

Rolando A. Gonzales[*,‡], Eddy A. Morris[*,§], and Tim Jacks[†,¶]

ESAN University, Santiago de Surco, Peru
†*Southern Illinois University Edwardsville, Edwardsville,*
IL, USA
‡*rgonzales@esan.edu.pe*
§*emorris@esan.edu.pe*
¶*tjacks@siue.edu*

Summary

This chapter about information technology (IT) issues in Peru addresses important attitudes and perceptions of IT workers in the context of a small Latin American country. The data were obtained through questionnaires completed by IT employees of 17 different organizations in Peru.

First, the chapter explains important elements of the country's background and history. Next, there is a preliminary explanation of the state of IT in Peru, especially compared with other countries in Latin America. Then, the methodology of the research is presented, and specific organizational, technological, and individual issues that IT employees face are explained. Overall, the IT workers feel satisfied with their jobs, do not feel excessive work pressure, and have a feeling of accomplishment. Top-ranked organizational IT issues include alignment between IT and business and IT strategic planning. For technology issues, the top-ranked concerns are business intelligence and analytics and enterprise application integration.

26.1 Introduction

Peru is a small developing country in South America that has been using information technology (IT) for many years. As a developing country, Peru does not have the best trained IT professionals and the best hardware, software, and other elements of information systems (IS). It is, however, making successful attempts to be competitive globally and to position itself for the future. There are different characteristics for each particular country such as the level of economic development, size, population, language, literacy rate, main industrial activities, international trade, and several others that have to be considered when analyzing the development of IT in a nation.

It is desirable to determine the main characteristics of both IT and the employees that work in this profession in the context of large enterprises and organizations in this country. By analyzing the results from the information collected in several organizations as part of the World IT Project, it will be possible to present a general idea of the state of IT in Peru.

26.2 Country Background and History

Peru, formally Republica del Peru, is a medium-size country (1,285 million km^2) in South America, in a tropical climate beside the Pacific Ocean. Peru has over 32 million inhabitants and has a democratic form of government. There are three main geographical areas: the coast, the mountains, and the jungles of the Amazon basin. Around 56% of the population lives on the coast. Lima, also on the coast, is the capital with about 10 million inhabitants, which comprise 31% of the national population. The Sierra in the high mountains (Andes Mountains) has 30% of the population. The Selva, the jungle area of the country (high jungle and low jungle including the Amazon basin) with 61% of the national area has just 15% of the total population. Geographically, Peru is a complicated country with advantages and disadvantages. With the Andes Mountains, there are abundant mineral resources, but the Sierra population lives at the altitude between 2,000 and 4,600 m above the sea level and has specific problems of communication. Transportation costs in the mountains and jungle are high which makes those areas less competitive for business purposes.

The heterogeneity of the population is another complicating factor, with 47% of mixed races, 32% native race, 18% European, 2% Africans, and 1% Asians, which is similar to other Latin American countries such as Mexico, Guatemala, Ecuador, and Bolivia. Before European colonization, Peru was part of the Incan Empire, one of the most advanced pre-Columbian cultures in the Americas. The literacy rate in Peru is 92.9% (80.3% in the rural areas), but the education system has received low scores on the PISA index for the last several years. The official language is Spanish, with two additional native languages: Quechua and Aymara. Proficiency in English is generally low.

The Global Competitiveness Report 2017–2018 indicates that Peru is in the 72nd position in the world (Schwab, 2017). The Global Competitive Index covers 137 nations and measures national competitiveness, as the set of institutions, policies and factors that determine the level of productivity. Peru has a nominal Gross Domestic Product (GDP) of US$232 billion, 42nd in the world, and per capita GDP of US$7,200, 82nd in the world. GDP PPP is US$450 billion and US$1,400 per capita, with a Gini index of 44.3 (medium). According to the report Doing Business 2018 of the World Bank Group, which researches governmental regulations that enhance business activity, Peru ranks 58th among 190 countries. While there is a problem with corruption, the larger economic concern is the lack of investment in innovation (0.08% of GDP, compared with Chile that invests 0.37% of their GDP) (Stunt, 2017). Today, Peru continues to be primarily a commodity producer, similar to most of the Latin American countries. The supply of minerals, such as copper, gold, zinc, tin, lead, and silver, accounts for 15% of the Domestic National Product and 40% of international trade.

26.3 Information Technology in Peru

In terms of the use of advanced hardware and software, IT in Peru has been growing steadily since the 1990s, on par with the other Latin American countries. Many large technology firms such as IBM, Microsoft, Apple, and Oracle have a significant presence in Peru.

The Global Information Technology Report 2016 (World Economic Forum, INSEAD, and Cornell University) regularly evaluates Information and Communication Technologies (ICT) and ranks 139 countries according to the Network Readiness Index (NRI) that uses four sub-indexes: Environment, Readiness, Usage, and Impact (Baller, 2016). The Network Readiness Index for Peru is 3.8 (values from 1 to 7, with 7 being the best value) in the 90th position. Thus, Peru can be considered an upper-middle economy, although slightly behind other Latin American countries: Chile (#38), Uruguay (#43), Costa Rica (#44), Colombia (#68), Brazil (#72), Mexico (#76), and Argentina (#89). The Environment sub-index (3.7, #97) indicates a weak Political and Regulatory Environment (3.1, #118). The other sub-indexes are Readiness (4.4, #89), Usage (3.5, #92), and Impact (3.5, #81). As can be seen, this position has to improve, especially in comparison to other Latin American countries, if Peru is going to stay competitive.

ICT penetration is similarly low, although it is better in the capital. The Instituto Nacional de Estadística e Informática of Peru reports that households with access to ICTs in 2017 were: 21.9% for landline phones, 90.2% for mobile phones, 37.4% for cable TV, 33.2% for PCs, and 28.2% for the Internet. For Lima, the capital, these numbers were: 46.3% for landline phones, 94.0% for mobile phones, 58.7% for cable TV, 51.7% for PCs, and 52.1% for Internet. For businesses, the same report says that out of 77,000 companies in 2016, 91.3% use computers (98.1% for big companies, 96.8% for medium-size, and 90.2% for the small ones), 88.5% use the Internet, 88.2% use landline telephones, and 94.3% use mobile telephones. Only 14.1% provide training in ICTs, 15.2% buy products and services on the Internet, and only 7.2% sell products and services on the Internet. Only 29.7% of companies have a home page. Many of Peru's ICT challenges are due to its geography as noted above.

On average, 14.4%, of companies invest in advertisement on the Internet. These are mainly in private education (29.6%), lodging, accommodation and restaurants (27.9%) and other services (24.2%). The main types of enterprise software used are accounting and taxes 81.9%, sales 39.5%, logistics 18.7%, human resources 15.4%, finance 10.9%, informatics support 10.3%, and production 7.3%. Sources of software are 36.9% from providers, 17.6% from free-access systems, and 10.9% developed by companies. The use of social networking is 72.3%.

Several countries in Latin America have implemented a Ministry of Electronics and Information Technology, which has significantly increased their digital connectivity. Some examples are Colombia (92% of digital connectivity), Mexico, Brazil, and Chile. Peru is not yet at this level (54% of digital connectivity). Smart Cities are another concept that Lima is trying to develop in order to manage and administer its resources and assets more efficiently.

26.4 Methodology

The questionnaire developed by the World IT Project (Palvia *et al.*, 2017; Palvia *et al.*, 2018) was in English and it had to be translated and back-translated three times to obtain a comparable instrument in Spanish. CIOs of large firms were approached to participate in the study and were contacted through ESAN University's Graduate School of Business which offers a Master's program in IT. CIOs were requested to have 10 to 20 questionnaires completed by their company IT employees. After the initial contact, one of the researchers visited each one of the CIOs in order to fully explain the purpose of the study and request their collaboration. The questionnaire could be received in two ways, either on paper or electronically. The data was obtained from January to October 2016. No incentive was provided to complete the questionnaire.

A total of 159 responses were received from 17 companies representing different industries in Peru. The Peruvian sample was mainly from ages 30–39 years (40.9%), 40–49 years (22.0%), and 21–29 years (20.1%). The most common IT roles were project management (22.3%), management and strategy (19.6%), analysis and design (12.5%), and programming, operations, and database administrators with the same percentage each (6.3%). 80.5% of respondents were male and 50.3% were not part of IT Management. More details are provided in Table 26.1.

26.5 Organizational IT Issues

Participants ranked 18 organizational IT issues according to their importance based on a 5-point Likert scale (1 being the most important and 5 being of no importance). The ranking, mean value, and standard deviation for each issue are shown in Table 26.2.

Alignment between IT and business is the most important organizational IT issue, with the mean value of 1.42. IT strategic planning is in the

Table 26.1: Descriptive Statistics

Characteristics	N	%	Characteristics	N	%
Education:			Years of work experience:		
High school or less	0	0	0–4 years	15	9.4
Associate degree	24	15.1	5–9 years	38	23.9
Bachelor's degree	74	46.5	10–19 years	58	36.5
Master's degree	61	38.4	20–29 years	34	21.4
Ph.D.	0	0	30+ years	14	8.8
Years of IT experience:			Organizational location:		
0–4 years	21	13.2	IT department employee	137	86.2
5–9 years	35	22.0	IT worker in non-IT department	13	8.2
10–19 years	61	38.4	Contract employee	5	3.1
20–29 years	31	19.5	Consultant	3	1.9
30+ years	11	6.92	Vendor employee	1	0.6
Work as:			Work position:		
Mostly full time	148	93.1	Not part of management	80	50.3
Mostly part time	0	0	In lower management	31	19.5
Mostly over time	11	6.9	In middle management	35	22.0
Been laid off from IT job:			In senior management	13	8.2
Yes	12	7.6			
No	147	92.5			

Table 26.2: Organizational IT Issues in Peru

Organizational IT Issues	Rank	Mean Rating*	Std. Deviation
Alignment between IT and business	1	1.42	0.64
IT strategic planning	2	1.42	0.65
IT reliability and efficiency	3	1.50	0.64
Revenue-generating IT innovations	4	1.53	0.70
Project management	5	1.55	0.65
Security and privacy	6	1.59	0.71
Knowledge management	7	1.61	0.72
Continuity planning and disaster recovery	8	1.65	0.77
IT service management (e.g., ITIL)	9	1.67	0.66
Business productivity and cost reduction	10	1.70	0.73
Business agility and speed to market	11	1.73	0.73
Attracting and retaining IT professionals	12	1.73	0.74
Business process reengineering	13	1.77	0.71
Enterprise architecture	14	1.82	0.71
IT cost reduction	15	1.87	0.84
Globalization	16	2.06	0.79
Outsourcing	17	2.51	0.97
Bring your own computing device (BYOD)	18	2.60	1.13

*Rating scale ranges from 1 to 5: 1 as most important and 5 as no importance.

second place, with the same mean value. IT reliability and efficiency is in the third place with the mean value of 1.50. The next places were for revenue generating IT innovations, project management, security and privacy, knowledge management, and continuity planning and disaster recovery. The less relevant items were BYOD, outsourcing, and globalization.

Alignment between IT and business, IT strategic planning, and IT reliability and efficiency are items related to the concept of a competitive enterprise that uses IT in order to behave efficiently (Kappelman *et al.*, 2017), and in the case of the Peruvian economy, they continue to be very relevant. Project management and revenue generating IT innovations are items related to specific activities that permit the teams of IT executives to be more efficient and competitive. Security and privacy and continuity planning and disaster recovery are important topics today in the IT world both nationally and internationally. With the great amount of data that we have now, and the big databases that each company has, it is very relevant to ensure the security and privacy of the customer, but at the same time to use all the information that it has and care for their customers at any moment.

Other middle-rated items are knowledge management, business productivity and cost reductions, and IT service management, which are relevant in the Peruvian case, for an enterprise that is improving its competitive edge. Other IT items, such as attracting and retaining IT professionals, business agility and speed to market, business process reengineering, IT cost reduction, enterprise architecture, globalization, outsourcing, and BYOD were all ranked lower, but they are, nevertheless, important. Some issues, for example, globalization and outsourcing, were likely more prominent several years ago.

26.6   Technology and Infrastructure Issues

About 16 technology issues were rated by the participants on a 5-point Likert scale, in which number 1 was of most importance and 5 of no importance. The overall ranking, mean, and respective standard deviation are reported in Table 26.3.

Business intelligence and analytics is a very relevant IT issue in the world (third most important IT management issue in 2017, (see Kappelman *et al.*, 2017), and in Peru it is also very important. Most IT executives are under pressure to get specific solutions that improve the effectiveness and

Table 26.3: Technology and Infrastructure Issues in Peru

IT-Related Issues	Rank	Mean Rating*	Std. Deviation
Business intelligence/analytics	1	1.62	0.63
Enterprise application integration	2	1.70	0.69
Networks/telecommunications	3	1.76	0.73
Mobile and wireless applications	4	1.76	0.73
Business process management systems	5	1.77	0.67
Mobile apps development	6	1.81	0.78
Customer relationship management (CRM) systems	7	1.87	0.81
Collaborative and workflow tools	8	1.87	0.68
Big data systems	9	1.88	0.80
Data mining	10	1.89	0.74
Enterprise resource planning (ERP) systems	11	1.93	0.79
Service-oriented architecture (SOA)	12	1.95	0.72
Virtualization (desktop or server)	13	2.00	0.72
Software as a Service	14	2.04	0.83
Cloud computing	15	2.04	0.81
Social networking/media	16	2.08	0.86

*Rating scale ranges from 1 to 5: 1 as most important and 5 as no importance.

efficiency of their organizations. There are several tools that enable improvement without very demanding effort. Enterprise application integration is another important IT issue that permits an organization to improve the use of sophisticated enterprise applications such as supply chain management systems, ERP systems, CRM systems, business intelligence systems, and other systems which may break the inefficient information silos within the organization. In Peru, this is important because historically the larger enterprises started with one of these systems, such as an ERP system, then developed a separate CRM system, and continuing in this way, created information silos.

Network and telecommunications is also very relevant because of the great amount of information from various sources and newer technologies that allow better ways of communicating. However, in Peru, a developing country, it is necessary to make considerable efforts in this regard. Business process management is a relevant IT Issue that permits improvements and automation of business processes and is related to the ideas of total quality management, continuous improvement of process methodologies, and others. In Peru, this is especially relevant due to the absence of high quality business processes and systems that already exist in developed countries.

There are several other IT issues that continue to be relevant, as they were several years ago, including collaborative and workflow tools, CRM, ERP, SOA, virtualization, software as a service, and social networking/media. Other issues, such as data mining, could be considered as part of business intelligence or analytics.

There are other issues that draw attention to the fact that they were not ranked higher for the Peruvian IT workers, such as big data systems and cloud computing. It could be, in the case of big data systems that the Peruvian enterprises are not big enough or are not up to date on this technology. In the case of cloud computing, several organizations are slowly starting to use this model of computing with good results. But it will likely take several years ahead to completely adopt this technology.

26.7   Individual IT Employee Issues

Individual IT employee issues were analyzed in seven areas and the details are shown in Table 26.4. For this analysis, we do not reverse any of the scales but present the raw scores.

Table 26.4:   Individual IT Employee Issues in Peru

Individual Issues	Mean Rating*	Std. Deviation
Job satisfaction		
In general, I like working here.	1.78	0.69
All in all, I am satisfied with my current job.	2.07	0.89
In general, I don't like my current job.	4.13	0.97
Perceived work overload		
I feel that the number of requests, problems or complaints that I deal with at work is more than expected.	3.30	1.22
I feel that the amount of work I do interferes with how well it is done.	3.37	1.13
I feel busy or rushed at work.	3.22	1.19
I feel pressured at work.	3.42	1.06
Work–home conflict		
There is a blurring of boundaries between my job and my home life.	3.74	1.03
My work-related responsibilities create conflicts with my home responsibilities.	3.95	0.99
I do not get everything done at home because I find myself completing job-related work.	3.97	1.04

(Continued)

Table 26.4:   (*Continued*)

Individual Issues	Mean Rating*	Std. Deviation
Work exhaustion/strain		
I feel drained from activities at work.	3.77	1.01
I feel tired from my work activities.	3.77	1.01
Working all day is a strain for me.	3.99	0.94
I feel burned out from my work activities.	3.82	1.05
Professional self-efficacy		
I feel I'm making an effective contribution to what this organization does.	1.88	0.92
In my opinion, I do a good job.	1.72	0.90
I have accomplished many worthwhile things in this job.	1.65	0.73
At my work, I feel confident that I am effective at getting things done.	1.75	0.82
Job insecurity		
I am worried that future technology advancements may pose a threat to my job.	3.82	1.11
I believe that other people may be able to perform my work activities.	2.80	1.10
I am concerned that my job may be eliminated soon.	4.01	0.98
I am concerned that my job may be outsourced soon.	3.81	1.06
Turnover intention		
I will be with this organization 1 year from now.	2.26	1.05
I will take steps during the next year to secure a job at a different organization.	3.24	1.15
I will be with this organization 5 years from now.	2.70	1.05
Turnover intention — IT profession		
I will be working in the IT field 1 year from now.	1.91	1.06
I will take steps during the next year to secure a job outside the IT field.	3.59	1.14
I will be working in the IT field 5 years from now.	2.13	1.13

*Rating scale ranges from 1 to 5: 1 as strongly agree and 5 as strongly disagree.

In relation to job satisfaction, we can state that the IT employees are generally satisfied with their current occupation and will continue working in IT. In terms of work pressure, it is observed that the participants do not feel excessive pressure in their work. In terms of work–life balance, they do not feel a problem between their jobs and their home life. For workload and burnout, they do not feel that the pressure is excessive. Of course, they need to work hard, but they can manage a good relationship between home life and work, and do not feel extreme pressure in their jobs. With respect to a sense of accomplishment and professional self-efficacy, the IT workers feel that they are doing a good job and accomplish good outcomes for themselves and their organizations. They do not feel a strong threat of losing their jobs. They think they will continue working in the IT sector, be with the same organization in the short term, and possibly continuing with the same organization in the next 5 years.

26.8 Conclusion

Peru is a medium-size country in South America that has a developed IT industry for a country of its characteristics, but it is in position #90 of 139 countries, according to the Global Information Technology Report 2016 of the World Economic Forum. There is a strong need to improve the current state of technology in order to better compete with other Latin American countries, such as Chile or Uruguay. Some Organizational IT issues that are important in Peru are alignment between IT and business, IT strategic planning, and IT reliability and efficiency. In terms of technology and infrastructure issues, the relevant ones are business intelligence and analytics, enterprise application integration, and networks and telecommunications. With respect to individual IT employee issues, the IT employees are generally satisfied with their jobs, do not feel too much pressure at work, get a feeling of accomplishment from work, and plan to be working in the IT field in the near future. The authors hope that the findings of this study will provide new insights that can be used by academics and practitioners both in Peru and globally.

References

Baller, S., Dutta, S., & Lanvin B. (2016). The Global Information Report 2016. *World Economic Forum, INSEAD, and Johnson Cornell University,* pp. 1–185.

Kappelman, L, Johnson, V., Maurer, C., McLean E., Torres, R., David, A., & Nguyen. Q. (2017). "The 2017 SIM IT Issues and Trend Study," *MIS Quarterly Executive,* 17(1), 53–88.

Instituto Nacional de Estadística e Informática (2018). *Estadísticas de la Tecnologías de Información y Comunicación en los Hogares.* Informe Técnico No 2. Junio 2018, pp. 1–55. www.inei.gob.pe.

Instituto Nacional de Estadística e Informática (2018). *Perú: Tecnologías de Información y Comunicación en las Empresas, 2015. Encuesta Económica Anual 2016,* pp. 1–133.

Palvia, P., Jacks, T., Ghosh, J., Licker, P., Romm-Livermore, C., Serenko, A., & Turan, A. H. (2017). The World IT Project: History, trials, tribulations, lessons, and recommendations. *Communications of the Association for Information Systems*, 41(18), 389–413.

Palvia, P., Ghosh, J., Jacks, T., Serenko, A., & Turan, A. (2018). Trekking the globe with the World IT Project. *Journal of Information Technology Case and Application Research*, 20(1), 3–8.

Schwab, K. (2017). The Global Competitiveness Report 2017-2018. *World Economic Forum*.

Stunt, V. (2017). Peru is a bid to catch up with its innovative Latin American neighbor. https://techcrunch.com/2017/08/10/.

World Bank Group (2018). *Doing Business 2018, Reforming to Create Jobs*, pp. 1–332.

Chapter 27

Information Technology Issues
in Poland

Stanislaw Wrycza[*,‡], Damian Gajda[*,§], and Prashant Palvia[*,¶]

*University of Gdansk, Gdańsk, Poland
†University of North Carolina at Greensboro,
Greensboro, NC, USA
‡swrycza@ug.edu.pl
§damian.gajda@ug.edu.pl
¶pcpalvia@uncg.edu

Summary

In this chapter, the top issues related to information technology (IT) professionals and organizations in Poland are reported. A survey was conducted with the instrument provided by the World IT Project (Palvia et al., 2017; Palvia et al., 2018). Included in this chapter are challenges of IT in Poland, specifically organizational IT issues, technological and

346 S. Wrycza, D. Gajda, & P. Palvia

infrastructural concerns and, finally, individual IT employee issues. The top three organizational issues for IT employees are: Knowledge management, IT reliability and efficiency, and Security and privacy. The top three technology issues are: networks/ telecommunications, software as a service, and mobile and wireless applications. Furthermore, the IT employees are highly satisfied with their jobs and feel secure in the profession.

27.1 Introduction

For the goals of this chapter, the World Information Technology Project (WITP) provided useful, professional, and deep insights into IT issues in Polish companies and institutions. The objective of WITP research (Palvia *et al.*, 2017) is the understanding and interpretation of the influence of contextual factors on important IT issues, with regards to a multi-dimensional ecosystem that includes cultural, economic, managerial and social factors in domestic and international environments of firms. In order to produce a scientifically valid survey, the local investigators (Wrycza *et al.*, 2016) fine-tuned and described the survey process in order to ensure the representativeness of the responses, and took methodological steps to verify that quality responses were received. While the instrument captured a variety of topics related to IT professionals, this chapter reports the organizational, technological, and individual issues in Poland.

27.2 Country Background and History

Poland is a country in Central Europe with about 38 million inhabitants and is a member of the European Union and NATO. Poland's history dates back over a 1,000 years and has produced many artists and scientists known both in Europe and throughout the world. Poland's most famous musician was Frédéric Chopin, who was a piano virtuoso and a great composer. In the field of literature, Poland boasts several Nobel Prize winners such as Henryk Sienkiewicz, Wladyslaw Reymont, Czeslaw Miłosz and Wisława Szymborska. Furthermore, Lech Walesa, the leader of Solidarity (Solidarnosc), independent trade union, won the Nobel Peace Prize. Prominent scientists from Poland include Nicolaus Copernicus (1473–1543); Jan Heweliusz (Johannes Hevelius, 1611–1687); Daniel Gabriel Fahrenheit (1686–1736); Maria Skłodowska-Curie (1867–1934); Stanislaw Ulam (1909–1984), among others.

27.3   Information Technology in Poland

The dynamic development of the information and communication technology (ICT) sector in Poland began in the early 1990s due to political transformation and the beginning of a market economy. Despite the difficulties in acquiring new technologies and attainment of capital, Polish companies in the ICT sector have maintained a rapid growth rate over the last 25 years (The Development Perspectives, 2017). The development of the sector was supported by the improving economic situation in the country, the demographic structure of Poland with a large population of young people born in the 1980s, a high level of science in the education system as well as good working conditions in the ICT sector in comparison with other sectors of the economy. For many years, the ICT industry has been at the forefront in terms of employee earnings, and an ICT specialist can count on a salary that exceeds 2–3 times the average national. Employment in the ICT sector in 2014 numbered 315,000 people, and for many years the employment has increased in this sector by an average of 5% per year. In connection with Poland's membership in the European Union since 2004, Poland is an important country for international IT companies to establish branches, including research and development centers. In recent years, a number of Polish IT companies have developed both startups, offering mobile applications or computer games as well as large IT companies specializing in financial services and systems.

27.4   Methodology

The proper preparation of a survey questionnaire has a significant impact on its success. At the design stage, many key issues were identified which, in turn, influenced the high quality and credibility of the research and the usefulness of its findings. The sampling frame selected from the Polish Central Statistical Office resources and used for the aim of this research project included over 60,000 small, 15,500 medium and 3,200 large organizations.

The next stage in the process was the questionnaire preparation. For the goals of this research, a purposeful questionnaire, constructed in English by the Core Research Team of the World IT Project (Palvia *et al.*, 2017; Palvia *et al.*, 2018), was used and it had over 150 questions. All responses required a selection from a menu of options, and most questions had to be evaluated on a 5-point Likert scale. An important concern was the appropriate translation of the original questionnaire in English into the local official Polish language. The translated questionnaire was subsequently subjected to a forward and back-translation to ensure consistency and accuracy. The use of the forward and back translation method was very effective while also taking into account that most Polish professionals are fluent in English. In particular, they constantly use professional English IT terminology in daily communication in an international working environment of global firms and the IT teams in which they are employed.

The direct interview method was selected to collect data from the respondents. This method generated a good number of quality responses, thereby reducing the errors resulting from a blind mail survey. In this way, access to the IT staff of targeted organizations was considerably facilitated. A high participation rate was achieved in the survey; 70% of the organizations which had been chosen answered the questionnaire. This can be considered a very good result among statistical surveys, in particular when taking into account the low-to-moderate responsiveness and cooperativeness of big companies and corporations, in general.

The interview method proved to be effective even in terms of the participation rate of the IT employees, working in the sampled firms and institutions. By visiting the companies, the interviewer could easily take the individual questionnaire to the IT staff and encourage them to participate by providing a direct presentation of the expected research outcomes. As a result, the participation rate of employees in the randomly selected organizations reached almost 100%. Only one IT team member refused to participate in the study, that too because of urgent work assignments. The execution of 300 interviews among the IT workers required the selection of 51 companies. The size and branches of the firms are presented in Table 27.1 (Wrycza *et al.*, 2016).

Table 27.2 shows the descriptive statistics of the IT professionals who took part in the study.

27.5   Organizational IT Issues

The identification of organizational IT issues and challenges is important and is a precursor to achieving the full potential of IT in Polish organizations. As indicated in Table 27.3, the five most important organizational IT issues in order in the Polish survey are: Knowledge management, IT reliability and efficiency, security and privacy, project management, and continuity planning and disaster recovery. Most of these issues are in the operational realm. More strategic issues like strategic planning and alignment between

Table 27.1:   Employees by Branch and Size of Firm

Branch	Total Employment			
	10–49	50–249	>250	Total
Production		1	2	3
Wholesale trade	1	1	1	3
Retail trade, services	1	1	1	3
Construction		1	2	3
Transport		1	2	3
Education		1	2	3
Finance			3	3
Tourism		3		3
Information technology	2	1		3
Public utility companies		3		3
Freelance services	2	1		3
Healthcare	1	2		3
Entertainment	1	2		3
Media		2	1	3
Governmental/Public institution		2	1	3
Nongovernmental (NGO) and non-profit	2	1		3
Others	1	1	1	3
Total number of organizations	**11**	**24**	**16**	**51**
Total number of interviews	**33**	**144**	**123**	**300**

Table 27.2:   Descriptive Statistics

Characteristics	N	%	Characteristics	N	%
Education:			Years of work experience:		
High school or less	13	4.3	0–4 years	52	17.3
Associate degree	42	14.0	5–9 years	81	27.0
Bachelor's degree	41	13.7	10–19 years	129	43.0
Master's degree	196	65.3	20–29 years	25	8.3
Ph.D.	8	2.7	30+ years	13	4.3
Years of IT experience:			Organizational location:		
0–4 years	85	28.3	IT department employee	200	66.7
5–9 years	104	34.7	IT worker in non-IT department	60	20.0
10–19 years	96	32.0	Contract employee	14	4.7
20–29 years	14	4.7	Consultant	13	4.3
30+ years	1	0.3	Vendor employee	13	4.3
Work as:			Work position:		
Mostly full time	237	79.0	Not part of management	210	70.0
Mostly part time	22	7.3	In lower management	43	14.3
Mostly over time	41	13.7	In middle management	35	11.7
			In senior management	12	4.0
Been laid off from IT job:					
Yes	67	22.3			
No	233	77.7			

Table 27.3:   Organizational IT Issues in Poland

Organizational IT Issues	Rank	Mean Rating*	Std. Deviation
Knowledge management	1	2.06	0.91
IT reliability and efficiency	2	2.11	0.86
Security and privacy	3	2.12	0.86
Project management	4	2.15	0.97
Continuity planning and disaster recovery	5	2.19	0.95
IT service management (e.g., ITIL)	6	2.25	0.90
Revenue-generating IT innovations	7	2.38	1.08
IT strategic planning	8	2.39	1.02
Business productivity and cost reduction	9	2.40	1.10
Business agility and speed to market	10	2.44	1.06
Alignment between IT and business	11	2.45	1.13
IT cost reduction	12	2.53	1.10
Attracting and retaining IT professionals	13	2.57	0.91
Enterprise architecture	14	2.63	1.03
Business process reengineering	15	2.68	1.10
Globalization	16	2.71	1.00
Outsourcing	17	2.78	1.00
BYOD	18	2.83	0.99

*Rating scale ranges from 1 to 5: 1 as most important and 5 as no importance.

IT and business were ranked in the middle, at eighth and 11th places, respectively. Also lower were issues like business agility, enterprise architecture and business process reengineering. As an emerging market economy, these issues suggest that Polish companies first have to master operational concerns before moving on to addressing longer-range and strategic goals.

Ranked lowest among the 18 issues were: globalization, outsourcing, and bring your own computing device (BYOD). It appears that the global concerns have not taken paramount importance in Polish companies. As an aside, only 20 years ago, Poland was not even a player and not known for its technological achievements. Today, it is one of the top destinations in the world for software development. However, Polish companies themselves may not engage in outsourcing or offshoring IT development or operations to outside companies.

27.6   Technology and Infrastructure Issues

Table 27.4 shows the ranked list of 16 technology issues. According to Polish IT professionals, networks and telecommunications is the most significant technology issue. The next four important issues, in order are: software as a service (SaaS), mobile and wireless applications, cloud computing, and

Table 27.4: Technology and Infrastructure Issues in Poland

IT-Related Issues	Rank	Mean Rating*	Std. Deviation
Networks/telecommunications	1	2.27	0.98
Software as a service	2	2.43	0.92
Mobile and wireless applications	3	2.44	0.91
Cloud computing	4	2.45	0.91
Enterprise application integration	5	2.46	0.92
Mobile apps development	6	2.50	1.04
Data mining	7	2.52	0.94
Social networking/media	8	2.54	0.96
Service-oriented architecture (SOA)	9	2.55	0.95
Virtualization (desktop or server)	10	2.58	0.98
Business process management systems	11	2.62	1.10
Collaborative and workflow tools	12	2.63	0.99
Big data systems	13	2.64	1.02
Business intelligence/analytics	14	2.67	1.15
Enterprise resource planning (ERP) systems	15	2.70	1.13
Customer relationship management (CRM) systems	16	2.70	1.11

*Rating scale ranges from 1 to 5: 1 as most important and 5 as no importance.

enterprise application integration. Networks, telecommunications and associated technologies such as mobile and wireless applications are extremely important in Poland as they can considerably improve the IT infrastructure, thus enabling and accelerating the use of IT in business.

Ranked somewhere in the middle are modern operational-level IT solutions recently applied in business, such as social networks, data mining, service-oriented architecture, and virtualization. Such technologies and solutions in business have been adopted in countries with a very high level of advancement in IT. Thus, these seem like aspirational goals for Polish companies. Surprisingly, technologies such as business intelligence/analytics, ERP systems, and CRM systems appeared at the bottom of the rankings. One possible explanation is that Poland may be behind the curve in the adoption of these technologies. Consistent with the organizational issues, these results suggest that Polish companies are more focused on operational efficiencies than strategic applications of technology.

27.7 Individual IT Employee Issues

Table 27.5 presents employee responses on individual issues related to such issues as job satisfaction, work pressure, work burnout, sense of

Table 27.5:   Individual IT Employee Issues in Poland

Individual Issues	Mean Rating*	Std. Deviation
Job satisfaction		
In general, I like working here	1.79	0.88
All in all, I am satisfied with my current job	1.83	0.90
In general, I don't like my current job	4.00	1.13
Work pressure		
I feel that the number of requests, problems or complaints that I deal with at work is more than expected	3.39	1.27
I feel that the amount of work I do interferes with how well it is done	3.54	1.19
I feel busy or rushed at work	3.55	1.22
I feel pressured at work	3.62	1.14
Work–life balance		
There is a blurring of boundaries between my job and my home life	3.64	1.10
My work-related responsibilities create conflicts with my home responsibilities	3.75	1.11
I do not get everything done at home because I find myself completing job-related work	3.77	1.12
Workload and burnout		
I feel drained from activities at work	3.77	1.10
I feel tired from my work activities	3.75	1.09
Working all day is a strain for me	3.49	1.18
I feel burned out from my work activities	3.79	1.14
Sense of accomplishment		
I feel I'm making an effective contribution to what this organization does	2.66	1.11
In my opinion, I do a good job	2.10	1.00
I have accomplished many worthwhile things in this job	2.23	1.02
At my work, I feel confident that I am effective at getting things done	2.24	1.05
Threats to one's job		
I am worried that future technology advancements may pose a threat to my job	3.25	1.10
I believe that other people may be able to perform my work activities	3.00	1.13
I am concerned that my job may be eliminated soon	3.56	1.10
I am concerned that my job may be outsourced soon	3.49	1.11
Career plans		
I will be with this organization 1 year from now	2.38	0.94
I will take steps during the next year to secure a job at a different organization	3.60	1.07

(*Continued*)

Table 27.5: (*Continued*)

Individual Issues	Mean Rating*	Std. Deviation
I will be with this organization 5 years from now	2.67	0.87
I will be working in the IT field 1 year from now	2.11	0.98
I will take steps during the next year to secure a job outside the IT field	3.61	1.00
I will be working in the IT field 5 years from now	2.26	0.98

*Rating scale ranges from 1 to 5: 1 as strongly agree, to 5 as strongly disagree.

accomplishment, treatment at work and career plans. Note that a lower score represents higher agreement with the listed statement. On a 5-point scale, an average below 3 would indicate agreement and above 3 would represent disagreement.

Based on the responses, it is noticed that employees in the IT sector in Poland are highly positive about their jobs and the profession. Their job satisfaction ratings are extremely high. Moreover, they do not feel overwhelmed by work or burned out professionally, and can maintain a good balance between work and home life. They have a high sense of accomplishment at work. Further, they do not perceive much job insecurity; however, they acknowledge that their jobs can be performed by others. With such positive aspects, no wonder, they plan to stay on the job and have no plans to change jobs or the profession in the near-term or the long-term. Such positive opinions are due to very good working conditions and salaries for IT employees in Poland. Reviews found on IT employment sites are generally positive and use such phrases as: career opportunities, good salary, atmosphere, work–life balance, supportive team leaders, flexible time options, and company perks, to describe employment conditions.

27.8 Conclusion

The study of IT professionals undertaken in Poland revealed that the top five organizational IT issues in business and administrative organizations are: Knowledge management, IT reliability and efficiency, security and privacy, project management, and continuity planning and disaster recovery. As for technology and infrastructural issues, the top five are: Knowledge management, IT reliability and efficiency, security and privacy, Project management, and continuity planning and disaster recovery. Both sets of issues confirm that Polish organizations are more concerned about operational issues related to technology and less focused on strategic issues. On a positive note, the Polish IT professionals are remarkably pleased with their jobs and the IT profession and intend to continue in the profession for a long time.

References

Palvia, P., Jacks, T., Gosh, J., Licker, P., Romm-Livermore, C., Serenko, A., & Turan, A. H. (2017). The World IT Project: History, trials, tribulations, lessons, and recommendations. *Communications of the Association for Information Systems*, 41, 18.

Palvia, P., Ghosh, J., Jacks, T., Serenko, A., & Turan, A. (2018). Trekking the globe with the World IT Project. *Journal of Information Technology Case and Application Research*, 20(1), 3–8.

The Development Perspectives of Poland's Development ICT Industries Until 2025 (original title: Perspektywy Rozwoju Polskiej Branży ICT do roku 2025) (2017), Ministry of Development, Warsaw: PARP.

Wrycza, S., Gajda D., Palvia, P., & Turan, A. H. (2016). Representativeness in the 'World IT Project' survey research. The methodological prerequisites and verification. In *17th Annual Global Information Technology Management Association (GITMA) World Conference 2016*, August 2016.

Chapter 28

Information Technology Issues
in Portugal

Rui Dinis Sousa[*,‡], João Álvaro Carvalho[*,§],
Luis Alfredo Martins do Amaral[*,¶], and Prashant Palvia[†,‖]

*University of Minho, Guimarães, Portugal
† The University of North Carolina,
Greensboro, NC, USA
‡ rds@dsi.uminho.pt
§ jac@dsi.uminho.pt
¶ amaral@dsi.uminho.pt
‖ pcpalvia@uncg.edu

Summary

Information technology (IT) professionals were surveyed in organizations across several industry sectors in Portugal to capture their concerns regarding organizational, technological and individual issues. Reliability and efficiency in IT, alignment between IT and the business, attracting and retaining IT professionals, and IT project planning were found to be the four most important organizational IT-related issues. Among technologies, the most important were business intelligence/analytics, enterprise application integration, ERP systems and business process management systems. Regarding individual issues, IT professionals seem quite satisfied with their jobs and their contribution to business goals, seem to have reasonable workloads, and have little or no intent of getting a job outside the IT sector.

28.1 Introduction

This is a study essentially descriptive of major issues for information technology (IT) professionals, a first deliverable from the team of Portuguese investigators working under the research framework of the World IT Project (Palvia *et al.*, 2017). Like the well-known IT issues studies from the Society of Information Management carried out since the 1980s, this study covers organizational IT-related issues as well as technologies of the most concern to IT professionals. However, this study clearly differs from those studies.

First, instead of only a senior IT leader in each organization as the respondent, our study involves at least 10 IT professionals in each organization, aiming for a more inclusive survey. Second, in addition to the usual IT issues, this study also addresses individual issues that include work pressure, life balance, satisfaction and sense of accomplishment at work. Third, as part of a major and coordinated effort around the world, instead of providing a single, US-centric view, the World IT Project provides a comprehensive view of IT issues.

28.2 Country Background and History

Portugal is a long-established country (in the 12th century), located in the Iberian Peninsula in southwestern Europe. It borders Spain in the North and East and has a coastline of 1800 km facing the Atlantic. The Portuguese Republic — its official designation — includes the mainland, two archipelagos located in North-Atlantic (Madeira and Azores), and an extensive maritime exclusive economic zone (1.7 million km^2). This maritime territory expanded to 2.1 million km^2 resulting from a claim for extending the country's jurisdiction over the neighboring continental shelf.

The ocean played an important role in Portuguese history. Viewing the ocean as its natural way for growth, in the 15th century Portugal started a systematic exploration of the west coast of Africa. The goal of reaching India by sea ended up on a global pursuit that took the Portuguese to Africa, Asia, Oceania and South America. By the 16th century, Portugal became a leading economic, political and military nation, and had established the first global empire.

Portuguese discoveries were built upon knowledge. The systematic nautical expeditions launched by the Portuguese led to important developments in nautical technology in areas such as ship-building, cartography, sailing and navigation, and relevant contributions in mathematics, astronomy, zoology and botany. Portugal played a major role on the inception of an early science-and-technology-grounded globalization wave that strongly influenced the modern world.

The Portuguese territory has a long history of cultural blending. It underwent a wide variety of influences by people coming from the Mediterranean, North and Central Europe. Among these, Romans and Arabs are well present in the country's memory. The former for the infrastructures they built and whose remains can be found all over the country. The latter because the country borders were established during the Christian Reconquista, which ended seven centuries of domination of most of the Iberian Peninsula by the Arabs (711–1491).

Later influences came from regions and countries that were once Portuguese colonies in Asia, Africa and South America. The decolonization that followed the 1974 revolution, which ended 48 years of dictatorship, brought to Portugal around 1 million people. Coming from Angola and Mozambique and, to a lesser extent, form Guinea-Bissau and East-Timor, these people led to a growth of the Portuguese population of about 10% and had a major influence in the country's culture and on its evolution after a period marked by censorship, ideological restrictions and limitations of civil freedom.

In 1985, together with Spain, Portugal joined the European Union endeavor adhering to the European Economic Community. This allowed Portugal to consolidate its young democracy and to leap into the level of social and economic development of the European countries already bonded together by the European project.

28.3   Information Technology in Portugal

Portugal made its way through the computer age with no special prominence. Early computers in the country can be traced back to the late 1950s.
They were mainly used for technical and scientific problems in engineering
areas. However, it did not take long for computers to start being used for
business purposes in larger corporations, and later on, in public administration. Universities were also important centers of computer usage and
promoters of its advantages. Early high education degrees in informatics
were launched in the "new universities", a set of higher education institutions created in the mid-1970s by an insightful Ministry of Education.
These degrees built upon classical education in systems and control (at
University of Minho) and electrotechnical engineering (at New University
of Lisbon), but quickly evolved to degree programs with a clear focus on
informatics.

In the 1980s, the country followed a promising attempt of creating
a personal computer. Despite its flexible architecture and other advantages in relation to the emerging IBM PC, the country didn't possess the
technological basis for a successful industry. A successful history can be
associated to the production of software. In late 1970s and early1980s,
there were already cases of development of computer applications for common business needs (accounting, inventory and invoicing, salaries payment, etc.). These computer applications were made available as minicomputers became affordable in medium-size companies. With the dissem-
ination of personal computers, these computer applications were rewritten
for MS-DOS and later for Windows. During the late 1980s and early 1990s,
in the vicinity of several universities it was possible to find many vibrant
start-up companies, eager to explore the opportunities brought upon by an
era of affordable computers. Along the years, the software industry consol-
idated. Several Portuguese companies are nowadays world leaders, developing IT products worldwide. Others, although operating internationally,
exploit the large market of the Portuguese former colonies that, besides
using the Portuguese language, keep their administrative and legal basis
similar to the Portuguese.

The liveliness of the IT sector and the availability of young IT talent,
together with other factors, have led Portugal to become a relevant hub for
the IT sector. Many international IT and consulting companies joined the
existing IT industry by establishing Portugal software development units
and nearshore centers. At the same time, the quality of Portuguese higher

education in this field led to the employment of many graduates into established companies worldwide.

28.4 Methodology

The instrument for data collection used in the study was developed and made available to the country investigators by the leaders of the World IT Project (Palvia *et al.*, 2017; Palvia *et al.*, 2018). We entered the project at the stage of data collection to carry out the survey in Portugal according to what was established in the memorandum of understanding signed between the project leader and the liaison, the lead investigator and country investigators.

The Portugal team decided to use the questionnaire in its original version, i.e., in English. English has been a mandatory foreign language at school for quite some time now. Educated people can therefore easily understand English. This is even more prominent in the IT sector. In fact, people working in IT make use of a mix of Portuguese and English words and expressions in their daily work activities. Thus, we avoided issues related to the translation of the questionnaire. In any case, to aid in the understanding of the questionnaire, we provided some explanations in Portuguese for some words and expressions in English.

The questionnaire was administered through a web tool, LimeSurvey. However, right at the beginning of data collection, in the case of four organizations (three from the IT sector and one from the manufacturing sector, given the geographical proximity and personal connections between the country investigators and the CEO or the IT manager), we took the opportunity to visit them, gather the IT professionals in a meeting, and have them answer the questionnaire in a paper version. It worked as a test of the questionnaire, to reassure that 30 minutes were enough to answer it, as indicated, and to get a sense of its understandability. No relevant issues were reported and that gave us the confidence to move to the online version.

Among the three co-country investigators, the data collection efforts were distributed proportionally according to the motivation and ease for the investigator to assure the commitment of a targeted organization to participate in the study, first taking in consideration the investigator's familiarity with the industry sector and then the number of organizations to involve. It was not easy to recruit organizations to fulfill the requirement of having a minimum of 10 employees in IT since 99.9% of the organizations in Portugal are micro, small and medium enterprises that do not require or cannot afford having so many IT professionals.

These organizations were contacted by an email sent to a CEO, CIO or IT manager, usually a person with whom the country investigator had already interacted in other situations. It was up to this contact to proceed with a call to the IT employees in the organization. The contact person was provided with the URL to access the survey online and a key code to associate the respondents to a particular organization. Twenty organizations were targeted across seven industry sectors: manufacturing, retail, transportation, education, IT, professional services, and government. While monitoring the data collection, reminders were sent to assure that a minimum of 10 IT professionals would respond to the questionnaire.

Table 28.1 shows the descriptive statistics for the sample of 224 IT professionals that participated in this study in Portugal.

Only data coming from four companies that got the survey in a paper version had to be manually checked and entered into the Excel spreadsheet. All the other data were automatically exported to Excel from LimeSurvey and, after checking and cleaning unfinished responses, sent to the project leader.

Table 28.1:   Characteristics of Portuguese IT Professionals

Characteristics	N	%	Characteristics	N	%
Education:			Work experience:		
High school or less	22	9.8	0–4 years	26	11.6
Associate degree	14	6.3	5–9 years	47	21.0
Bachelor's degree	50	22.3	10–19 years	93	41.5
Master's degree	137	61.2	20–29 years	45	20.1
Ph.D.	1	0.4	30+ years	13	5.8
IT experience:			Reporting relationship:		
0–4 years	33	14.7	IT department employee	148	66.1
5–9 years	51	22.8	IT worker in non-IT department	7	3.1
10–19 years	94	42.0	Contract employee	11	4.9
20–29 years	38	17.0	Consultant	58	25.9
30+ years	8	3.6	Vendor employee	0	0.0
Work as:			Management level:		
Mostly full time	213	95.1	Not part of management	85	37.9
Mostly part time	0	0.0	In lower management	49	21.9
Mostly over time	11	4.9	In middle management	57	25.4
Been laid off from IT job:			In senior management	33	14.7
Yes	17	7.6			
No	207	92.4			

28.5   Organizational IT Issues

The participants in the study were asked to consider 18 organizational IT-related issues and rate each one on a 5-point Likert scale, from 1 as the most important to 5 as no important. Table 28.2 shows the ranking for the 18 issues according to the average rating of each one.

Among this set of organizational issues, IT reliability and efficiency was the most important issue for this sample of IT professionals. To understand this result, we should bear in mind that data collection took place right at the end of a severe economic crisis that Portugal experienced between 2009 and 2016, shortly after the end of a foreign intervention in the country by the "Troika", a group formed by the European Commission, the European Central Bank and the International Monetary Fund. A key message from the Troika intervention, that permeated organizations and society, was that we need to be more efficient. IT professionals may have acknowledged that message to contribute to business efficiency and to the importance of delivering reliable, accurate and timely data for effective decision-making.

The alignment between IT and business comes as no surprise as the second highest-ranked issue. This is something that has been consistently

Table 28.2:   Organizational IT Issues in Portugal

Organizational IT Issue	Rank	Mean Rating*	Std. Deviation
IT reliability and efficiency	1	1.70	0.68
Alignment between IT and business	2	1.71	0.66
Attracting and retaining IT professionals	3	1.75	0.71
IT strategic planning	4	1.81	0.67
Security and privacy	5	1.84	0.73
Project management	6	1.86	0.68
Knowledge management	7	1.91	0.74
Business agility and speed to market	8	2.02	0.77
Continuity planning and disaster recovery	9	2.04	0.83
Business productivity and cost reduction	10	2.07	0.67
Enterprise architecture	11	2.17	0.78
Revenue-generating IT innovations	12	2.18	0.74
Business process reengineering	13	2.21	0.73
IT service management (e.g., ITIL)	14	2.23	0.82
Globalization	15	2.51	0.93
IT cost reduction	16	2.54	0.92
Outsourcing	17	3.19	0.95
BYOD (Bring Your Own Computing Device)	18	3.37	1.06

*Rating scale ranges from 1 to 5: 1 as most important and 5 as no importance.

shown as a top concern for IT management. At the time of our data collection, another well-known annual study, the 2016 SIM IT issues and trends study (Kappelman *et al.*, 2017), presented this issue as the first top concern 4 years in a row for IT managers. Aligning business and IT continues to be elusive since not only alignment has to be achieved but also kept in a fast-changing environment making it a moving target to pursue.

Attracting and retaining IT professionals, as the third highest-ranked organizational IT issue, is understandable since IT professionals have been experiencing a high demand in Portugal and internationally. The excellence of higher education to churn out highly qualified tech professionals together with policy changes from Portuguese government to flow foreign business investment in Portugal have been leading big and small companies to open new technology focused hubs in Portugal raising the competition for attracting and retaining IT professionals.

IT strategic planning ranked as the fourth most important organizational IT issue. It is certainly important in periods of economic growth but also in periods of economic downturn or even recession for an organization to succeed. More than looking at IT as a supporting function and the first one to suffer budget cuts, namely, during an economic crisis, this concern may be more related to help the organization not only on how to leverage IT to increase efficiency but also on how to identify strategic, business opportunities to increase revenues and profits. In fact, the line between IT strategy and business strategy is disappearing since there should be no IT strategy separate from the business strategy. For example, in the annual surveys of the Society for Information Management, IT strategic planning was the top concern in the 1980s and since then was ranked continuously in the top 10 management concerns till 2012. We no longer see it at the very top, but the concern has surfaced embedded in Strategic Planning for Business, ranked in the ninth and 11th positions, respectively, in the 2016 and 2017 surveys.

28.6   Technology and Infrastructure Issues

IT professionals were also asked to rate a set of 16 issues, according to their importance, in a 5-point Likert scale ranging from 1 as the most important to 5 as not important. Table 28.3 shows the ranking with Business Intelligence/Analytics as the most important IT issue for the Portuguese IT professionals in our sample.

Table 28.3: Technology and Infrastructure Issues in Portugal

IT Issue	Rank	Mean Rating*	Std. Deviation
Business intelligence/analytics	1	1.96	0.83
Enterprise application integration	2	2.08	0.76
Enterprise resource planning (ERP) systems	3	2.18	0.85
Business process management systems	4	2.18	0.77
Collaborative and workflow tools	5	2.19	0.80
Mobile and wireless applications	6	2.24	0.81
Customer relationship management (CRM) systems	7	2.29	0.82
Networks/telecommunications	8	2.32	0.97
Service-oriented architecture (SOA)	9	2.34	0.80
Software as a service	10	2.37	0.83
Data mining	11	2.45	0.87
Mobile apps development	12	2.48	0.99
Virtualization (desktop or server)	13	2.49	0.96
Cloud computing	14	2.52	0.84
Big data systems	15	2.52	0.89
Social networking/media	16	2.91	1.00

*Rating scale ranges from 1 to 5: 1 as most important and 5 as no importance.

Business intelligence/analytics comes, with no surprise, right at the top of the concerns for the IT professionals. This could be the result of the increasing recognition that applying business intelligence/analytics to business data repositories leads to effective decision-making. This finding is consistent with findings from other studies showing Analytics as the largest IT investment for nine consecutive years and that it should get more investment in the coming years (Kappelman *et al.*, 2018a).

Enterprise Application Integration and ERP systems follow business intelligence/analytics, respectively, as the second and third top IT issues. It is understandable since we need them to make data available in a consistent, timely and accurate way. Without enterprise application integration to assure an integrated IT infrastructure, we can easily see IT islands emerging as new technologies make their way in the organization. Without ERP systems that provide comprehensive and integrated data, analytics will not be able to deliver the expected benefits. The combination of analytics with applications designed to provide transactional data that can be trusted should be a powerful one to deliver benefits of another magnitude.

Business Process Management Systems is the fourth-ranked IT issue. These systems that allow modeling and design as well as execution and tracking of business processes also incorporate components as Analytics for performance management. Business processes need to be managed and optimized. Process and data are just two sides of the same coin.

28.7 Individual IT Employee Issues

The general profile of the IT professional depicted in Table 28.4 can be summarized as follows: a person satisfied with her/his job, who perceives his/her work as contributing to the purpose/goals of the employer organization, sees being relevant to the organization and society, has a sense of self-efficacy and confidence in what concerns the security of her/his job, has ambivalent feelings regarding the burden of the working dimension of her/his life, is loyal to the employer organization but also open to explore new opportunities in other organizations, and has no plans to exit the IT field.

Although this profile might be criticized for being simplistic and failing to account for some nuances of the collected responses to the questionnaire, in most aspects it is consistent with the authors' understanding of the nature of IT-related jobs and their perception of the current situation of the IT job market in Portugal.

IT is a recent and lively work field. On the one hand, new technologies, new techniques, methods and tools become available regularly, requiring that IT professionals continuously pay attention to the news of the field. On the other hand, the field is prolific in what concerns new products and services. Their impact (positive and negative) is often a hot topic in the news, raising the awareness of the responsibility associated with the IT sector. The field has, therefore, a good fit for people that appreciate a varied and challenging job. At the same time, the IT sector relies heavily on talent and creativity. Companies in the IT sector are aware of the need to set up a working environment that is intellectually stimulating and agreeable in general. They often compete to be perceived as providing the best working environment or for being recognized as the best-place-to-work. Furthermore, the current shortage of IT professionals in the global job market pushes companies in the IT sector to be particularly active in order to attract talented people and retain their employees. As a result, working environments in the IT sector often exhibit exotic and exuberant facilities, and provide flexible working conditions and competitive wages. Job satisfaction and loyalty to the employer are therefore easy to understand.

The IT sector in Portugal is facing a major problem related to the shortage of IT professionals. Schools in general and higher education institutions in particular are being pressed to increase the number of graduates

Table 28.4: Individual IT Employee Issues in Portugal

Individual Issue	Mean Rating*	Std. Deviation
Job satisfaction		
In general, I like working here	1.88	0.793
All in all, I am satisfied with my current job	2.03	0.808
In general, I don't like my current job	4.22	0.901
Work pressure		
I feel that the number of requests, problems or complaints that I deal with at work is more than expected	2.94	1.014
I feel that the amount of work I do interferes with how well it is done	2.68	1.015
I feel busy or rushed at work	2.74	0.933
Work–life balance		
There is a blurring of boundaries between my job and my home life	3.32	1.058
My work-related responsibilities create conflicts with my home responsibilities	3.70	1.104
I do not get everything done at home because I find myself completing job-related work	3.62	1.094
Workload and burnout		
I feel drained from activities at work	3.27	1.040
I feel tired from my work activities	3.14	1.019
Working all day is a strain for me	3.70	1.003
I feel burned out from my work activities	3.54	1.025
Sense of accomplishment		
I feel I'm making an effective contribution to what this organization does	2.02	0.813
In my opinion, I do a good job	1.79	0.586
I have accomplished many worthwhile things in this job	2.04	0.756
At my work, I feel confident that I am effective at getting things done	1.96	0.661
Threats to one's job		
I am worried that future technology advancements may pose a threat to my job	3.77	1.039
I believe that other people may be able to perform my work activities	2.46	0.983
I am concerned that my job may be eliminated soon	4.13	0.916
I am concerned that my job may be outsourced soon	4.04	0.934
Career plans		
I will be with this organization 1 year from now	2.26	1.040
I will take steps during the next year to secure a job at a different organization	3.65	1.056
I will be with this organization 5 years from now	2.75	1.027
I will be working in the IT field 1 year from now	1.68	0.832
I will take steps during the next year to secure a job outside the IT field	3.98	1.130
I will be working in the IT field 5 years from now	1.93	0.883

*Rating scale ranges from 1 to 5: 1 as strongly agree, to 5 as strongly disagree.

they produce. The response is normally slow, as the balance between quantity and quality is not easy to maintain. Professionals are often enticed by their employers' competitors and job switching is frequent. It should be noted that this study was carried out in a period were the effects of a major crisis that Portugal faced were still present. This led to a significant exodus of qualified young Portuguese professionals, including many IT professionals. The temptation of career progression, better conditions and the will to try new experiences is something difficult to keep away from. We believe that the responses received in this study reflect a possible blend of these pressures with the traditional posture of staying on in the current job.

The responses to the items under work pressure, work–life balance, and workload and burnout reveal some ambivalence regarding the burden of the working dimension on personal life. In most cases, the computed scores are in the vicinity of the neutral point (3). This raises a question that might be worth pursuing: is the shortage of IT professionals putting extra demands on the existing work force or is it just a manifestation of the trends in modern life?

28.8 Conclusion

In this survey of Portuguese IT professionals, the four most important IT-related organizational issues are IT reliability and efficiency, alignment between IT and business, attracting and retaining IT professionals, and IT strategic planning. The top four technological issues for these professionals are business/analytics, these professionals place as the most important, are business intelligence/analytics, enterprise application integration, ERP systems and business process management systems.

Compared with a similar survey conducted with 276 European IT executives in the summer of 2017 (Kappelman et al., 2018b), and looking at the top five IT-related organizational issues, our study shows two issues in common: business-IT alignment, and security and privacy. When looking at the top five technological issues, there were three in common: business intelligence/analytics, ERP systems, and business process management. Therefore, we may infer that Portuguese and European IT professionals share similar concerns with some minor differences.

These similarities and differences notwithstanding, there is at least one noteworthy difference regarding IT organizational issues: Attracting and retaining IT professionals, something that is quite important for Portuguese organizations, but not so important for European organizations. This may be explained by the shortage of IT professionals in Portugal since the country is facing an increasing demand from national and international organizations that choose Portugal to open tech hubs while there is also high international mobility among Portuguese IT professionals. Nevertheless, Portuguese IT professionals seem quite satisfied with their jobs and their contribution to business goals, perceive to have reasonable workloads, and do not intend to leave the IT profession.

References

Kappelman, L., McLean Ephraim, Vess, J., Torres, R., Nguyen, Q., Maurer, C., & David, A. (2018a). The 2017 SIM IT Issues and Trends Survey. *MIS Quarterly Executive*, 17(1), 53–88.

Kappelman, L., Johnson, V., Torres, R., Maurer, C., & McLean, E. (2018b). A study of information systems issues, practices, and leadership in Europe. *European Journal of Information Systems*, 28(1), 26–42. https://doi.org/10.1080/0960085X.2018.1497929

Kappelman, L., McLean, E., Johnson, V., Torres, R., Nguyen, Q., Maurer, C., & Snyder, M. (2017). The 2016 SIM IT Issues and Trends Study. *MIS Quarterly Executive*, 16(1), 47–80.

Palvia, P., Jacks, T., Gosh, J., Licker, P., Romm-Livermore, C., Serenko, A., & Turan, A. H. (2017). The world IT project: History, trials, tribulations, lessons, and recommendations. *Communications of the Association for Information Systems*, 41(1), 389–413.

Palvia, P., Ghosh, J., Jacks, T., Serenko, A., & Turan, A. (2018). Trekking the globe with the World IT Project. *Journal of Information Technology Case and Application Research*, 20(1), 3–8.

Chapter 29

Information Technology Issues in Romania

Doina Fotache[*,||], Vasile-Daniel Pavaloaia[*,**], Luminiţa Hurbean[†,††],
Octavian Dospinescu[*,‡‡], Alexander Serenko[‡,§,§§], and Tim Jacks[¶,¶¶]

Alexandru Ioan Cuza University of Iasi, Romania
†West University of Timisoara, Timişoara, Romania
‡University of Toronto, Toronto, Canada,
§University of Ontario Institute of Technology, Oshawa, Canada
¶Southern Illinois University Edwardsville,
Edwardsville, IL, USA
||doina.fotache@feaa.uaic.ro
***danpav@uaic.ro*
††luminita.hurbean@e-uvt.ro
‡‡doctav@uaic.ro
§§a.serenko@utoronto.ca
¶¶tjacks@siue.edu

Summary

Nowadays, companies in Romania and abroad compete intensely for attracting employees, but the demand in Romania for qualified information technology (IT) personnel exceeds the supply. Previous studies have also reported that the IT market is quite "choosy". The World IT Project contributes to a better understanding of perceptions of Romanian IT workers towards organizational, technological, and individual issues. It was discovered that IT workers have different preferences, taking into account that the average scores indicate all issues as being important. Continuity planning and disaster recovery, security and privacy, and IT reliability and efficiency are the most important organizational issues. The key technology issues are customer relationship management systems, collaborative and workflow tools, and networks/telecommunications. Overall, Romanian IT professionals are knowledgeable, hard-working, satisfied with their jobs, and willing to develop themselves in their current or other organizations.

29.1   Introduction

The information technology (IT) industry has become the most dynamic field in Romania. It has been supported by entrepreneurial experience accumulated in the last decades, high quality of higher education, academic research in the relevant technical disciplines, and the presence of major multinational companies. Considering the level of technical proficiency and soft skills (common sense, positive and flexible attitude, high responsibility, and a good sense of teamwork), Romania is highly competitive being considered "superior to what is typically found in other outsourcing locations" (Fiscutean, 2015). Young and motivated professionals having European values and a good command of English, German, French, and Italian are the key elements of Romania's appeal.

Unemployment in the Romanian IT sector is close to zero mainly due to high qualification of the workforce, the presence of global companies, and the development of numerous startups. In 2017, over 100,000 Romanians worked in IT. This domain of activity, together with communication technologies, exceeded 6% of the 2017 GDP growth rate, and the IT market was valued at over 5 billion Euros. The sector is quickly growing, and the number of IT firms has increased by 50% in the last 8 years, according to the National Study of the IT market (Aries Transilvania, 2017) conducted by the IT Cluster of ARIES Transylvania. The competition for the title of the Best Performing IT Center in Romania included such cities as Cluj-Napoca,

Timisoara, and Iasi (Florescu, 2017). The number of national IT companies increased between 2011 and 2016 from 9,823 to 14,339, reaching 17,000 by the end of 2017. The number of startups has doubled from 2010 to 2018, indicating the presence of young entrepreneurs, promoters of new ideas and solutions, along with a new business attitude. However, their capacity to face strong competition is actually an issue that should be observed for continued IT market development.

29.2   Country Background and History

With an area of 238,400 km^2, Romania is the 12th largest country in Europe. The name "România" was first used when the two Principalities of the country were united in 1859, and it reflects the influence of ancient Rome on the nation's language and culture. The Romanian people are Europeans by birth (Romanian Academy, 2007), as the Carpathian-Danubian-Pontic area where they live has always been a part of Europe. The Geto-Dacians, the Thracians, and their ancestors did not come to this territory from other parts of the world. Romanian is an Indo-European language, belonging to the Romance group of languages (the Eastern part). Among Romance Languages, Romanian ranks fifth according to the number of speakers, following Spanish, Portuguese, French and Italian. The Romanian state was founded by Alexandru Ioan Cuza in 1859 through the union of Moldova and Walachia. It was officially recognized as an independent country in 1866. The third province, Transylvania, joined the other two in 1918 when the Great Union was completed. Officially, the constitutional monarchy governed Romania between 1881 (the coronation year of King Carol I) and 1947. In this period, Romania was ruled by four monarchs of German origin, the Hohenzollern-Sigmaringen — a branch of an old German imperial family. Starting with King Ferdinand I (the Unificator), the Romanian monarchs became a distinctive royal family. The monarchy was unconstitutionally abolished by *coup d'état* on December 30, 1947, during the military occupation of the country by the Soviet Union. This was the moment when the Romanian Communist Party seized power. Communist Romania (1947–1989) is an unofficial name used sometimes with reference to the Communist period in the history, when the country was known by the official names of the People's Republic of Romania and the Socialist Republic of Romania.

The Romanian Revolution of December 1989 was part of the revolutions in Eastern Europe in that year which led to the abolishment of the communist regime. This Revolution marked a significant change. The Adherence of

Romania to the European Union (EU) took place on January 1, 2007 (Filip, 2014). Romanian has become one of the official languages of the Union (seventh by the number of speakers, in strong competition with Dutch). Today, Romania ranks seventh in size in the EU, based on its population (22.19 million inhabitants according to the National Statistics Institute, INS, 2018) and is a semi-presidential republic. The Romanian population is made up of 70% Christian Orthodox, 6% Roman Catholics, 6% Protestants, and 18% with no religious affiliation.

29.3   Information Technology in Romania

According to the European Commission (European Commission, 2017), "the main challenge for Romania is its low competitiveness". The data (Dăianu, 2015) also indicate the country's low capacity to commercialize research output and innovation. To support the economic development of all countries and to ensure sustainable economic growth, the EU published the Europe Digital Agenda 2020 aimed to develop a Digital Single Market (European Commission, 2016). This Agenda has been adopted in Romania (Romanian Government, 2016) in order to develop the IT industry and to increase the level of IT creativity.

While in the western European countries the economic impact of IT sector is 5% of GDP growth rate, in Romania the percentage was 4.1 in 2011, in decline compared to 2010. Investments in the IT sector in 2011 reached 720 million Euros, much lower than in other EU countries. In 2020, Romania aims to reach 250,000 IT employees. The most recent study (published in 2015 by iTech Transylvania Cluster, https://itech.aries-transilv ania.ro/) shows that now there are almost 17,000 IT companies operating in Romania, half of which are located in Bucharest and Cluj-Napoca. As for the IT workforce, 75% of the total number of IT employees are located in Bucharest, Cluj-Napoca, Iasi, and Timisoara. The capital city of Bucharest is by far the place with the highest number of employers, providing almost half of the IT jobs in Romania. In 2016, Romania had a shortage of 50,000 IT workers. The IT staff crisis has been marked by a low number of graduates (universities supply around 7,000 IT graduates a year, while the demand is at least twice as high), resulting in a significant increase of salaries in the IT industry. In 2018, the net average salary reached the amount of 1,325 Euros a month, 2.4 times higher than the average salary in the country. Practitioners speak about IT cannibalization, given the specialist hunt and theft from one firm to another. The sad reality is that only one in four IT specialists is senior, and the migration from one job to another has become very common.

To find a strategic solution to all these problems, Romania has adopted the National Digital Strategy by 2020 with four major directions of action:

1. E-governance, Interoperability, Cybernetic Security, Cloud Computing, Open Data, Big Data and Social Media;
2. IT in Education, Health and Culture;
3. E-commerce, Research, Development and IT Innovation;
4. Broadband and Digital Services Infrastructure.

A complete implementation of this strategic vision will require a total investment of over 3.9 billion Euros. A direct and indirect impact on the economy, calculated considering the best practices in the other European countries, can generate an increase of GDP and jobs by 13% and 11%, respectively, and the reduction of administrative costs by 12%.

As shown above, Romania has become an incredibly attractive country for IT investments (Fotache *et al.*, 2016). Tholon's Top 100 Outsourcing Destinations 2016 Report ranks the Romanian capital of Bucharest 41st on the list compared to position 44 held in 2012. Located at the Eastern border of the European Union, only one hour away from Berlin and at a 3-hour flight to London, the country has turned into a hub of flourishing technology. The Romanian government supports the IT industry development by exempting software developers from the payment of the 16% income tax.

29.4 Methodology

The standard instrument of the World IT Project (Palvia *et al.*, 2017; Palvia *et al.*, 2018) was translated into Romanian as the team considered that it would help the respondents understand and complete it better. To ensure the quality of translation, the instrument was translated back (from Romanian into English) by a different interpreter, and it was compared with the original. The translated instrument was tested, and its completion required from 20 to 30 minutes. In order to reach as many respondents as possible, the survey was posted and distributed online; 350 potential respondents received the survey request. They were contacted online by email and also through online groups (Yahoo and Facebook based). Many of the respondents were graduates of the Business Computer Science study program of the two main Faculties of Economics and Business Administration in Romania, and most were employed in large cities, such as Iasi

Table 29.1: Descriptive Statistics

Characteristics	N	%	Characteristics	N	%
Education:			Years of work experience:		
High school or less	28	8.5	0–4 years	115	35.1
Associate degree	5	1.5	5–9 years	55	16.8
Bachelor's degree	151	46.0	10–19 years	99	30.2
Master's degree	126	38.4	20–29 years	46	14.0
Ph.D.	18	5.5	30+ years	13	4.0
Years of IT experience:			Organizational location:		
0–4 years	129	39.3	IT department employee	88	26.8
5–9 years	76	23.2	IT worker in non-IT department	38	11.6
10–19 years	103	31.4	Contract employee	187	57.0
20–29 years	17	5.2	Consultant	10	3.0
30+ years	3	0.9	Vendor employee	5	1.5
Work as:			Are you:		
Mostly full time	296	90.2	Not part of management	128	39.0
Mostly part time	14	4.3	In lower management	113	34.5
Mostly over time	18	5.5	In middle management	57	17.4
Have you ever been dismissed from an IT job?			In senior management	30	9.1
Yes	4	1.2			
No	324	98.8			

and Timisoara. No face-to-face meetings were arranged. In some instances, graduates that reached managerial IT positions were contacted and invited to support the survey in order to attain a reasonable number of responses. As a result, 332 IT workers completed the survey instrument. The collected responses were imported to Excel and validated, and 328 responses were confirmed as usable.

Table 29.1 displays descriptive statistics, such as respondents education, professional experience, organization location, type of employment contract, management level, and whether the respondents had been ever dismissed from an IT job. The results show that the majority (46%) have a bachelor's degree, 38.4% hold a master's degree, while 5.5% have a Ph.D. A majority of respondents fall into the first category of 0–4 years for both years of all work experience (35.1%) and years of IT experience (39.3%). Most of them (39%) do not hold managerial positions or work at lower managerial levels (34.5%), 90.2% are employed full time, and 57% are contract employees. Among the 328 respondents, only four had been previously dismissed from an IT job. A low level of dismissal is justified by a high demand for IT specialists, generating migration from one job to another in search of a better position.

29.5 Organizational IT Issues

Respondents were asked to grade 18 organizational IT-related issues in terms of their importance based on a 5-point Likert scale. The average rating for each issue was computed and a ranking list was developed (see Table 29.2). If we compare the Romanian results with those of the 2017 SIM IT Issues and Trends Report (Kappelman *et al.*, 2017), it reveals some similarities and differences. For instance, in both studies, security and privacy and alignment between IT and business ranks are within the top 10 issues. There are also differences between the two studies, namely, IT cost reduction is ranked 15th in Romania while in the SIM Report it is ranked sixth.

Romanian IT workers assign the highest importance to continuity planning and disaster recovery and security and privacy, both issues being linked to importance of security for all applications and platforms taking into consideration recent cyber attacks in Europe and worldwide. It is possible that continuity planning and disaster recovery is listed first due to both global

Table 29.2: Organizational IT Issues in Romania

Organizational IT Issues	Rank	Mean Rating*	Std. Deviation
Continuity planning and disaster recovery	1	1.43	0.75
Security and privacy	2	1.51	0.81
IT reliability and efficiency	3	1.57	0.76
Knowledge management	4	1.61	0.82
Project management	5	1.66	0.82
Business productivity and cost reduction	6	1.76	0.84
Business agility and speed to market	7	1.77	0.82
Alignment between IT and business	8	1.80	0.83
Attracting and retaining IT professionals	9	1.86	0.99
Bring your own computing device (BYOD)	10	1.87	0.99
Revenue-generating IT innovations	11	1.89	0.93
IT service management (e.g., ITIL)	12	1.90	0.93
IT strategic planning	13	1.91	0.88
Enterprise architecture	14	2.05	0.97
IT cost reduction	15	2.10	0.98
Business process reengineering	16	2.23	0.89
Outsourcing	17	2.33	1.08
Globalization	18	2.36	1.03

*Rating scale ranges from 1 to 5: 1 is for most important and 5 for no importance.

political and economic instability and more recent political unpredictability in Romania. Moreover, in 2018, the country tightened the legislation on the importance of personal data, and all organizations have been required to apply the new General Data Protection Regulation (GDPR) rules. GDPR 2016/679 is a regulation in the European Union (EU) law on data protection and privacy for all individuals in the EU and the European Economic Area (EEA). This regulation also addresses the issues of export of personal data outside of the EU and the EEA. The GDPR aims primarily to give control to citizens and residents over their personal data and simplifies the regulatory environment for international business by unifying the regulation across the EU.

IT reliability and efficiency was ranked as the third most important issue due to the fact that consistent delivery of IT services to support business operations is a key concern of IT leaders. Efficiency is a fundamental issue underlying any investment decision, and IT efficiency is one of the most important aspects when evaluating project opportunity. The relationship between new IT investments and efficiency and performance is compressed in the famous productivity paradox (Solow paradox) (Orlikowski, 1993): "We can see computers being used everywhere, except for productivity statistics." Even though technology is one of the most important factors influencing productivity and, at least in theory, all countries have equal access to technological innovation, in fact, productivity is influenced by many other factors: the accumulation of physical and human capital, infrastructure, market structure, demographic developments, competition, etc. The main difficulty lies in the ability to create teams made of efficient and reliable specialists. Flexibility is also becoming an imperative objective in organizations that must cope with the constant migration of IT staff. The value of IT services is given by the quality of human resources and service-level satisfaction. These are embedded in the brand and reputation of the IT service provider, generating added value and competitive advantage.

The last two options that round up the top five organizational IT issues in Romania are knowledge management and project management. Because we live in a knowledge-based economy, people and their knowledge make the difference in the competitiveness race. The value of IT products and services is generated by the amount of data and information modeled within them and the performance gained from their implementation. Knowledge management is designed as an umbrella concept that aggregates information on current needs and customer aspirations, market and macroeconomic environment, competition, new technologies and developments in other industries, etc. In current IT projects, classical organizational modes (based on value channels) have been eclipsed by project-based organization. Managers revise their style and strike a balance between ensuring the best execution and the autonomy of the working groups. The presence of knowledge management and project management is logical and interconnected with IT reliability and efficiency.

Overall, when browsing the calculated means in Table 29.2, one may observe the close values, the interval ranging from 1.43 to 2.36. We may say that our respondents consider all the 18 issues of some importance, as if they did not want to neglect any of the listed issues.

29.6   Technology and Infrastructure Issues

Table 29.3 shows the technology and infrastructure issues in Romania ranked in order of importance.

Major interest in marketing effectiveness exists because many companies realize that they need to "win the battle" for valuable customers. The large amount of customer data that needs to be analyzed and processed requires customer relationship management (CRM) solution support. Web 2.0 and social platforms have revolutionized the way organizations treat

Table 29.3:   Technology and Infrastructure Issues in Romania

IT-Related Issues	Rank	Mean Rating*	Std. Deviation
Customer relationship management (CRM) systems	1	1.76	0.91
Collaborative and workflow tools	2	1.85	0.91
Networks/telecommunications	3	1.87	0.91
Enterprise resource planning (ERP) systems	4	1.88	1.02
Business intelligence/analytics	5	1.89	0.96
Business process management systems	6	1.91	0.86
Mobile and wireless applications	7	1.92	0.89
Enterprise application integration	8	1.93	0.93
Mobile apps development	9	2.01	0.97
Service-oriented architecture (SOA)	10	2.02	1.03
Virtualization (desktop or server)	11	2.03	0.99
Cloud computing	12	2.06	1.08
Software as a service	13	2.07	1.01
Data mining	14	2.16	1.12
Big data systems	15	2.17	1.00
Social networking/media	16	2.19	1.04

*Rating scale ranges from 1 to 5: 1 as most important and 5 as no importance.

their customers. Four out of five Romanian companies do not provide a prompt response to customer interactions on social media, such as Facebook, LinkedIn, or Twitter. Response time is unacceptably long, more than 48 hours on average, due to manual management of social media accounts. Public analysis shows that in Romanian companies, social media is mainly used for promotion and less for sales and support (Zamfir, 2016). As can be seen in Table 29.3, CRM ranks first, while social media has the bottom position.

Collaborative and workflow tools provide the infrastructure for automating and improving workflows and assigning tasks to processes within the organization. While workflow designates a set of organizational maps of tasks and activities carried out by a number of individuals or groups, collaborative business systems (Groupware) are synonymous with providing tools for good communication at the organizational level (horizontally and vertically) to assist with the management and automation of business processes. Their significance led to their incorporation into integrated enterprise systems (Hurbean *et al.*, 2016), such as ERPs. In this study, collaborative and workflow and ERP systems hold positions two and four.

Networks/telecommunications is closely related to infrastructure, and it was ranked third. Overall, Romania has a very good telecommunications and network infrastructure, ensuring the technical conditions required for the implementation of sophisticated projects, including big data. However, big data appears at the bottom of the list, indicating that organizations do not currently consider it an important issue.

Given the current business environment, the quality and timeliness of information is not a choice between profit and loss for the organization but is rather a matter of survival. The benefits of business intelligence (BI) systems are obvious — all analysts and professionals are convinced of the importance of visual analysis tools and BI. The market is already mature in terms of offering diverse analytical applications capable of providing a wide range of analyses needed to support decision-making at all levels of an organization. Again, respondents graded all the sixteen issues within a small interval, from 1.76 to 2.19, showing that organizations perceive all issues as important or at least of some importance.

29.7 Individual IT Employee Issues

From the analysis of results presented in Table 29.4, which pertain to individual employee issues, several important findings emerged.

Table 29.4: Individual IT Employee Issues in Romania

Individual Issues	Mean Rating*	Std. Deviation
Job satisfaction		
In general, I like working here	1.67	0.80
All in all, I am satisfied with my current job	2.12	1.21
In general, I don't like my current job	3.88	1.21
Work pressure		
I feel that the number of requests, problems or complaints that I deal with at work is more than expected	3.28	1.22
I feel that the amount of work I do interferes with how well it is done	3.19	1.26
I feel busy or rushed at work	3.13	1.17
I feel pressured at work	3.31	1.22
Work–life balance		
There is a blurring of boundaries between my job and my home life	3.36	1.34
My work-related responsibilities create conflicts with my home responsibilities	3.74	1.31
I do not get everything done at home because I find myself completing job-related work	3.68	1.32
Workload and burnout		
I feel drained from activities at work	3.54	1.25
I feel tired from my work activities	3.51	1.24
Working all day is a strain for me	3.86	1.27
I feel burned out from my work activities	3.22	1.39
Sense of accomplishment		
I feel I'm making an effective contribution to what this organization does	1.85	0.86
In my opinion, I do a good job	1.67	0.69
I have accomplished many worthwhile things in this job	1.88	0.80
At my work, I feel confident that I am effective at getting things done	2.29	1.35
Threats to one's job		
I am worried that future technology advancements may pose a threat to my job	3.66	1.27
I believe that other people may be able to perform my work activities	3.32	1.27
I am concerned that my job may be eliminated soon	4.08	1.06
I am concerned that my job may be outsourced soon	3.58	1.38
Career plans		
I will be with this organization 1 year from now	2.48	1.45
I will take steps during the next year to secure a job at a different organization	3.38	1.43

(*Continued*)

Table 29.4: (*Continued*)

Individual Issues	Mean Rating*	Std. Deviation
I will be with this organization 5 years from now	2.61	1.39
I will be working in the IT field 1 year from now	2.32	1.48
I will take steps during the next year to secure a job outside the IT field	3.57	1.53
I will be working in the IT field 5 years from now	2.41	1.55

*Rating scale ranges from 1 to 5: 1 as strongly agree to 5 as strongly disagree.

An overwhelming majority of the IT workers responded that they like their job (job satisfaction/In general, I like working here) which means that they feel that they perform their duties well and there is a positive correlation with sense of accomplishment/In my opinion, I do a good job, rated as 1.67. Overall, the employees experience an acceptable degree of work pressure and workload and burnout, which means that although they perform their tasks with high responsibility, they perceive that they are not burdened with workloads and feel an acceptable level of pressure. Since scores are all greater than 3, it is assumed that organizations manage to have an optimal load of positions and working hours for their employees.

Considering their sense of accomplishment, the IT employees rated themselves positively, showing good confidence in their skills — calculated means are between 1.67 and 2.29. They seem to be aware of the high demand for their IT skills in the job market. They also consider their professional skills valuable to their organizations. They perceive little threat to their jobs, and this confirms the existence of little competition for the entry-level positions where minimum IT experience is required. At the same time, a minority of IT workers may potentially leave their organizations and the entire IT profession.

29.8 Conclusion

Overall, the study reveals that professionals in the Romanian IT industry are well positioned in their organizations, in line with their qualifications and knowledge, where they can continuously grow, diversifying their IT skills (over the next 5 years, most of them believe they will continue to be working in this area).

While the investigation of individual IT employee issues indicated specific differences among respondents, the organizational and technology/infrastructure sections show that the means of all issues are above the average, indicating that respondents consider all issues as being important or very important.

In Romania, government programs support the development of a strategy for bridging the gap between supply/demand for labor in the IT industry (digital strategies, tax exemptions, increased number of state-funded places at universities for IT study programs). The status of the IT role in Eastern Europe gives Romania attractiveness in terms of IT competences and outsourcing. The individual profile of Romanians can be summarized as follows: professional trust, competence orientation, and mistrust for collaboration and teamwork. Their excellent IT skills, good command of foreign languages and the so-called soft skills, and the cultural proximity to the West make them attractive and enable them to compete with IT professionals from India — a popular country for outsourcing.

Future studies on Romania should look at how IT employees could be retained in organizations by other means than benefits or incentives (wage increases, private health insurance, workplace relaxation, and other similar benefits). The Romanian IT industry is on the rise and it is expected to develop further. For the first time in its recent history, Romania desires to be not just an IT support country, but also a country that develops value-added IT products.

References

Aries Transilvania (2017). IT Market Study — Comparative Analysis 2017. Aries Transilvania. Retrieved from https://www.itstudy.ro/.

Dăianu, D. (2015). *Marele impas în Europa.* Iasi. Romania: Polirom Publishing House.

European Commission (2016). Digitising European Industry. Reaping the full benefits of a Digital Single Market COM 180 final. Brussel. Retrieved from https://ec.europa.eu/digital-single-market/en/news/communication-digitising-european-industry-reaping-full-benefits-digital-single-market [Accessed April 17, 2018].

European Commission (2017). Europe's Digital Progress Report (EDPR) 2017 Country Profile Romania. Retrieved from https://ec.europa.eu/digital-sin gle-market/en/news/europes-digitalprogress-report-2017.

Filip, P. (2014). *România și experiența aderării.* Bucuresti. Romania: Editura Economica.

Fiscutean, A. (2015). IT outsourcing: As Romania vies to be the new India, can the country keep up? Retrieved from Zdnet.com:

https://www.zdnet.com/article/it-outsourcing-as-romania-vies-to-be-the-new-india-can-the-country-keep-up/ [Accessed March 12, 2018].

Florescu, R. (2017). Capitala IT-ului din România: oraşul în care 1 din 11 angajaţi lucrează în domeniu. Adevarul.ro. Retrieved from https://adevarul.ro/loc ale/cluj-napoca/capitala-it-ului-romania-orasul-1-11-angajati-lucreaza-do meniu-1_5a2be6b35ab6550cb804b8b6/index.html [Accessed April 19, 2018].

Fotache, M., Dumitriu, F., & Greavu-Serban, V. (2016). Information system skills differences between offshoring source and destination markets. A Romanian perspective. *Transformations in Business and Economics*, 15(3C), 452–477.

Hurbean, L., Fotache, D., & Pavaloaia V.D. (2016). *Modern Business Information Systems: The Enterprise Resource Planning and its Functions*, Volume 1. LAMBERT Publishing.

Orlikowski, W. (1993). Learning from notes: Organizational issues in groupware implementation. *The Information Society*, 237–250.

Palvia, P., Jacks, T., Ghosh, J., Licker, P., Romm-Livermore, C., Serenko, A., & Turan, A. H. (2017). The World IT Project: History, trials, tribulations, lessons, and recommendations. *Communications of the Association for Information Systems*, 41(18), 389–413.

Palvia, P., Ghosh, J., Jacks, T., Serenko, A., & Turan, A. (2018). Trekking the globe with the World IT Project. *Journal of Information Technology Case and Application Research*, 20(1), 3–8.

Romanian Academy (2007). *Cunoaste Romania, membra a Uniunii Europene.* Bucuresti. Romania: Editura Economica.

Romanian Government (2016). National Strategy on the Digital Agenda for Romania 2020. Bucharest: Romanian Government. Retrieved from https ://www.comunicatii.gov.ro/?page_id=3496 [Accessed February 9, 2018].

Zamfir, C. (2016). Infografic Studiu: 74% din firmele romanesti se promoveaza pe retele sociale. Retrieved from https://www.startupcafe.ro/stiri-marketing-20821833-studiu-74-din-firmele-romanesti-promoau-retele-sociale-2015.htm [Accessed February 3, 2018].

Chapter 30

Information Technology Issues in Russia

Tim Jacks[*,||], Nikolay Kazantsev[†,‡,**], and Alexander Serenko[§,¶,††]

*Southern Illinois University Edwardsville, Edwardsville,
IL, USA
†Alliance Manchester Business School,
Manchester, UK
‡Higher School of Economics, National Research University,
Moscow, Russia
§University of Toronto, Toronto, Canada
¶University of Ontario Institute of Technology, Oshawa, Canada
||tjacks@siue.edu
**nicolay.kazantsev@gmail.com
††a.serenko@utoronto.ca

Summary

This chapter presents the top organizational, technological, and individual issues facing information technology (IT) workers in Russia. Due to its political and economic history, Russia has a unique combination of issues and concerns not seen in Western countries. The top organizational IT issues in Russia were IT reliability and efficiency, security and privacy, and revenue-generating IT innovations. While the first two are similar to Western countries, the third is particular to the Russian context. The top-technology issues included business intelligence/analytics, business process management systems, and enterprise application integration. Surprisingly, networks/telecommunications was at the bottom of the list. While levels of job satisfaction, perceived work overload, work/home conflict, and work exhaustion/strain, were similar to those reported in other countries, professional self-efficacy was markedly lower as was turnover intention due to the likelihood of finding better work in other countries. Such differences highlight the importance of conducting research in non-Western countries.

30.1 Introduction

Approximately 37% of information technology (IT) projects, which include the implementation of information systems, are not completed successfully (Standish Group, 2015). The implementation of American and European information systems in Russia causes difficulties that have not been encountered in the countries of the West. This is due to the fact that information systems were developed by the United States for American companies and companies located in Western Europe, taking into account the economic and social benefits of these countries. In Russia, these systems are not operating and being used as planned. The reasons can also be related to the style of management, the development of IT architecture, IT culture, public beliefs, and various intercultural aspects that must be taken into account for the successful implementation of foreign information systems. A better understanding of the top organizational, technological, and individual issues facing IT workers in Russia should help in addressing such difficulties.

30.2 Country Background and History

The Russian Federation is the largest country on Earth by area (more than 17 million km^2), with the population of more than 140 million people. Russia lies in Europe and Asia and has common borders with the Scandinavian countries, Eastern Europe, South Caucasus, Central Asia, China, North Korea, Japan, and the USA. This geographical location divides its multicultural society into eleven time zones, having the central regions closer to the West and the rest of the country closer to the East. Economic and political development of Russia is therefore two-dimensional.

Russian history goes back more than 1100 years and includes periods of *openness* and *closeness* to the influences from the West. After the adoption of Orthodox Christianity in 988, the period of openness was followed by the invasion of Mongols in 13th–14th centuries that had closed the country for almost 200 years. Openness to the West flourished again in the beginning of the 18th century with the Europe-style reforms of Peter the Great, which influenced the development of science, music, art, and technology till the first decade of 20th century. After the Second World War, the country was locked again behind the famous "iron curtain" due to the confrontation between the communist block countries and the Euro-Atlantic part of the world.

Modern history starts in 1991 with the dissolution of the former Soviet Union and again its openness to the West. Despite a difficult start, the real GDP growth reached 8.3% in 2000 (Johnston, 2018). After this initial growth, Russia suffered badly during the recession of 2008, losing over $1 trillion in market capital (Faulconbridge, 2008). Russia rebounded and, by 2016, its economy was the sixth largest in the world by Purchasing Power Parity and twelfth largest at market exchange rates (Excif, n.d.). Despite periodic instability of the Russian economy, its industrial output grows rapidly. Russia joined the World Trade Organization in 2012. It is the part of the UN, the Council of Europe, the G-20 countries, the World Trade Organization, and other major economic and political organizations. Russia is also a member of BRICS, i.e., one of five major emerging economies (including Brazil, Russia, India, China, and South Africa). Oil accounts for 30% of the country's GDP and 50% of the government's budget (Candau, 2018). Weak diversification of the Russian economy leads to overdependence of the country on the energy sector and unstable energy prices.

Russia has experienced a number of political and economic challenges in recent years. Today, the deployment of economic sanctions against Russia, announced by the US and the EU, has triggered another loop of closeness for the Russian economy. With the fall in oil prices, the Russian ruble fell to a historic low (Rutland, 2014) and the negative economic outlook impeded foreign investment. Any further weakening of the national currency may lead to dire consequences for the entire economy, and the situation in the IT market will correspond to the general economic situation in the country.

30.3 Information Technology in The Country

The Russian IT sector drew very little from the Soviet-era institutions. The first Russian IT companies were launched in the early 1990s by founders with an academic background seeking to find a place in the new market economy. Software piracy was widespread in the country, with an estimated 90% of all software in Russia being unlicensed in 1997 (Associated Press, 1997). In the 1990s, companies such as Vist began assembling computers out of foreign-made components, targeting small businesses and families who could not afford major foreign brands such as IBM and Compaq. DVM Computer gained some traction in the laptop market with its Rover-Book brand. The Russian Computer was the trade association representing the sector. In 1997, Yandex, a multinational corporation specializing in Internet-based services, was established in Moscow. In 1999, MCST developed the E2K processor, which was initially hyped as an Itanium killer, but the project was hampered by a chronic lack of funding.

Over time, Russian companies moved to software development, an activity which enjoyed higher margins. Exports of software and IT services from Russia reached $7 billion in 2015, up from $2.8 billion in 2009 (East-West, 2016). Local companies cater to the specific needs of the Russian market, such as ERP software developed by 1C Company with a focus on Russian accounting rules. Kaspersky Labs is described as the flagship company of the Russian IT industry. Currently, the biggest companies are Yandex, Mail.ru Group, and Avito.

Worsening relations between the United States and Russia have led some to advocate for a ban of Russian software, such as Kaspersky's antivirus software. In response, the Russian government passed a law in 2015 to give preference to software developed in Russia for all state institutions (EY, 2015). In the aftermath of the armed conflict in the Donbass region of Ukraine, the Ukrainian government banned a number of Russian IT companies from conducting business in the country.

Another negative factor in the development of the IT market in Russia is the inefficiency of large government projects. Poor cooperation between federal and regional authorities leads to an increase in the cost of IT projects, unsuccessful IT implementations, and bloated budgets. The surge of inflation and a rise in prices for imported products also significantly hamper the development of the IT market, forcing Russian and foreign entrepreneurs to discontinue investment in the country. High inflation undermines consumer confidence and adversely affects the entire market.

The reality is that the economic condition of the country affects the IT market much more than the labor market does. IT specialists are ready to work in Russia, but the situation in the entire market is far from prosperous. Therefore, IT workers tend to relocate to other countries with more opportunities and higher salaries, such as the United States, Germany, and the United Kingdom. Therefore, the context of IT workers in Russia is significantly different from that of Western countries.

30.4 Methodology

The data from 206 IT professionals in Russia were collected through an online survey in 2016. Respondents represented a variety of companies, industries, and IT roles. The items were measured on a 5-point Likert scale. The English survey developed for the World IT Project (Palvia *et al.*, 2017; Palvia *et al.*, 2018) was translated into Russian and then back-translated to verify that the meaning had not changed. No incentives were offered to complete the survey. However, one difference with data collection in Russia is that the respondents typically wanted to know (1) why the research was being performed and (2) what the expected outcomes would be. Many of their questions and concerns had to be addressed before IT workers would participate in the survey (an issue not reported for any other country in the World IT Project) and this highlights important cultural differences in Russia. All calculations for the analysis were performed in IBM SPSS software. Table 30.1 shows the overall demographics of the respondents.

30.5 Organizational IT Issues

The top organizational IT issues in Russia were IT reliability and efficiency, security and privacy, and revenue-generating IT innovations. The presence of IT reliability as well as security are in sync with the findings of other countries in the project and, in fact, match precisely with the top two issues in the US. However, the high ranking of revenue-generating IT innovations and business productivity and cost reduction is further evidence of the particular economic conditions facing Russia highlighted above. Bring your own device (BYOD) computing was at the bottom of the list and is similar to findings in other countries. The relatively lower ranking of attracting and retaining IT professionals is slightly surprising given the increased likelihood of Russian IT workers immigrating to other countries (see Table 30.2).

30.6 Technology and Infrastructure Issues

The top technology issues for IT workers in Russia were business intelligence/analytics, business process management (BPM) systems, and enterprise application integration. The presence of BPM at the top of the list may reflect a lower level of maturity of IT organizations given that this

Table 30.1:   Descriptive Statistics

Characteristics	N	%	Characteristics	N	%
Education:			Years of work experience:		
High school or less	7	4.7	0–4 years	125	84.5
Associate degree	70	47.3	5–9 years	12	8.1
Bachelor's degree	48	32.4	10–19 years	8	5.4
Master's degree	22	14.9	20–29 years	1	0.7
Ph.D.	1	0.7	30+ years	2	1.4
Years of IT experience:			Organizational location:		
0–4 years	133	89.9	IT department employee	62	41.9
5–9 years	8	5.4	IT worker in non-IT	43	29.1
10–19 years	4	2.7	department		
20–29 years	1	0.7	Contract employee	15	10.1
30+ years	2	1.4	Consultant	28	18.9
Work as:			Work position:		
Mostly full time	54	36.5	Not part of management	81	54.7
Mostly part time	90	60.8	In lower management	37	25.0
Mostly over time	4	2.7	In middle management	22	14.9
Been laid off from IT job:			In senior management	8	5.4
Yes	7	4.7			
No	141	95.3			

Table 30.2:   Organizational IT Issues in Russia

Organizational IT Issues	Rank	Mean Rating*	Std. Deviation
IT reliability and efficiency	1	1.90	0.917
Security and privacy	2	2.05	1.087
Revenue-generating IT innovations	3	2.09	0.971
Business productivity and cost reduction	3	2.09	0.936
Business agility and speed to market	5	2.11	0.877
Project management	6	2.14	1.060
IT strategic planning	7	2.15	0.928
Knowledge management	8	2.17	0.958
Enterprise architecture	9	2.22	1.042
Alignment between IT and business	10	2.28	0.896
Business process reengineering	11	2.32	1.018
Attracting and retaining IT professionals	12	2.34	1.041
IT service management (e.g., ITIL)	13	2.41	1.068
Continuity planning and disaster recovery	14	2.55	1.058
Globalization	15	2.57	1.107
IT cost reduction	16	2.61	1.134
Outsourcing	17	2.78	1.085
BYOD	18	2.94	1.108

*Rating scale ranges from 1 to 5: 1 as most important and 5 as no importance.

Table 30.3: Technology and Infrastructure Issues in Russia

IT-Related Issues	Rank	Mean Rating*	Std. Deviation
Business intelligence/analytics	1	1.86	0.940
Business process management systems	2	2.11	1.040
Enterprise application integration	3	2.14	1.043
Big data systems	4	2.15	1.078
Enterprise resource planning (ERP) systems	5	2.21	0.957
Mobile and wireless applications	6	2.27	1.092
Data mining	7	2.30	1.123
Customer relationship management (CRM) systems	8	2.34	1.027
Cloud computing	9	2.37	1.052
Virtualization (desktop or server)	10	2.39	1.092
Collaborative and workflow tools	11	2.43	1.057
Mobile apps development	12	2.44	1.213
Service-oriented architecture (SOA)	13	2.49	1.128
Software as a service	14	2.53	1.078
Networks/telecommunications	15	2.60	1.153
Social networking/media	16	2.65	1.148

*Rating scale ranges from 1 to 5: 1 as most important and 5 as no importance.

is one of the older issues on the list. Social networking/media appearing at the bottom of the list is not surprising and in fact is the same as the US ranking. What is much more noteworthy is networks/telecommunications appearing in the second-to-last position when it is in the top position for the US. Further research will need to address the reasons for this vast disparity (see Table 30.3).

30.7 Individual IT Employee Issues

Mean ratings for job satisfaction were high but not as high as in the US. There is a moderate amount of perceived work overload and low amount of work–home conflict. Likewise, strain due to work exhaustion is similar to levels reported in other countries. These results are not surprising given that the work being done in the different countries is basically the same. Professional self-efficacy was markedly lower than in the US and other countries and may reflect lower levels of morale and perceived contribution to the firm due to economic conditions. Surprisingly, there was no high perception of job insecurity, perhaps because Russian firms do attempt to retain local talent because highly qualified workers are likely to emigrate to other

more prosperous countries. What is not surprising for Russia is the higher turnover intention. Most IT workers did not think they would still be at the same firm in 5 years' time. Similarly, intention to potentially leave the IT field altogether was higher than in other countries (see Table 30.4).

Table 30.4: Individual IT Employee Issues in Russia

Individual Issues	Mean Rating*	Std. Deviation
Job satisfaction		
In general, I like working here	1.88	0.976
All in all, I am satisfied with my current job	2.05	0.935
In general, I don't like my current job	3.82	1.129
Perceived work overload		
I feel that the number of requests, problems or complaints that I deal with at work is more than expected	3.33	1.166
I feel that the amount of work I do interferes with how well it is done	3.17	1.184
I feel busy or rushed at work	2.03	0.971
I feel pressured at work	3.22	1.134
Work–home conflict		
There is a blurring of boundaries between my job and my home life	3.31	1.187
My work-related responsibilities create conflicts with my home responsibilities	3.34	1.252
I do not get everything done at home because I find myself completing job-related work	3.48	1.246
Work exhaustion/strain		
I feel drained from activities at work	3.24	1.213
I feel tired from my work activities	2.93	1.203
Working all day is a strain for me	2.99	1.185
I feel burned out from my work activities	3.67	1.067
Professional self-efficacy		
I feel I'm making an effective contribution to what this organization does	2.60	1.099
In my opinion, I do a good job	2.00	0.860
I have accomplished many worthwhile things in this job	2.64	1.066
At my work, I feel confident that I am effective at getting things done	2.28	0.949
Job insecurity		
I am worried that future technology advancements may pose a threat to my job	3.85	0.989
I believe that other people may be able to perform my work activities	2.71	1.118

(*Continued*)

Table 30.4: (*Continued*)

Individual Issues	Mean Rating*	Std. Deviation
I am concerned that my job may be eliminated soon	3.85	1.143
I am concerned that my job may be outsourced soon	3.87	0.946
Turnover intention		
I will be with this organization 1 year from now	2.48	1.087
I will take steps during the next year to secure a job at a different organization	3.22	1.189
I will be with this organization 5 years from now	3.13	1.186
Turnover intention — IT profession		
I will be working in the IT field 1 year from now	2.24	1.178
I will take steps during the next year to secure a job outside the IT field	3.53	1.234
I will be working in the IT field 5 years from now	2.33	1.206

*Rating scale ranges from 1 to 5: 1 as strongly agree and 5 as strongly disagree.

30.8   Conclusion

The top-organizational IT issues in Russia were IT reliability and efficiency, security and privacy, and revenue-generating IT innovations. While the first two are similar to Western countries, the third is particular to the Russian context. The top technology issues included business intelligence/analytics, business process management (BPM) systems, and enterprise application integration. Surprisingly, networks/telecommunications was at the bottom of the list.

While levels of job satisfaction, perceived work overload, work–home conflict, and work exhaustion/strain were similar to those reported in other countries, professional self-efficacy was markedly lower as was turnover intention due to the likelihood of finding better work in other countries. Russian IT workers face a unique combination of issues and concerns reflecting both the commonality of the IT occupation and the severity of the economic situation in Russia that is not seen in Western countries. These findings serve to underscore the importance of the World IT Project.

References

Associated Press (1997). Gates Urges Russia to Stop Software Piracy. *New York Times*, October 11, 1997.

Candau, M. (2018). Russia's Economy: Still Dependent on Oil. Euractiv. Retrieved from https://www.euractiv.com/section/energy-environment/news/russias-economy-still-dependent-on-oil/.

East-West (2016). 2015 in Review: Russia's Software Export Kept Growing While Authorities Developed Import Substitution Strategy. *East-West Digital News*, January 26, 2016. Retrieved from http://www.ewdn.com/2016/01/26/2015-in-review-the-it-service-market-kept-growing-while-software-publishers-expanded-internationally/.

Excif (n.d.). Russian Federation — Return to Growth. Excif.de. Retrieved from https://excif.de/en/portfolio_page/russland/.

EY (2015). Restrictions on Foreign Software for State Procurements. Ernst & Young Tax Alert. Retrieved from https://web.archive.org/web/20171117003215/ http://www.ey.com/Publication/vwLUAssets/EY-Tax-Alert-06-July-2015-ENG/$FILE/EY-Tax-Alert-06-July-2015-ENG.pdf.

Faulconbridge, G. Russian Stocks Shed Over $1 trillion in crisis. Reuters. Retrieved from https://www.reuters.com/article/us-markets-russia-trillion/russian-stocks-shed-over-1-trillion-in-crisis-idUSTRE4AC5M020081113.

Johnston, M. (2018). The Post-Soviet Union Russian Economy. Investopedia. Retrieved from https://www.investopedia.com/articles/investing/012116/russian-economy-collapse-soviet-union.asp.

Palvia, P., Jacks, T., Ghosh, J., Licker, P., Romm-Livermore, C., Serenko, A., & Turan, A. H. (2017). The World IT Project: History, trials, tribulations, lessons, and recommendations. *Communications of the Association for Information Systems*, 41(18), 389–413.

Palvia, P., Ghosh, J., Jacks, T., Serenko, A., & Turan, A. (2018). Trekking the globe with the World IT Project. *Journal of Information Technology Case and Application Research*, 20(1), 3–8.

Rutland, P. (2014). The Impact of Sanctions on Russia. *Russian Analytical Digest*, 157(17), 2–7.

Standish Group (2015). CHAOS Report. Retrieved from https://www.standishgroup.com/sample_research_files/CHAOSReport2015-Final.pdf.

Chapter 31

Information Technology Issues
in South Africa

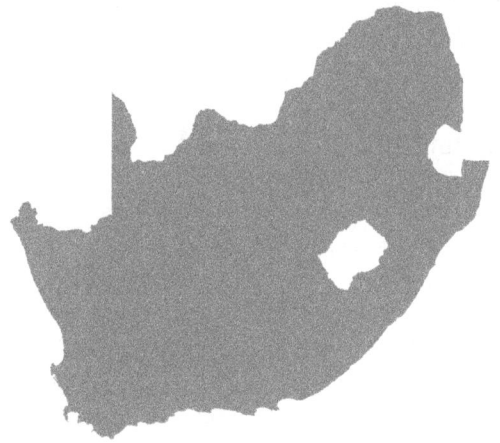

Kennedy Njenga[*,||], Brenda Scholtz[†,**], Jean Paul van Belle[‡,††],
and Alexander Serenko[§,¶,‡‡]

*University of Johannesburg, Johannesburg, South Africa
†Nelson Mandela University, Summerstrand,
Port Elizabeth, South Africa
‡University of Cape Town, Rondebosch,
Cape Town, South Africa
§University of Toronto, Toronto, Canada, ON M5S 1A1
¶University of Ontario Institute of Technology, Oshawa, Canada
||knjenga@uj.ac.za
**brenda.scholtz@mandela.ac.za
††jean-paul.vanbelle@uct.ac.za
‡‡a.serenko@utoronto.ca

Summary

Many skilled information technology (IT) professionals working in South Africa understand that they work in a fast-changing environment. The emergent nature of IT, coupled with growing demands for new sets of IT skills in South Africa, has fostered a miscellany of employment issues in the IT profession at a higher percentage relative to other professions. Arguably, the antecedents that shape these issues in the profession are critical for managers in the industry. This chapter presents findings that are based on an investigation carried out in South Africa regarding the miscellany of organizational, technological, and individual issues that IT professionals face. Having a reliable and efficient IT infrastructure was by far the most highly ranked organizational issue. The most important technology issue as perceived by respondents was networking and telecommunications. Insights to these findings are discussed in this chapter.

31.1   Introduction

South Africa's demand for skilled professionals in information technology (IT) has been growing, but many in the profession understand that they work in a fast-changing environment characterized by a miscellany of issues that are increasingly influencing the demand and supply of the IT professional. The South African government's renewed focus on skills generation, technological advancements, and the designating of the IT profession as a 'critical skill' are crucial examples of issues directly affecting the profession. The country has experienced a notable exit of workers out of the IT profession due to skills made obsolete by technological advancements, and the attractiveness of the profession to more advanced economies (Lambert *et al.*, 2001). In order to stabilize IT employment in South Africa, many organizations are pursuing strategies such as engineering *on-the-job* transitions for IT employees who work with them and presenting IT employees the opportunity to extend skills into in-demand areas within the organization rather than allowing them to leave. Not all organizations are doing this, and many in the IT profession are progressively facing career uncertainty. Understanding the relationship between technology, employment and skills has been critical in the current dispensation because of the impact this understanding has to a country's economic wellbeing (Silva and Lima, 2017). It is noted that a majority of studies that relate to technology and employment issues have been carried out in more technologically advanced countries outside of South Africa and these studies usually carry views that may be biased towards their own countries. Studies conducted in South Africa regarding employment issues have targeted a diverse range of industries and professions (Stoermer *et al.*, 2017). Investigations focusing on the miscellany of issues that specifically face the IT profession in South Africa were not identified at the time of this research undertaking. South Africa, as a developing country, may not face the same challenges developed countries do, and research relating to this country's context is therefore timely.

The research thereof that has been carried out presents the unique contexts that address the myriad of issues facing the IT profession in South Africa and forms part of a larger research project, namely, the World IT Project (Palvia *et al.*, 2017; Palvia *et al.*, 2018). The World IT Project's goal is to present a global perspective of IT issues and it is one of the largest projects undertaken within the IT field. While the wider project's purpose is to examine the role of personal and organizational factors in a global IT work environment, this chapter presents issues that specifically address the context in South Africa. The following pertinent questions are answered in this chapter:

- What are the miscellany of IT employment issues that face the IT professional in South Africa?
- How do these issues differ from those reported in the literature?

The answers to the above questions and reported in this chapter are based on data obtained from a survey of 301 IT professionals in South Africa and are reported in the penultimate sections. The section that follows presents the background of South Africa's employment contexts. A section on the research methodology adopted in this study follows, and then an analysis of the findings is provided. The sections thereafter provide some discussion and conclusions.

31.2   Country Background

South Africa is a country on the southernmost tip of the African continent, characterized by a unique social stratification of economic inequality among its populace in the labor market (Horwitz, 2017). By the end of 2017, South Africa's gross domestic product (GDP) was estimated at US$355 Billion with an annual economic growth of 1.3%. It is one of the most industrialized countries in Africa. There are several significant scientific and technological developments that have originated in South Africa, such as the first human-to-human heart transplant performed at Groote Schuur Hospital on December 1967 by cardiac surgeon Christiaan Barnard. Another South African by the name of Allan McLeod Cormack also pioneered the X-ray computed tomography, while Aaron Klug developed crystallographic electron microscopy techniques. Notwithstanding the above, South Africa has an over-represented but semi-skilled labor market coupled with a shortage of high level skills in the IT sector (Horwitz, 2017).

31.3   Information Technology in South Africa

One of the possible reasons for the IT skills shortage in South Africa is the growth in the IT service sector with IT service exports as a percentage of total exports increasing from 12.8% to 16.9% in 2016 (World Dev, 2017). In addition, there has been a notable decline in the number of skilled IT professionals in this growing sector, from 2001 till 2016 (World Dev, 2017). Another issue influencing the skills shortage is overall job satisfaction, which has been reported as a critical contributor to decline in the IT profession (Joo and Park, 2010). Studies suggest that the IT professional in South Africa, particularly those in lower levels (e.g., helpdesk tech support, IT technicians), do not feel adequately compensated nor respected for the work they perform (Jackson, 2014).

The issues affecting IT employees' skills and performance have become popular research topics due to high levels of career abandonment in this industry (Colomo-Palacios *et al.*, 2014; Palvia *et al.*, 2017). Knowledge of these issues is important for managing IT staff and for organizational survival and competitiveness (Jiang and Klein, 2002). The rapidly changing technology landscape also requires organizations to have employees who are conversant with the latest technologies, particularly with those that can support readiness for the fourth industrial age. For those moving into the IT profession, understanding the critical IT issues being faced is important in helping to position themselves effectively in the profession. Employers are therefore demanding qualified IT specialists with high expertise in networking, designing, programming, and deploying pervasive computing systems and communication architectures for business sustainability (Lotriet, Matthee and Alexander, 2010). These demands for skills also put pressure on higher education institutions to provide qualified IT workers.

New technologies and new ways of carrying out business and the pressures that accompany these are forcing more IT professionals who cannot cope out of the profession (Altbach *et al.*, 2010). Lotriet *et al.* (2010) raised the concern that IT employers are still constantly faced with the task of recruiting skilled graduates into their organizations with up-to-date skills. The changing technology landscape has made those with obsolete skills irrelevant to the demands of organizations. Lotriet *et al.* (2010) have thus proposed a rethink of how both universities and IT employers can work together towards solving the skills' shortage issue that South Africa faces or reskilling those in employment to avoid them being made redundant. In this regard, South African universities face challenges of generating, accessing and disseminating information, thus making it even more difficult for such universities to respond to business challenges (Popescu, 2015).

31.4 Methodology

This research was positivistic and aimed at explaining the miscellany of issues facing the IT profession in South Africa. Quantitative data were collected from IT employees by means of a survey using a cross-sectional time-frame. The sampling approach used stratified, convenience and purposive sampling. While the broader World IT Project has a global application, this work is primarily centered in South Africa. Three South African-based researchers were each allocated a geographic region of the country based on the location of their home university to collect data. A standardized questionnaire (5-point Likert scale) designed and prescribed by the World IT Project was used for purposes of allowing international comparability (Palvia *et al.*, 2017).

A total of 310 responses (205 online responses and 105 paper-based format) were obtained. Nine responses were excluded because of insufficient quality. Respondents were generally young (38% aged 30–39 years and a further 32% aged 21–29 years), mostly male (72%) and well-educated (48% has a bachelor's degree, 16% a master's or Ph.D., with only 11% having a high-school diploma or less). Regarding sample representativeness, this corresponds well with the South African IT industry's overall demographic profile, albeit with a slight bias towards better educated employees. Respondents were fairly experienced with one-third having between 10 and 19 years of IT work experience. Most work for large organizations, i.e., those with more than 1000 employees. This represents a bias towards the larger organizations in line with the intended focus of the world IT survey. Table 31.1 shows some descriptive statistics of the IT professionals who took part in the study.

Table 31.1:   Descriptive Statistics

Characteristics	N	%	Characteristics	N	%
Education:			Years of work experience:		
High school or less	32	10.8	0–4 years	62	20.9
Associate degree	74	24.9	5–9 years	68	22.9
Bachelor's degree	141	47.5	10–19 years	99	33.3
Master's degree	37	12.4	20–29 years	46	15.5
Ph.D.	13	4.4	30+ years	22	7.4
Years of IT Experience:			Organizational location:		
0–4 years	83	27.7	IT department employee	216	72.0
5–9 years	72	24.0	IT worker in non-IT department	14	4.7
10–19 years	95	31.6	Contract employee	25	8.3
20–29 years	36	12.0	Consultant	39	13.0
30+ years	14	4.7	Vendor employee	6	2.0
Work as:			Work position:		
Mostly full time	282	94.0	Not part of management	154	51.9
Mostly part time	6	2.0	In lower management	41	13.7
Mostly over time	12	4.0	In middle management	57	19.2
Been laid off from IT job:			In senior management	45	15.2
Yes	20	6.7			
No	279	93.3			

31.5   Organizational IT Issues

Respondents were asked to rate 18 organizational issues arising from the
organizations' IT engagement according to their perceived relative importance. Table 31.2 presents the summary of results of organizational issues
sorted on the level of importance, starting by the most important issue and
ending with the least important. IT reliability and efficiency was the clear
top priority issue that faced organizations. IT/business alignment was second while security/privacy concerns emerged as the third most important
issue of concern to organizations. Interestingly, security and privacy has
remained the second most important issue according to the findings of the
Society for Information Management's 38th Anniversary IT Trends Study
(Kappelman et al., 2017).

Further, despite the acclaimed shortage of IT skills in the country,
attracting and retaining IT staff is a prominent issue but ranks sixth. At
the opposite end of the spectrum, globalization and outsourcing are not
seen as major issues and the often-hyped BYOD issue is not even on the
radar. In the surveys conducted by the Society for Information Management (SIM), the issues of alignment, security and privacy were also rated

Table 31.2:   Organizational IT Issues in South Africa

Organizational IT Issues	Ranking	Mean Rating*	Std. Deviation
IT reliability and efficiency	1	1.49	0.61
Alignment between IT and business	2	1.61	0.64
Security and privacy	3	1.67	0.76
IT strategic planning	4	1.72	0.74
Knowledge management	5	1.78	0.72
Attracting and retaining IT professionals	6	1.80	0.74
Continuity planning and disaster recovery	7	1.85	0.85
Business agility and speed to market	8	1.86	0.76
Project management	9	1.89	0.80
Revenue-generating IT Innovation	10	1.93	0.79
Business productivity and cost reduction	11	1.96	0.78
Enterprise architecture	12	2.03	0.83
IT Service management	13	2.05	0.90
Business process reengineering	14	2.15	0.88
IT cost reduction	15	2.22	0.89
Globalization	16	2.39	0.91
Outsourcing	17	3.15	0.99
Bring your own device (BYOD)	18	3.30	1.16

*Rating scale ranges from 1 to 5: 1 as most important and 5 as no importance.

as very important — usually featuring in the top three issues both in the US as well as internationally (Europe, China). However, IT reliability and efficiency, the most important issue in South Africa, did not even make the top five issues in any of the other international surveys.

By contrast, Johnston *et al.* (2007), who used a different list of issues, found in their admittedly small sample that building a responsive IT infrastructure, IT value management and service delivery were three of the top four issues; these can arguably be said to align with what we called IT reliability and efficiency in our survey. Similarly, most of the issues rated as less important in South Africa were also not ranked highly overseas. We carried out an exploratory principal component analysis on these issues to possibly group higher-order clusters. Four larger factors emerged. The factor accounting for the biggest variance in the data groups is made up of four items: IT/business alignment, agility, innovation, and IT skills attraction/retention. They can thus be conceptualized as issues relating to the organization reacting to or engaging with its market environment. The factor with the second-largest explained variance refers to distinct internal management abilities or concerns, loading the issues of disaster planning, project management, and knowledge management. The third factor loads security/privacy with reliability/efficiency and service management, which appear to group the operational business concerns (two of which feature in the top-three concerns). A fourth factor loads IT cost reduction, productivity as well as IT strategy planning (with business process re-design and enterprise architecture loading more than 0.40). These seem like more IT-internal specific concerns, although a more natural descriptor is not so evident. A fifth factor, low in importance, combines globalization with outsourcing and can thus be seen as the international concerns.

The implications for the South African IS curriculum appear to be relatively minor. The issues are generally all covered quite well by senior or capstone IS management courses at most universities. Of note is that the BYOD, globalization and outsourcing issues appear to be less important in the industry. The relatively high ranking of continuity/disaster recovery planning (seventh) belies its sometimes low visibility in many university curricula. The largest surprise, perhaps, is that the rather mundane issue of ensuring a reliable and efficient IT infrastructure is still the foremost concern as seen by IT practitioners, ranked quite distinctly above all other, sometimes much more glamorous IT concerns.

31.6   Technology and Infrastructure Issues

Respondents were also asked to rate 16 contemporary technology issues in terms of their importance using a 5-point Likert scale (most importance to no importance). Table 31.3 also presents the summary of results of technology issues sorted on the level of importance, starting with the most important issue and ending with the least important one. Business intelligence/analytics was seen as the most important technology issue. The often undervalued network/telecoms infrastructure is seen as the second most important technical issue.

Mobile/wireless applications, partly related to the telecoms issue, was considered the third most important issue while, enterprise application integration (EAI) followed this issue. Noteworthy is that EAI is more visible in academic curricula. Interestingly, although software-as-a-service (SaaS) was ranked fifth, cloud computing, which is technically more related to SaaS, was ranked as the ninth most important issue. Less than two-thirds of respondents ranked service oriented architecture (SOA), data mining and social networking systems/media as the least important technology issues. When we compare the international SIM studies with the above findings, business intelligence (BI) was also ranked as a top concern. However, our second most important networking/telecoms issue was ranked only eighth

Table 31.3: Technology and Infrastructure Issues in South Africa

Technology Issues	Ranking	Mean Rating*	Std. Deviation
Business intelligence/analytics	1	1.81	0.78
Networks/telecommunications	2	1.94	0.87
Mobile and wireless applications	3	1.99	0.86
Enterprise application integration	4	1.99	0.82
Software as a service	5	2.08	0.86
Business process management systems	6	2.09	0.89
Virtualization (desktop or server)	7	2.10	0.92
Collaboration and workflow tools	8	2.12	0.85
Cloud computing	9	2.19	0.85
Big data systems	10	2.22	0.91
Customer relationship management (CRM)	11	2.23	1.01
Mobile apps development	12	2.23	1.01
ERP systems	13	2.25	0.98
Data mining	14	2.25	0.91
Service-oriented architecture	15	2.25	0.88
SNS/Media	16	2.62	1.06

*Rating scale ranges from 1 to 5: 1 as most important and 5 as no importance.

and ninth in the last two surveys, respectively, and not evident at all in the prior SIM surveys. Only one of our top six concerns appears in the top six concerns in any of the other SIM surveys, namely, EAI (fourth in South Africa) was the number five issue in the 2009 international SIM survey. That year, virtualization (our number six), was the second most important issue in the same SIM survey. Perhaps this marks South Africa as lagging in some respects or issues, although the issues of mobile applications, SaaS and cloud computing are definitely recent global developments.

We also conducted an exploratory principal component analysis to see which of these issues could be bundled together based on their variability. Interestingly, the factor accounting for most of the explained variance loads business process management (BPM), ENTERPRISE RESOURCE PLAN-NING (ERP), customer relationship management (CRM) with EAI and workflow/collaboration. These are indeed logically and conceptually linked technology issues, which are often grouped together in a single academic course; perhaps those teaching this course can appeal to this fact as support for the importance of their course in the undergraduate major curriculum. The second largest (in terms of variance) factor groups mobile/wireless, app development, networking and social networking/media. The third factor groups not only BI, Big Data and data mining (which are often taught together), but somehow also SOA (this may just be an accidental data artefact). Finally, virtualization and SaaS load on the same factor, although cloud computing loaded separately on its own factor.

31.7 Individual IT Employee Issues

Table 31.4 summarizes the individual IT employee issues reported in South Africa. The responses indicate that the group of workers face significant work pressure with many stating that they experience burnout as well. Generally, those liking their work environment rated low.

The individual issues affecting IT staff were analyzed according to seven broad themes. These are: job satisfaction; work pressure; work–life balance; workload and burnout; sense of accomplishment; threat to one's job and career plans. According to the scale used, any mean score below 3 shows an agreement with the stated item and above 3 shows disagreement. With regards to job satisfaction, respondents are generally satisfied with their jobs since all three items in this theme had a mean rating of less than 3. The standard deviation is also low, thus indicating general agreement on this issue.

With regards to the theme of work pressure, for the first two items, the mean ratings were very close to the mid-point. Thus respondents did not seem to have significant issues related to the number of requests/problems/complaints dealt with as compared with what is expected and how the amount of work interfered with the quality of their work. For the third and fourth items related to work pressure, the respondents generally felt that they were busy/rushed and pressured at work.

All items in the work–life balance theme had mean ratings of above 3; therefore, it can be deduced that respondents disagreed that there is a problem with their work–life balance. Similarly, respondents disagreed with all four items in the workload and burnout theme. This is unexpected, since several studies cite high workloads and burnout among IT staff.

In the sense of accomplishment theme, all four items were below 2, showing that respondents strongly agreed. It can therefore be said that South African IT staff generally feel that they are making a contribution to the organization, feel that they are doing a good job and are confident about their effectiveness on the job.

One item in the threat to one's job theme had validity issues and could not load. This is the item stating "I believe that other people may be able to perform my work activities". For this reason, the item is not reported on and is marked as not applicable (N/A). The other three items in this theme were

Table 31.4: Individual IT Employee Issues in South Africa

Individual Issues	Mean Rating*	Std. Deviation
Job satisfaction		
In general, I like working here	1.98	0.84
All in all, I am satisfied with my current job	2.21	0.91
In general, I don't like my current job	2.03	0.97
Work pressure		
I feel that the number of requests, problems or complaints that I deal with at work is more than expected	3.14	1.10
I feel that the amount of work I do interferes with how well it is done	3.00	1.16
I feel busy or rushed at work	2.94	1.13
I feel pressured at work	2.89	1.12
Work–life balance		
There is a blurring of boundaries between my job and my home life	3.18	1.18
My work-related responsibilities create conflicts with my home responsibilities	3.47	1.16
I do not get everything done at home because I find myself completing job-related work	3.45	1.15
Workload and burnout		
I feel drained from activities at work	3.12	1.19
I feel tired from my work activities	3.03	1.19
Working all day is a strain for me	3.34	1.12
I feel burned out from my work activities	3.38	1.12
Sense of accomplishment		
I feel I'm making an effective contribution to what this organization does	1.99	0.88
In my opinion, I do a good job	1.76	0.70
I have accomplished many worthwhile things in this job	1.92	0.82
At my work, I feel confident that I am effective at getting things done	1.84	0.75
Threat to one's job		
I am worried that future technology advancements may pose a threat to my job	3.73	0.94
I believe that other people may be able to perform my work activities	4.02	0.92
I am concerned that my job may be eliminated soon	4.02	0.92
I am concerned that my job may be outsourced soon	4.03	0.96

(*Continued*)

Table 31.4:   (*Continued*)

Individual Issues	Mean Rating*	Std. Deviation
Career plans		
I will be with this organization 1 year from now	3.70	1.11
I will take steps during the next year to secure a job at a different organization	3.40	1.22
I will be with this organization 5 years from now	3.03	1.13
I will be working in the IT field 1 year from now	4.18	0.92
I will take steps during the next year to secure a job outside the IT field	3.96	1.10
I will be working in the IT field 1 years from now	2.06	1.04

*Rating scale ranges from 1 to 5: 1 as strongly agree to 5 as strongly disagree.

all above 3, showing that respondents disagreed that they were concerned about job security. From this it can be inferred that South African IT staff are generally not concerned about the security or future of their jobs. One of the reasons is that IT skills are scarce globally and in South Africa; therefore, staff realize that they have more security in their jobs than in other professions.

The item "I will be working in the IT field one year from now" was rated highly ($\mu = 4.18$) indicating that the respondents disagreed with this statement and had a strong tendency towards turnover intention for leaving the industry. Respondents also disagreed with the statement "I will be with this organization one year from now", thus indicating turnover intention among the South African IT staff. Turnover intention (of both organization and industry) is therefore strong among IT staff in South Africa. The reasons for this are not clear, since workload and burnout were not rated as issues, nor was work–life balance. In addition, respondents agreed with all the job satisfaction items. One possible reason for organization turnover intention could be the high demand for IT skills, since staff could be searching for higher salaries or other opportunities within the field. The reasons for respondents agreeing to industry turnover intention are not clear however, and should be explored in future analysis.

31.8   Conclusions

The chapter has provided insights to the context of South Africa and presented from the miscellany of issues facing the IT profession, i.e., the most urgent and pressing organizational, technological, and individual issues as perceived by IT professionals and managers. The issue rated as most important by South African organizations interestingly did not match what the vendors or academics perceive as popular, urgent or current. For instance, a reliable and efficient IT infrastructure was by far the most highly ranked organizational issue. This issue is hardly ever foregrounded in contemporary MIS or ISM courses. The most important technical issues as perceived by respondents was the networking/telecommunication infrastructure and EAI (often absent from academic curricula). Conversely, issues such as globalization, outsourcing, BYOD (organizational) or SNS/Social Media and data mining (technical) were not seen as major issues. Hopefully, insights provided by these findings will lead to some interesting discussions, such as the need to plow additional resources into IT education to address the IT skills shortage. We anticipate these discussions to be of benefit to both academics and practitioners.

References

Altbach, P., Reisberg, L., & Rumbley L. (2010). *Trends in Global Higher Education: Tracking an Academic Revolution Sense.* Rotterdam: UNESCO.

Colomo-Palacios, R., Casado-Lumbreras, C., Misra, S., & Soto-Acosta, P. (2014). Career abandonment intentions among software workers. *Human Factors and Ergonomics in Manufacturing & Service Industries*, 24, 641–655 (doi:10.1002/hfm.20509).

Horwitz, F.M., & Mellahi, K. (2009). Human resource management in emerging markets. In *Human Resource Management: A Critical Approach*, D.G. Collings and G. Wood (eds.). London: Routledge, 263–295.

Jackson, T. (2014). Employment in Chinese MNEs: Appraising the Dragon's gift to sub-Saharan Africa. *Human. Resource. Management*, 53(6), 897–919.

Jiang, J. J., & Klein, G. (2002). A discrepancy model of information system personnel turnover. *Journal of Management Information Systems*, 19(2), 249–272.

Johnston, K., Muganda, N., & Theys, K. (2007). Key issues for CIOs in South Africa. *The Electronic Journal of Information Systems in Developing Countries*, 30(1), 1–11.

Joo, B., & Park, S. (2010). Career satisfaction, organizational commitment, and turnover intention. *Leadership & Organization Development Journal*, 31(6), 482–500 (doi: 10.1108/01437731011069999).

Kappelman, L., Johnson, V., Maurer, C., McLean, E., Torres, R., David, A., & Nguyen, Q. (2018). The 2017 SIM IT issues and trends study. *MISQ Exec*, 17(1), 53–88.

Lambert, E.G., Hogan, N. L., & Barton, S. M. (2001). The impact of job satisfaction on turnover intent: a test of a structural measurement model using a national sample of workers. *The Social Science Journal*, 38(2), 233–250 (doi: 10.1016/S0362-3319(01)00110-0).

Lotriet, H. H., Matthee, M. C., & Alexander, P. M. (2010). Challenges in ascertaining ICT skills requirements in South Africa. *South African Computer Journal*, 2010(46), 38–48.

Palvia, P., Jacks, T., Gosh, J. Licker, P. Romm-Livermore, C., Serenko, A., & Turan, A. H. (2017). The World IT Project: History, trials, tribulations, lessons, and recommendations. *Communications of the Association for Information Systems*, 41(18), 389–413.

Palvia, P., Ghosh, J., Jacks, T., Serenko, A., & Turan, A. (2018). Trekking the globe with the World IT Project. *Journal of Information Technology Case and Application Research*, 20(1), 3–8.

Popescu, F. (2015). South African globalization strategies and higher education. *Procedia — Social and Behavioral Sciences*, 209, 411–418.

Silva, H. C., & Lima, F. (2017) Technology, employment and skills: A look into job duration. *Research Policy*, 46, 1519–1530.

Stoermer, S., Hitotsuyanagi-Hansel, A., & Froese, F. J. (2017). Racial harassment and job satisfaction in South Africa: the moderating effects of career orientations and managerial rank. *The International Journal of Human Resource Management* (doi: 10.1080/09585192.2016.1278254).

World Development Indicators Database. (2017). Aavailable at http://databank.w orldbank.org/data/reports.aspx?source=world-development-indicators#.

Chapter 32

Information Technology Issues
in South Korea

Kyootai Lee[*,§], Youngkyun Kim[†,¶], and Prashant Palvia[‡,∥]

*Sogang University, Seoul, South Korea
†Incheon National University,
Incheon, South Korea
‡University of North Carolina at Greensboro,
Greensboro, NC, USA
§kyootai@sogang.ac.kr
¶kkart1@incheon.ac.kr
∥pcpalvia@uncg.edu

Summary

The central theme that arises from the study of information technology/information systems (IT/IS) industries in South Korea is that IT/IS service companies have strategically placed a greater emphasis on domestic markets and the clients engaged in their conglomerates. Hence, they have considered reliable support and business continuance as key organizational issues. In addition, these firms have acutely reacted to the current social issues pertaining to IT/IS in South Korea. These are reflected in the fact that mobile apps, security and privacy, and networks are highly ranked among the organizational and technological issues. Employees in the IT/IS service industry sectors are generally satisfied with their current jobs, although they experience some levels of work pressure and burnout.

32.1   Introduction

Information technology (IT) services were initially developed to support information systems (IS) functions in traditional business operations of firms in South Korea. Around 1980s and 1990s, independent IT service firms were spun off from their parent companies. IT service industry firms experienced a major boom in the late 1990s, as the needs for up-to-dated IS rapidly rose across industries. As firms did not possess capabilities to develop and implement systems, they made strategic alliances with global firms. For instance, Samsung Data Systems (SDS) made alliances with Microsoft and HP in 1997 and 1998, respectively. LG started IS services with EDS in 1995. SK also had a relationship with TELUS. They also played agent roles to implement global IT service firms' packages such as Oracle, SAP, and HP. Then, the impetus of IT services moved to the public sectors around 2000. The South Korea government ministries and agencies initiated a vast number of projects, and the IT/IS service companies actively participated in e-government projects. Now, South Korea is one of the countries with the most ambitious targets for Internet usage across the population, where citizens commonly obtain certified copies of real estate registration, birth, residency, and other certificates right in their homes through government websites (Lee and Joshi, 2015). Some of the IT service packages have started to be exported to developing and/or under-developing countries as part of official development assistance (ODA) programs.

However, IT service firms are facing major challenges due to stagnated demand in the domestic markets and technological changes such as Internet of things (IOT) and artificial intelligence (AI). IDC reported that the domestic IT service sector has grown at the rate of 1.5%, and such a growth pattern will be sustained until 2021. IT service firms are overcoming such industry downturns by actively seeking new markets by collaborating with firms in traditional businesses. That is, although they had focused on systems integrations (SI) based on users' and/or customers' needs, they are now offering new IT-based business concepts to their potential partners. As such, many industries are experiencing software and platform-based transformation in South Korea. For instance, IT-based companies are closely working with banks to support and implement fin-tech and/or blockchain-based operations.

32.2 Country Background and History

Korea has a long history, as several ancient dynasties have been established and have flourished in the Korean peninsula for about couple of thousands of years. It became a Japanese colony in 1910, but earned independence in 1945 right after the Second World War. Then, Korea was divided in two, South and North Korea. The Korean Civil War occurred in 1950 and ceased in 1952. South Korea was a historical recipient of ODA (official development assistance) from OECD about 50 years ago.

South Korea has shown remarkable rise from being one of the poorest countries in the world to a developed, high-income country within just a generation since the 1960s. This economic miracle, commonly called as the Miracle on the Han River, brought South Korea to the ranks of elite countries in the OECD and the G-20. However, in the 1997 Asian financial crisis, the South Korean economy suffered a liquidity crisis and relied on the bailout by the IMF that restructured and modernized the South Korean economy, including the national development of the ICT industry. The country recovered from the crisis within 5 years, which is also known as another miracle by the economists.

South Korea was one of the few developed countries that was able to escape from a recession during the global financial crisis, and its economic growth rate reached 6.2% in 2010, a sharp recovery from the economic growth rates of 2.3% in 2008 and 0.2% in 2009 when the global financial crisis hit. South Korea has the fourth largest economy in Asia and the 11th largest in the world as of 2016. It has market structures open to the world economy. South Korea has made free trade agreements with 52 countries, to name a few, the United States of America in 2007, the European Union in 2009, Canada in 2014, and China in 2014. South Korea's economy is dominated by family-owned conglomerates called *chaebols*, which may influence the development of IS industries in a quite unique pattern, to be discussed in Section 32.3.

32.3   Information Technology in South Korea

South Korea has emerged from a low technology base to become one of the frontline nations in advanced IT. Recognizing this emergence, the International Telecommunication Union (ITU) has ranked South Korea as the top country in the world in terms of an IT development index (ITU, 2013). According to Akamai's 2014 State of the Internet Report, the Internet speed in South Korea is on average 24.6 Mbps, which is the fastest in the world (Akamai, 2014). The status of South Korea as one of the advanced IT-based countries has been further strengthened as the IT usage trend has migrated to mobile devices from desktop devices.

Unlike the rapid development of IT hardware and diffusion of IT infrastructure, software and IS development are still relatively weak points in the industry sectors. As mentioned at the outset, although government and enterprise information systems have been actively adopted and diffused in South Korea, these systems were not developed in South Korea. Instead, system integration (SI) companies focused on customizing the systems based on the platforms developed mainly by foreign companies such as Microsoft, SAP, and Oracle.

One of the major reasons for the above situation is that the IT/IS service firms were established and developed to support their conglomerates. Also, these firms have been actively leveraged to transfer profits within their conglomerate by (a group of) owners. That is, an IT/IS service firm has obtained revenues from other sister companies within its conglomerate. Based on enormous profits, the owners of the IT/IS service firms, who are generally a group of family members, have increased their assets. A substantial portion of the revenues of the larger IT service companies are still based on projects with other companies under the same umbrella. As system integration firms have grown at relatively low levels of competition but based on their kinships with parent firms, it is not surprising that South Korea does not have IT services and/or software firms with global leadership.

Another reason may be that employment in the IT and other industry sectors is not considered to be a very attractive career path for young people. South Korean firms and even government and its agencies have not properly managed their project schedules and expenses. Hence, software developers have experienced work–life imbalance. Thus, they are likely to perceive that their efforts have not been properly rewarded. Top talented students wish to become medical doctors rather than software programmers.

32.4 Methodology

The first author who was fluent in Korean and English translated the World IT Project (Palvia *et al.*, 2017; Palvia *et al.*, 2018) questionnaire from English to Korean, and then another author with proficiency in Korean and English back-translated the Korean version to English. A third expert confirmed that there were no semantic differences between the two versions (Brislin, 1980).

The survey was conducted in South Korea with the assistance of a regional Chamber of Commerce and Industry in one of the metropolitan cities in 2015 and 2016. The local research team leveraged the chamber because it is not easy for researchers to collect responses from IT workers. Very low response rate is quite common in South Korea. One of the primary investigators directly approached senior managers in the chamber based on his personal relationships. Based on their approval, official letters of introduction and reference were sent, in which the research objectives and the data collection plan were clearly explained. Then, the Chamber of Commerce and Industry sent an official participation request to member firms. The request outlined research objectives and procedures. The Chamber of Commerce and Industry sent a web-based survey link to the HR departments of the firms. The HR departments then sent a survey link via email to their respective employees. Employees were assured that answers would be anonymous and the researchers (not the firms) would directly collect and handle the responses. Web-based systems generated data in an Excel-based format.

A total of 338 responses were initially received; 37 were deleted for being incomplete and unreliable. For instance, responses were considered incomplete if more than 30% of the questions were not answered. Thus, 301 responses were used in the analyses. Table 32.1 shows the descriptive statistics of the IT professionals who took part in the study. Note that the totals may be less than 301 due to some missing data.

Table 32.1:   Descriptive Statistics

Characteristics	N	%	Characteristics	N	%
Education:			Years of work experience:		
High school or less	3	1.0	0–4 years	36	12.0
Associate degree	29	9.7	5–9 years	89	29.8
Bachelor's degree	177	59.4	10–19 years	133	44.5
Master's degree	76	25.5	20–29 years	41	13.7
Ph.D.	13	4.4	30+ years	0	0
Years of IT experience:			Organizational location:		
0–4 years	57	18.9	IT department employee	185	61.7
5–9 years	98	32.6	IT worker in non-IT department	61	20.3
10–19 years	124	41.2	Contract employee	13	4.3
20–29 years	22	7.3	Consultant	37	12.3
30+ years	0	0	Vendor employee	4	1.3
Work as:			Work position:		
Mostly full time	257	87.7	Not part of management	84	28.1
Mostly part time	36	12.3	In lower management	59	19.7
Mostly over time			In middle management	127	42.5
Been laid off from IT job:			In senior management	29	9.7
Yes	9	3.0			
No	292	97.0			

32.5   Organizational IT Issues

Table 32.2 shows the organizational IT issues from the respondents from South Korea ranked in order of importance. Only two items in the top five, namely, security and privacy and alignment between IT and business, match the items in the top five of the 2017 SIM IT Key Issues and Trends study (Kappelman *et al.*, 2018). Further, four items in the top 10 match the items in the top 10 of the 2017 SIM issues list.

The highest-ranked item is security and privacy which also appears first in the 2017 SIM list. This issue is faced by most companies across the world and may not be specific to South Korea. Particularly, South Korean companies such as banks and securities have recently experienced customer information leaks several times. Many South Koreans have also experienced voice phishing. Thus, South Korean government requires private and public organizations to sustain the highest level of privacy and security when they use personal information. Hence, it is not odd for South Korean IT firms to consider security and privacy as the most important issue.

The second highest-ranked issue is IT reliability and efficiency, followed by continuity planning and disaster recovery, business agility and speed to

Table 32.2: Organizational IT Issue in South Korea

Organizational IT Issues	Rank	Mean Rating*	Std. Deviation
Security and privacy	1	1.80	0.67
IT reliability and efficiency	2	1.87	0.62
Continuity planning and disaster recovery	3	1.90	0.73
Business agility and speed to market	4	1.91	0.64
Business productivity and cost reduction	5	1.92	0.61
Alignment between IT and business	6	1.96	0.63
Revenue-generating IT innovations	7	1.97	0.62
Project management	8	1.98	0.59
IT service management (e.g., ITIL)	9	1.98	0.65
IT strategic planning	10	1.98	0.68
Attracting and retaining IT professionals	11	2.05	0.70
Enterprise architecture	12	2.06	0.69
Knowledge management	13	2.09	0.65
Globalization	14	2.12	0.75
Business process reengineering	15	2.16	0.69
IT cost reduction	16	2.25	0.82
Outsourcing	17	2.31	0.85
Bring your own computing device (BYOD)	18	2.35	0.97

*Rating scale ranges from 1 to 5: 1 as most important and 5 as no importance.

market, and business productivity and cost reduction in the third, fourth, and fifth places. Their high rankings seem to indicate the South Korean context. As discussed earlier, the major purpose of IS/IT service firms is to seamlessly support business operations of their client organizations that are involved in the same conglomerate. Thus, it is not surprising that the most critical issues would be reliable and efficient operations of IS for their client companies, which should not be disrupted by natural or artificial disasters. Ministries, government agencies, and banks have been major clients in this market; these institutions tend to lay further emphasis on reliability and continuity planning and disaster recovery. Thus, IS service firms naturally follow their clients' perspectives. Finally, the critical value being appreciated by clients is the completion of IS projects within a deadline. Therefore, IT service companies are required to build capabilities to promptly adapt to customer needs and swiftly deliver their products and services.

However, business process reengineering, globalization, and IT cost reduction, to name a few, have the lowest importance. These ranks would also reflect South Korean IT/IS organizational contexts well. As discussed earlier, South Korean IT/IS service firms are less likely to have global competence. Compared to IT hardware firms that have global competence, a very few IT/IS service organizations are serving global markets and are globally well known. Still, major IT/IS firms have a relatively large portion of revenues from their parent conglomerates that are owned by a handful of family members. Thus, they are less likely to recognize the needs for globalization and cost reduction. Their major goals would rather serve their parent and/or sister companies. Such situations may not be applicable to small to medium-sized IT/IS companies that are not engaged in conglomerates.

From 2013, the South Korean government has discouraged major big IT/IS companies to participate in government IS projects that comprise a large percentage of domestic markets. This regulation aims in part to resolve the inappropriate ownership structure of conglomerates. It also has a purpose to enhance capabilities of IT/IS service firms, particularly small to medium sized companies. Therefore, big IT/IS firms started to build their own capabilities and look for overseas opportunities based on their relationships with companies in their conglomerates. For instance, an IT/IS service company has recently expanded its global businesses based on its kinship with construction and petrochemical companies. The company has participated in refinery upgrade projects that are operated by construction companies, and built IT operating systems. The company has made efforts to build competences of IT services specific to traditional industry sectors.

In summary, IT/IS services companies in South Korea have had to place a relative emphasis on domestic markets and clients engaged in their conglomerates. In addition, they have focused on information security issues as part of reactions against government regulations and accidents. However, the companies may be less likely to build their own capabilities to manage IS/IT service projects and plan overall IT strategies.

32.6   Technology and Infrastructure Issues

Table 32.3 shows the technology and infrastructure issues from South Korea ranked in order of importance. Two of the top five responses from South Korea seem to match those in the top five of the 2017 SIM IT Key Issues and Trends study (Kappelman *et al.*, 2018). Seven of the top 10 in the South Korea list match the top 10 in the US list. Overall, it can be surmised that the technology and infrastructure issues identified in South Korea generally indicate similarity to the 2017 SIM list. We should note that 'analytics/business intelligence/data mining/forecasting/big data' in the US list is separately grouped into three areas in this study: big data systems, data mining, and business intelligence/analytics.

Table 32.3: Technology and Infrastructure Issues in South Korea

IT-Related Issues	Rank	Mean Rating*	Std. Deviation
Big Data systems	1	1.91	0.78
Mobile apps development	2	2.02	0.78
Networks/telecommunications	3	2.05	0.73
Customer relationship management (CRM)	4	2.07	0.69
Business process management system	5	2.08	0.72
Mobile and wireless applications	6	2.10	0.69
Enterprise resource planning (ERP) systems	7	2.11	0.74
Software as a service	8	2.11	0.72
Service-oriented architecture	8	2.11	0.70
Collaborative and workflow tools	10	2.11	0.71
Business intelligence/analytics	11	2.12	0.74
Data mining	12	2.16	0.65
Cloud computing	13	2.22	0.75
Social networking/media	14	2.24	0.83
Enterprise application integration	15	2.25	0.78
Virtualization (desktop or server)	16	2.27	0.76

*Rating scale ranges from 1 to 5: 1 as most important and 5 as no importance.

The top issue identified in South Korea is "big data systems". This is an area of expertise where the demand for services has begun to grow recently. Moreover, the focus is shifting to artificial intelligence. In addition, most conglomerates in South Korea generally own credit card companies, securities, and/or telecommunication companies, where big data systems are heavily used. Thus, big data systems development has been a critical issue for the industries. In the related area of artificial intelligence (AI), applications are being developed since 2016.

Mobile apps development and mobile and wireless applications were second and sixth ranked technology issues. Mobile apps have gained importance in two ways in the South Korean society. First, private and public organizations have seen trends to develop their apps and to increase the number of apps. Mobile apps by themselves have been considered as examples of innovation in many organizations. But, strategic considerations did not seem to be emphasized when developing the apps. Second, mobile app development is considered as being equivalent to venture business. Many low technology-based entrepreneurs have developed mobile apps to change the industry value chains and to increase the number of communication and marketing channels. For instance, food delivery apps may be the best examples. However, it must be said that among numerous mobile apps, a very few have been successful.

"Networking and telecommunications" was the third highest rated technology issue. This has been a major area of concern for the IT industry because several big companies have been attacked by hackers and experienced financial and reputational damages. For instance, a nuclear power plant company was threatened and attacked by hackers. Some other government agencies have also experienced such threats. Since then, one of the trends in network and telecommunication design has been to separate internal networks from outside ones and to build virtual private network (VPN). Incidentally, South Korea is one of the countries that is ranked at the top with regard to network speed. As per 2017, the average Internet speed in South Korea was 28.6 Mbps, which ranked on the top, followed by Norway (23.5 Mbps) and Sweden (22.5 Mbps). In addition, public WiFi has been widely diffused, and is known to be at the top position in the world. These are the results of the efforts of the government and the private companies to diffuse mobile networks and communications.

CRM and ERP were on the fourth and the seventh positions in the rankings, respectively. CRM and ERP have been widely adopted by companies in South Korea since the late 1990s. However, industry experts have recently reported that there are almost no new CRM and ERP development projects in South Korea. Most projects are about maintaining and upgrading existing systems. Thus, though the two systems are widely diffused, their importance may not be highly ranked now.

In summary, the technology and infrastructure issues ranked highly by the respondents in South Korea reflect the organizational contexts. They also clearly indicate the societal context which can be characterized as high levels of usages of smart phones and other mobile gadgets.

32.7   Individual IT Employee Issues

Table 32.4 shows the summary of the responses about individual IT employee issues. A lower average score means higher agreement with each statement. On a scale of 1–5, any average less than 3 reflects more agreement than disagreement. Thus, the responses indicate that the IT workers in general are satisfied with their current jobs. The responses also show that the IT service workers experience high work pressure, workload and burnout, although this does not seem to affect their home life significantly.

The responses related to work pressure and burnout seem to be real. As news media have reported, IT/IS workers do not have weekends, particularly when they are participating in development projects. One of the

Table 32.4: Individual IT Employee Issues in South Korea

Individual Issues	Mean Rating*	Std. Deviation
Job satisfaction		
In general, I like working here	2.14	0.75
All in all, I am satisfied with my current job	2.23	0.71
In general, I don't like my current job	2.34	0.80
Work pressure		
I feel that the number of requests, problems or complaints that I deal with at work is more than expected	2.92	0.83
I feel that the amount of work I do interferes with how well it is done	2.86	0.82
I feel busy or rushed at work	2.76	0.79
I feel pressured at work	2.75	0.85
Work–life balance		
There is a blurring of boundaries between my job and my home life	2.95	0.95
My work-related responsibilities create conflicts with my home responsibilities	3.21	0.96
I do not get everything done at home because I find myself completing job-related work	3.14	1.02
Work overload and burnout		
I feel drained from activities at work	2.80	0.98
I feel tired from my work activities	2.85	0.87
Working all day is a strain for me	2.85	0.91
I feel burned out from my work activities	2.82	0.83
Sense of accomplishment		
I feel I'm making an effective contribution to what this organization does	2.51	0.75
In my opinion, I do a good job	2.46	0.70
I have accomplished many worthwhile things in this job	2.44	0.72
At my work, I feel confident that I am effective at getting things done	2.44	0.79
Threats to one's job		
I am worried that future technology advancements may pose a threat to my job	2.72	0.97
I believe that other people may be able to perform my work activities	2.54	0.84
I am concerned that my job may be eliminated soon	3.22	0.92
I am concerned that my job may be outsourced soon	3.24	0.98

(Continued)

Table 32.4: (*Continued*)

Individual Issues	Mean Rating*	Std. Deviation
Career plans		
I will be with this organization 1 year from now	2.22	0.79
I will take steps during the next year to secure a job at a different organization	3.52	0.97
I will be with this organization 5 years from now	2.61	0.86
I will be working in the IT field 1 year from now	2.32	0.85
I will take steps during the next year to secure a job outside the IT field	2.97	1.06
I will be working in the IT field 5 years from now	2.37	0.92

*Rating scale ranges from 1 to 5: 1 as strongly agree, to 5 as strongly disagree.

tragic expressions that describe this industry in South Korea is "Monday, Tuesday, Wednesday, Thursday, Friday, Friday, Friday ..." Particularly, the salaries in the software, IS, and IT service fields tend to be based on the project, rather than work-hours. Many workers complain that they have not been appropriately compensated for their overtime. In spite of such perceived unfairness between workload and economic compensation, the IT workers seem to have a sense of accomplishment. They are also not worried too much about job security although they realize that technology advancements and other people may be able to take away their jobs.

Interestingly, despite high workload, work pressure and burnout, IT workers have high levels of intention to remain in their jobs in the current organizations, as well as in the IT profession. There may be three issues related to these results: labor market rigidity, job market situation, and career path rigidity. First, compared to other countries, labor market in South Korea is considered to be relatively rigid. That means that the rate of employee movement across organizations is still low though employees in the IT service sector are more likely to transfer compared to other industry sectors. IT employees may not have enough motivation to move to other organizations, knowing that the working environment is not much different across organizations. Second, the job market has not been very good since 2010. IT services is not an exception. Third, career path rigidity is also quite high in South Korea. This third issue is in part related to the first and the second issues. Culturally speaking, it is not common for employees to change their careers. As employees get older, they have little success finding a new job, particularly in a new area.

32.8 Conclusion

This chapter has reported the organizational, technological, and individual issues of IT employees in South Korea. In many ways, South Korea is unique in that the IT/IS service companies have strategically placed a greater emphasis on domestic markets and on clients engaged in their conglomerates. As a consequence, the top five organization-related issues of IT employees are: Security and privacy, IT reliability and efficiency, Continuity planning and disaster recovery, Business agility and speed to market, and Business productivity and cost reduction. The top five technology issues — big data systems, mobile apps development, networks/telecommunications, customer relationship management (CRM) systems, and Business process management system; they also reflect the organizational and social contexts of South Korea. IT employees are generally satisfied with their current jobs and have a sense of accomplishment at work, in spite of reporting moderate levels of work pressure and burnout. Finally, for a variety of reasons, they wish to remain in their jobs and in the IT profession not only in the short term but also the long term.

References

Akamai (2014). Q3 2014 state of the internet security report. Retrieved from http://www.stateoftheinternet.com/resources-eb-security-2014-q3-internet -security-report.html.

Brislin, R. W. (1970). Back-translation for cross-cultural research. *Journal of Cross-Cultural Psychology*, 1, 185–216.

Kappelman, L., Johnson, V., McLean, E., & Maurer, C. (2018). The 2017 SIM IT Issues and Trends Study. *MISQ Executives*, 17(1), 53–88.

Lee, K., & Joshi, K. (2015). Information technology (IT) development in Korea: Past, present, and future. *Journal of Information Technology Case and Application Research*, 17(2), 68–73.

Palvia, S. C. J., & Palvia, P. (2017). *Global Sourcing of Services: Strategies, Issues and Challenges*. Singapore: World Scientific.

Palvia, P., Jacks, T., Ghosh, J., Licker, P., Romm-Livermore, C., Serenko, A., & Turan, A. H. (2017). The World IT Project: History, trials, tribulations, lessons, and recommendations. *Communications of the Association for Information Systems*, 41(18), 389–413.

Palvia, P., Ghosh, J., Jacks, T., Serenko, A., & Turan, A. (2018). Trekking the globe with the World IT Project. *Journal of Information Technology Case and Application Research*, 20(1), 3–8.

Chapter 33

Information Technology Issues
in Taiwan

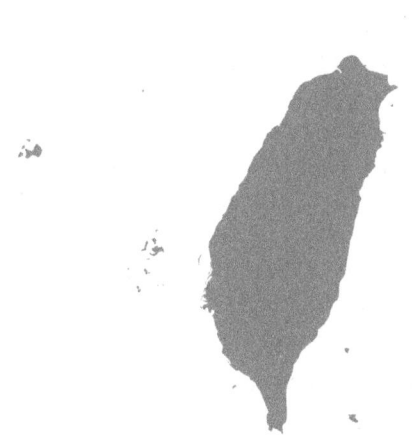

Benjamin Yeo[*,¶], Alexander Serenko[†,‡,||], and Tim Jacks[§,**]

[*]*Seattle University, Seattle, WA, USA*
[†]*University of Toronto, Toronto, Canada*
[‡]*University of Ontario Institute of Technology, Oshawa, Canada*
[§]*Southern Illinois University Edwardsville,*
Edwardsville, IL, USA
[¶]*byeo@seattleu.edu*
[||]*a.serenko@utoronto.ca*
[**]*tjacks@siue.edu*

Summary

This chapter covers the organizational, technological, and individual information technology (IT) issues among IT workers in Taiwan. The results were obtained from a survey of 303 IT workers in Taiwan. Top-organizational issues identified by IT workers include security and privacy, and IT reliability and efficiency. Network/telecommunications is the top-ranked technology and infrastructure issue, which is expected, given Taiwan's IT industry's strengths in hardware manufacturing. IT workers in Taiwan seem to be moderately satisfied with their jobs and intend to stay. However, they may not be confident about the sustainability of their companies, given the challenges facing the Taiwanese IT industry and economy.

33.1 Introduction

Studies on IT work/workers and related issues tend to be Western-centric (Palvia, 2013). However, IT work is increasingly global, as countries from different continents compete for the global market share. One such country is Taiwan, whose IT industry has propelled its economy since the 1980s (Fuller, 2015; Lin and Wong, 2016), even though it currently faces numerous challenges from both the developed and developing world. There are few studies on IT work/workers in Taiwan; and this chapter provides a descriptive overview of IT workers in Taiwan, as well as the organizational, technology and infrastructure, and individual employee issues.

33.2 Country Background and History

Taiwan is a Mandarin-speaking country in East Asia. It is important to note that Taiwan is sometimes not politically recognized as an independent country. The World Bank does not provide Taiwan's country data and does not list Taiwan as a separate country. However, its data are included in high income countries' aggregates (cf. World Bank, n.d.). According to its official statistics bureau, Taiwan is officially called "Republic of China (Taiwan)." Its population as of August 2018 stands at 23,577,271 people (National Statistics, Republic of China (Taiwan), n.d.). It is an island in the Pacific Ocean, about 100 miles southeast of China. Its geographical size is approximately 245 miles (395 km) long (north-south) and 90 miles (145 km) across at its widest point. The city of Taipei is its national capital.

Taiwan is listed among the four Asian Tigers, together with Singapore, Hong Kong, and South Korea, due to its strong economic growth prior to the 1990s. Indeed, Taiwan was among the fastest growing economies in Asia until the 1990s when its growth slowed down. Nonetheless, through its development of high tech industries, Taiwan has become an entrepreneurial democracy, whose economy is driven by its high tech industries. However, compared to the other three Asian Tigers, Taiwan's growth has been weak since the turn of the millennium. Taiwan's relatively weaker economic performance could be due to its relatively poor innovation and branding of high tech exports, which are vital for the country's overall economic performance (Lin and Wong, 2016).

In the late 1970s, Mainland China opened up its economy. This led Taiwanese manufacturers to shift their operations to China, so as to lower costs (Rigger, 2015). Taiwanese investments in Mainland China continue today. Whether or not this has eroded the strengths of the Taiwanese industries or boosted their competitiveness remains to be seen (Berger and Lester, 2005).

33.3 Information Technology in Taiwan

High tech industries are a main staple of Taiwan's economy (Fuller, 2015; Lin and Wong, 2016). According to the 2016 Global Information Technology Report, Taiwan, ranked 19th on the Network Readiness Index — which comprises indicators related to the environment, readiness, usage, and impact — sits among the top countries that are effectively harnessing IT for innovation (Baller *et al.*, 2016).

Taiwan has invested heavily in its IT industry, stemming from its government's industrial policy efforts on building science and technology capabilities. In 1973, the Industrial Technology Research Institute was founded to support applied research and facilitate technology transfer to the private sector. In 1980, Hsinchu Science Park was built as a hotbed for high tech industries. As a result, leading semiconductor firms, such as United Microelectronics Corp. and Taiwan Semiconductor Manufacturing Company (TSMC), have established operations in Taiwan (Lin and Wong, 2016).

Since the mid-1990s, Taiwan's IT industry developed rapidly and shifted the economy from labor-intensive to technology-driven (Schive and Chyn, 2001). In Taiwan, much of the IT work is located in the IT cluster of Hsinchu district (Hu *et al.*, 2005). Its broad IT industry is manufacturing-oriented and is made up of semiconductors, hardware, software, and various peripherals, and it stands in contrast to the US IT industry, which focuses on finished products such as communication services (Bridwell and Kuo, 2005).

Taiwan is a major player in the original equipment manufacturer (OEM) and original design manufacturer (ODM) IT product market. These products include laptops, tablets, smartphones, servers, network devices, and other forms of computer hardware. These are seen as brandless products, but they successfully compete with products of large IT companies such as Apple, Dell, and HP (Canada Trade Office, n.d.). Although the OEM/ODM business model has served the Taiwanese IT industry and economy well in the past, the industry faces competition from high tech firms in developing countries (Lin and Wong, 2016), as well as the innovation capabilities from technologically advanced nations, including the US (Berger and Lester, 2005).

The economic downturn from 2007 to 2009 affected Taiwan's economy. Coupled with increased competition, small- and medium-sized enterprises (SMEs) in Taiwan, which make up the majority of Taiwan's private sector (Lin and Wong, 2016), have identified the IT industry as a source of competitive advantage (Chang *et al.*, 2012). However, Taiwan's inflexible administrative regulations have deterred global firms from investing in the country (Lin and Wong, 2016). In addition, Taiwanese firms tend to rely on narrow markets. Innovative Taiwanese IT firms, such as Asus, have placed emphasis on research and development (R&D), but not on global after-sales services. This deters potential customers, and hence, investments, as competitive brands have a better global service network (Lin and Wong, 2016).

33.4   Methodology

A survey instrument that was developed as part of the World IT Project and included 160 questions (cf. Palvia *et al.*, 2018; Palvia *et al.*, 2017) was used in the present study. The survey was translated from its original English version to Mandarin Chinese, and localized to the Taiwanese context. Even though Taiwan, like Mainland China, is a Mandarin Chinese-speaking country, where Mandarin Chinese is the first language, there are subtle differences in their use of the language, which requires localization. It was administered online to IT workers in Taiwan in 2016. A competitive University Research Council (URC) research grant of $3,500 was obtained from DePaul University to support data collection. Graduate students in Taiwan from the National Chiao Tung University were recruited and trained

Table 33.1: Descriptive Statistics

Characteristics	N	%	Characteristics	N	%
Education:			Years of work experience:		
High school or less	9	3.0	0–4 years	115	38.0
Associate degree	33	10.9	5–9 years	45	14.9
Bachelor's degree	132	43.6	10–19 years	70	23.1
Master's degree	125	41.3	20–29 years	52	17.2
Ph.D.	4	1.3	30+ years	21	6.9
Years of IT experience:			Organization location:		
0–4 years	140	46.2	IT department employee	171	56.4
5–9 years	38	12.5	IT worker in a non-IT department	60	19.8
10–19 years	69	22.8	Contract employee	13	4.3
20–29 years	44	14.5	Consultant	27	8.9
30+ years	12	4.0	Vendor employee	32	10.6
Work as:			Work position:		
Mostly full time	284	93.7	Not part of management	172	56.8
Mostly part time	12	4.0	In lower management	49	16.2
Mostly over time	7	2.3	In middle management	56	18.5
Been laid off from a job:			In senior management	26	8.6
Yes	19	6.3			
No	284	93.7			

to conduct the survey. A total of 303 responses were obtained. Table 33.1 shows the frequencies of demographic variables from the survey.

A large majority of the respondents (86.2%) have at least a bachelor's degree. This is not surprising because IT jobs in Taiwan are likely to be white collar jobs. Although there appears to be a mismatch between university learning and job expectations (Chang, 2015), the IT field changes rapidly and IT workers are often expected to update their skills while on the job. In terms of work experience, 62.0% of them have worked for 5 years or more, while slightly less (53.8%) have worked as IT professionals for at least 5 years. This suggests that workers in general are inclined to move into IT work, possibly given its prestige and compensation compared to non-IT work. In their organizations, more than half of the respondents (56.4%) work in IT departments. This is also not surprising, given that IT work spans across occupations. It is plausible for an IT worker to be affiliated with the marketing department for instance, to oversee e-commerce systems.

33.5   Organizational IT Issues

The organizational IT issues perceived by respondents are given in Table 33.2. Given the tendency of modern firms to utilize connected IT systems, such as internet of things (IoT) and cloud computing, security is identified as a primary concern (Drew, 2017). It is therefore not surprising that security and privacy is ranked as the top-organizational issue in both Taiwan and the Society for Information Management (SIM) 2017 study of top issues.

Among the top five ranked issues, two are similar to the list of IT management issues identified by members of SIM (Kappelman et al., 2017). These are security and privacy, and continuity planning and disaster recovery, ranked first and third, respectively, in Taiwan, and ranked first and 18th, respectively, in the SIM study. Among the top 10 issues identified by respondents, alignment between IT and business, and business agility and speed to market, ranked seventh and eighth, respectively, in Taiwan, coinciding with those identified in the SIM study, where they were ranked second and ninth, respectively (Kappelman et al., 2017).

Cost reduction issues — business productivity and cost reduction, and IT cost reduction — were not ranked highly among IT workers in Taiwan.

Table 33.2:   Organizational IT issues in Taiwan

Organizational IT Issues	Rank	Mean Rating*	Std. Deviation
Security and privacy	1	1.69	0.68
IT reliability and efficiency	2	1.74	0.60
Continuity planning and disaster recovery	3	1.86	0.63
IT strategic planning	4	1.89	0.68
Project management	5	1.95	0.66
Knowledge management	6	1.97	0.67
Alignment between IT and business	7	2.00	0.56
Business agility and speed to market	8	2.04	0.67
Revenue-generating IT innovations	9	2.08	0.71
IT service management	10	2.10	0.68
Business productivity and cost reduction	11	2.13	0.68
Attracting and retaining IT professionals	11	2.13	0.75
Enterprise architecture	13	2.16	0.68
Business process reengineering	14	2.21	0.71
Globalization	15	2.26	0.82
IT cost reduction	16	2.39	0.83
Bring your own computing device	17	2.59	0.90
Outsourcing	18	2.89	0.88

*Rating scale ranges from 1 to 5: 1 as most important and 5 as no importance.

Engineers in Taiwanese IT firms focus on quality control in the manufacturing process, rather than innovation and R&D like in the US, that can yield innovative products and services that are more cost effective (Lin and Wong, 2016). The findings reflect the current mindset of IT workers in Taiwan, and can be addressed as Taiwan seeks to use its IT industry as a springboard towards a stronger economy.

Outsourcing was ranked as the least important issue among IT workers in Taiwan. It has been found that small and medium enterprises (SMEs) in Taiwan do not yet understand the importance of outsourcing, even though they have limited capital (Chang *et al.*, 2012), and the necessary skill sets from their employees usually lag behind large enterprises (Cragg *et al.*, 2011). In view of this challenge, a government agency can be set up to initiate and promote IT outsourcing activities to the SMEs and develop new opportunities for Taiwan's IT industry (Chang *et al.*, 2012).

33.6 Technology and Infrastructure Issues

Table 33.3 summarizes the technology and infrastructure issues ranked by IT workers in Taiwan. The fairly low standard deviations suggest that there is not much variation in their responses. These differ from recent US findings in 2017. While network/telecommunications is the highest-ranked IT

Table 33.3: Technology and infrastructure issues in Taiwan

IT Related Issues	Rank	Mean Rating*	Std. Deviation
Network/telecommunications	1	1.93	0.72
Customer relationship management systems	2	2.03	0.72
Business intelligence/analytics	3	2.06	0.69
Collaborative and workflow tools	3	2.06	0.69
Software as a service	5	2.08	0.65
Data mining	6	2.10	0.76
Enterprise resource planning systems	7	2.11	0.76
Mobile and wireless applications	8	2.13	0.81
Big data systems	9	2.14	0.84
Service-oriented architecture	9	2.14	0.68
Mobile apps development	9	2.14	0.82
Enterprise application integration	12	2.17	0.69
Business process management systems	13	2.20	0.73
Virtualization (desktop or server)	14	2.24	0.80
Social networking/media	15	2.29	0.87
Cloud computing	16	2.32	0.80

*Rating scale ranges from 1 to 5: 1 as most important and 5 as no importance.

issue perceived by IT workers in Taiwan, it is the eighth of the largest
IT investments made and is one of the least important areas that should
receive more investments in the US in 2017. In contrast, cloud computing
is ranked the lowest among technology and infrastructure issues in Taiwan,
but is the third largest IT investment and deemed as the third most important area that should get more investments in the US in 2017 (Kappelman
et al., 2017).

The top two technology and infrastructure issues ranked by IT workers
in Taiwan — network/telecommunications and customer relationship management systems — are hardware-related. Given that Taiwan's IT industry
is largely hardware manufacturing-based (Bridwell and Kuo, 2005), these
rankings are expected. Customer relationship management systems and
business intelligence/analytics, the second and third on the list, can be
argued to be service-related. It is plausible that IT workers are seeing the
need to diversify Taiwan's predominantly manufacturing base towards innovative service orientation as advocated by Lin and Wong (2016). Interestingly, social networking/media and cloud computing, both staples of the IT
world today, are the lowest-ranked issues.

33.7   Individual IT Employee Issues

Table 33.4 summarizes the issues faced by individual IT employees in Taiwan. They are categorized into job satisfaction, work pressure, work–life
balance, workload and burnout, sense of accomplishment, threat to one's
job, and career plans. Overall, IT workers in Taiwan appear to be moderately happy with their jobs and industry, and are successful in managing
job expectations, among other commitments, at least in the short term.

In terms of job satisfaction, IT workers in Taiwan appear to be moderately satisfied. Studies have attempted to explain job satisfaction. Although
Kuo and Chen (2004) found that IT managers in Taiwan are significantly
more satisfied than their lower level counterparts, East Asian cultures value
hierarchy and harmony, which implies that job positions do not matter
(Zhang et al., 2005). In general, workers who are highly educated may have
higher expectations, leading to disappointment (Ross and Reskin, 1992)
and, therefore, lower job satisfaction (Clark, 1997). The findings show that
slightly less than half the respondents (43.3%) are part of management.
At the same time, more than half (58.7%) of the respondents have less
than 10 years of experience in IT work, and less than half (42.6%) of
the respondents have a graduate degree. These demographic breakdowns

Table 33.4: Individual IT employee issues in Taiwan

Individual Issues	Mean Rating*	Std. Deviation
Job satisfaction		
In general, I like working here	2.09	0.79
All in all, I am satisfied with my current job	2.26	0.87
In general, I don't like my current job	3.55	1.00
Work pressure		
I feel that the number of requests, problems or complaints that I deal with at work is more than expected	3.24	0.98
I feel that the amount of work I do interferes with how well it is done	3.15	0.98
I feel busy or rushed at work	2.82	1.03
I feel pressured at work	2.55	0.91
Work–life balance		
There is a blurring of boundaries between my job and my home life	3.24	1.00
My work-related responsibilities create conflicts with my home responsibilities	3.42	0.98
I do not get everything done at home because I find myself completing job-related work	3.38	1.04
Workload and burnout		
I feel drained from activities at work	3.09	1.02
I feel tired from my work activities	2.79	0.99
Working all day is a strain for me	2.78	1.01
I feel burned out from my work activities	3.11	1.01
Sense of accomplishment		
I feel I'm making an effective contribution to what this organization does	2.04	0.65
In my opinion, I do a good job	2.08	0.66
I have accomplished many worthwhile things in this job	2.03	0.64
At my work, I feel confident that I am effective at getting things done	1.91	0.59
Threat to one's job		
I am worried that future technology advancements may pose a threat to my job	2.92	1.07
I believe that other people may be able to perform my work activities	2.53	0.92
I am concerned that my job may be eliminated soon	3.33	1.03
I am concerned that my job may be outsourced soon	3.54	1.00

(Continued)

Table 33.4:   (*Continued*)

Individual Issues	Mean Rating*	Std. Deviation
Career plans		
I will be with this organization 1 year from now	2.31	0.94
I will take steps during the next year to secure a job at a different organization	3.38	1.05
I will be with this organization 5 years from now	2.95	1.00
I will be working in the IT field 1 year from now	2.07	0.83
I will take steps during the next year to secure a job outside the IT field	3.63	0.91
I will be working in the IT field 5 years from now	2.34	0.91

*Rating scale ranges from 1 to 5: 1 as strongly agree and 5 as strongly disagree.

provide support for the literature in explaining the moderate job satisfaction levels among IT workers in Taiwan.

Looking at work pressure, IT workers in Taiwan seem to have more work than expected, but nonetheless, manage their work pressure well. It is important to note that the moderately high standard deviations may suggest that this varies across organizations or industries. Taiwanese work culture is characterized by Confucian values, which value hierarchy, and are different from the Western concept of egalitarianism. Recognition by superiors is an important motivator (Lu *et al.*, 2003). Furthermore, Taiwanese academic leaders have been found to have higher emotional intelligence than those in the US (Tang *et al.*, 2010), which has a negative relationship with perceptions of work pressure among workers in the semiconductor industry in Taiwan (Chang and Chang, 2010). It is plausible that IT workers in Taiwan exhibit sufficient levels of emotional intelligence, which helps them manage their workload and pressure. Thus, the finding on work pressure is not surprising, especially given that the employees are moderately satisfied with their jobs.

IT workers in Taiwan do not appear to have difficulties with balancing work and life commitments. These findings are consistent with those on work pressure. Again, the moderately high standard deviations may suggest variations across organizations and/or industries. In Taiwanese culture, the family is an important source of support for white collar workers (Luo, 1999), which includes IT workers. Given the hierarchical structures of Taiwanese organizations, it is plausible that IT workers in Taiwan recognize their responsibilities to their employers and families. This allows them to manage their professional and personal lives.

The findings on workload and burnout corroborate earlier findings on work pressure and work–life balance. The respondents moderately disagree that they are burned out, thus appearing to manage their work commitments well. This can be explained in two ways. First, they may be accustomed to the work culture in Taiwan, characterized by hierarchical organizations (Lu *et al.*, 2003) and the power superiors have over subordinates (Luo, 1999). Their expectations on their work commitment spur them onwards in completing their responsibilities. Second, recognition by superiors is important in Taiwanese culture (Lu *et al.*, 2003). This may be another source of motivation for IT workers in Taiwan.

As a related theme, IT workers in Taiwan have a strong sense of accomplishment, especially given the low standard deviations in the four items. This could be due to the recognition that they receive from superiors, which is a key motivator (Lu *et al.*, 2003). Their strong sense of accomplishment explains their high confidence that there are few or no threats to their jobs. They are not concerned that their jobs will be eliminated or outsourced, attesting to the value they bring to their organizations and faith in their professional skills. Their perceived threats stem more from technological advancements and their competitors. In the Taiwanese work culture, organizations put an emphasis on competitiveness (Lu *et al.*, 2003). Hence, IT workers may strive to outperform their peers to gain recognition and promotions. Incidentally, the moderately high standard deviations suggest some variations across organizations and/or industries.

Generally, IT workers in Taiwan seem moderately intent to stay at their jobs and the IT field in the short term. This is not surprising, given that they are moderately satisfied with their jobs, can handle the pressure, manage work and life commitments well, have a strong sense of accomplishment, and are confident about their skills. However, there appears to be a moderate intent to leave the organization in the longer term: they disagree that they will take steps to secure a job at a different organization or different field in the following year, but do not see themselves at this same organization in 5 years. These findings suggest that there may be a lack of confidence about the sustainability of the companies they are working for, which is not surprising, given the challenges facing the Taiwanese economy that is driven by its IT industry, compared to the other three Asian Tigers (cf. Lin and Wong, 2016).

33.8   Conclusion

The IT industry in Taiwan has played a major role in driving Taiwan's economy. However, it faces challenges today from both the developed and developing worlds. The findings suggest that IT workers in Taiwan are confident about their IT skills, but are concerned about the competition. They may perceive the limitations of Taiwan's IT industry and the sustainability of their companies and recognize the importance of a service-orientation model, given their identification of customer relationship management systems and business intelligence/analytics as top-organizational IT issues. Moving forward, Taiwan can leverage its strong IT base to diversify its IT industry from one that is largely manufacturing-based to an innovation-driven, service-based industry.

References

Baller, S., Dutta, S., & Lanvin, B. (2016). *Global Information Technology Report 2016*. Ouranos.

Berger, S., & Lester, R. K. (2005). Globalization and the future of the Taiwan miracle. *Global Taiwan: Building Competitive Strengths in a New International Economy*.

Bridwell, L., & Kuo, C.-J. (2005). An analysis of the computer industry in China and Taiwan using Michael Porter's determinants of national competitive advantage. *Competitiveness Review: An International Business Journal*, 15(2), 116–120.

Canada Trade Office (n.d.). *Information Communicaion Technology Sector Profile — Taiwan*. Taipei, Taiwan: Canada Trade Office. Retrieved from https://www.enterprisecanadanetwork.ca/_uploads/resources/Information-Communication-Technology-Sector-Profile-Taiwan.pdf.

Chang, C.-P., & Chang, F.-J. (2010). Relationships among traditional Chinese personality traits, work stress, and emotional intelligence in workers in the semiconductor industry in Taiwan. *Quality & Quantity*, 44(4), 733–748.

Chang, F. (2015). Educational Resources, Job Match, and Employment Outcomes in Taiwan. *TEPS-B Working Paper*, (2015–02). Retrieved from http://tepsb.nccu.edu.tw/download/TEPS-B_working_paper_201502.

Chang, S.-I., Yen, D. C., Ng, C. S.-P., & Chang, W.-T. (2012). An analysis of IT/IS outsourcing provider selection for small- and medium-sized enterprises in Taiwan. *Information & Management*, 49(5), 199–209.

Clark, A. E. (1997). Job satisfaction and gender: Why are women so happy at work? *Labour Economics*, 4(4), 341–372.

Cragg, P., Caldeira, M., & Ward, J. (2011). Organizational information systems competences in small and medium-sized enterprises. *Information & Management*, 48(8), 353–363.

Drew, J. (2017). The single factor of most concern in the cloud. *Journal of Accountancy*, 223(6), 52.

Fuller, D. B. (2015). Moving along the electronics value chain: Taiwan in the global economy. In *Global Taiwan: Building Competitive Strengths in a New International Economy* (pp. 159–187). Routledge.

Hu, T.-S., Lin, C.-Y., & Chang, S.-L. (2005). Technology-based regional development strategies and the emergence of technological communities: A case study of HSIP, Taiwan. *Technovation*, 25(4), 367–380.

Kappelman, L., Nguyen, Q., McLean, E., Maurer, C., Johnson, V., Snyder, M., & Torres, R. (2017). The 2016 SIM IT Issues and Trends Study. *MIS Quarterly Executive*, 16(1).

Kuo, Y.-F., & Chen, L.-S. (2004). Individual demographic differences and job satisfaction among information technology personnel: An empirical study in Taiwan. *International Journal of Management*, 21(2), 221–231.

Lin, M. C., & Wong, P. (2016). *Recapturing the Taiwan Miracle*. Santa Monica, CA: Milken Institute.

Lu, L., Cooper, C. L., Kao, S., & Zhou, Y. (2003). Work stress, control beliefs and well-being in Greater China: An exploration of sub-cultural differences between the PRC and Taiwan. *Journal of Managerial Psychology*, 18(6), 479–510.

Luo, L. (1999). Work motivation, job stress and employees' well-being. *Journal of Applied Management Studies*, 8, 61–72.

National Statistics, Republic of China (Taiwan). (2018). Latest indicators. Retrieved from https://eng.stat.gov.tw/np.asp?CtNode=1524 [Accessed October 9, 2018].

Palvia, P. (2013). The World IT Project: A program on international research and call for participation. *Journal of Global Information Technology Management*, 16(2), 1–5.

Palvia, P., Jacks, T., Gosh, J., Licker, P., Romm-Livermore, C., Serenko, A., & Turan, A. H. (2017). The World IT Project: History, trials, tribulations, lessons, and recommendations. *Communications of the Association for Information Systems*, 41(18), 389–413.

Palvia, P., Ghosh, J., Jacks, T., Serenko, A., & Turan, A. (2018). Trekking the globe with the World IT Project. *Journal of Information Technology Case and Application Research*, 20(1), 3–8.

Rigger, S. (2015). Taiwanese business in Mainland China: From domination to marginalization? In *Cross-Taiwan Strait Relations in an Era of Technological Change*. Springer, pp. 61–76.

Ross, C. E., & Reskin, B. F. (1992). Education, control at work, and job satisfaction. *Social Science Research*, 21(2), 134–148.

Schive, C., & Chyn, R. Y. S. (2001). Taiwan's high-tech industries. In L. K. Cheng & H. Kierzkowski (Eds.), *Global Production and Trade in East Asia*. Boston, MA: Springer, pp. 181–205.

Vivian Tang, H., Yin, M., & Nelson, D. B. (2010). The relationship between emotional intelligence and leadership practices: A cross-cultural study of academic leaders in Taiwan and the USA. *Journal of Managerial Psychology*, 25(8), 899–926.

World Bank. (n.d.). Where are your data on Taiwan? Retrieved from https://datahelpdesk.worldbank.org/knowledgebase/articles/114933-where-are-your-data-on-taiwan [Accessed August 24, 2018].

Zhang, Y. B., Lin, M.-C., Nonaka, A., & Beom, K. (2005). Harmony, hierarchy and conservatism: A cross-cultural comparison of Confucian values in China, Korea, Japan, and Taiwan. *Communication Research Reports*, 22(2), 107–115.

Chapter 34

Information Technology Issues in Thailand

Savanid Vatanasakdakul*,‡, Chadi Aoun*,§,
and Wachara Chantatub†,¶

*Carnegie Mellon University, Ar-Rayyan, Qatar
†Chulalongkorn University, Bangkok, Thailand
‡savanid@cmu.edu
§chadi@cmu.edu
¶wachara@cbs.chula.ac.th

Summary

Thailand is repositioning itself to reap the benefits of a digital knowledge-based economy. At a national level, the government is implementing Thailand 4.0, its strategic vision towards a digital future. Can Thailand achieve its vision? What challenges lie ahead? In addressing these fundamental questions, this study adopts a positivist epistemology, deploying a quantitative survey instrument. The results shed important light on organizational, infrastructural, and workforce-related issues in the information technology (IT) industry. The results highlight potential strengths and challenges that the industry faces, and provide in-depth and integrated analysis of these issues, along with avenues for future research.

34.1   Introduction

Thailand is currently undergoing many developments and positioning itself as an investment hub for the region by employing strategies to make starting businesses in Thailand easier and more attractive (KPMG, 2018). Over the past decade, the digital divide in Thailand has been reducing at a fast speed due to the country's significant improvements in workforce education and skills building (EU Gateway, 2018). The country's vision of Thailand 4.0 aims to release Thailand from a middle-income trap of over 20 years, to become a high-income country. Nonetheless, Thailand has many challenges to face. One of the key strategies is to drive digital transformation to transition into an innovation-driven economy in order to allow the country to overcome several economic challenges. The combination of good transport connections, national reforms and strong relationships with ASEAN countries will enable Thailand to be an attractive regional hub (KPMG, 2018). In terms of human resource development to serve the next-generation industries, according to the Thailand 4.0 vision the nation requires to upskill the labor force and recruit more foreign workers to help develop existing and new industries such as robotics, aviation and logistics, biofuels and biochemical, digitization, and healthcare (KPMG, 2018).

34.2   Country Background and History

Thailand is located in Southeast Asia, bordering Myanmar, Laos, Cambodia and Malaysia. Thailand is in the center of the Greater Mekong Subregion (GMS) and has become a regional logistics and trading hub. Thailand was known as Siam until 1939 and is the only Southeast Asian country never to have been colonized by European nations. Today, the current population of Thailand is 69.2 million and the country ranks 20th in the list of countries by population with a total land area of $510,890 \, km^2$ (197,256 sq. miles) (Worldometers, 2018). The GDP Annual Growth Rate in Thailand averaged 3.73% from 1994 until 2018 and Thailand's GDP grew by 4.6% year-on-year in the second quarter of 2018 (Trading Economic, 2018). Thailand today has a complex, multifaceted economy embracing industries that employ the latest and most sophisticated technologies (CountryWatch, 2018). Growth and diversification into new industrial areas have largely been initiated by the dynamic private sector (CountryWatch, 2018). In 2018, Thailand was ranked 26th out of 190 economies on ease of doing business and has more than US\$121 billion in assets under management and a market capitalization of US\$349 billion (KPMG, 2018).

At present, Thailand has a national vision called "Thailand 4.0". Thailand 4.0 aims to unlock the country from several economic challenges resulting from past economic development models which place emphasis on agriculture development (Thailand 1.0), light industry (Thailand 2.0), and advanced industry (Thailand 3.0). These challenges include "a middle income trap", "an inequality trap", and "an imbalanced trap" (Thailand Board of Investment, 2017). Thailand 4.0 plan is focused on 10 targeted industries, which are Next-Generation Automotive; Smart Electronics; High-Income Tourism and Medical Tourism; Efficient Agriculture and Biotechnology; Food Innovation; Automation and Robotics; Aerospace; Bio-Energy and Bio-chemicals; Digital; and Medical and Healthcare (Thailand Board of Investment, 2017).

34.3   Information Technology in Thailand

The development of information and communication technology (ICT) in Thailand started around 55 years ago when the first computer, IBM 1620, was brought into Thailand in 1963 (Kantabutra, 2001). It was installed at the Department of Statistics, Chulalongkorn University, for teaching computer programming classes. It was made available not only to students but also to the public (Kantabutra, 2001). According to IBM archive, IBM representative office was established in Bangkok in 1948 and became IBM Thailand Co., Ltd. in 1952 (IBM, 2013). Since then, ICT has revolutionized Thailand immensely. The remarkable events that are worth mentioning are as follows.

The Internet usage in Thailand for academics began in 1987 and it took until 1995 for the Internet to become commercialized and expand outside the academic realm (Palasri *et al.*, 1998). According to The Economist Group's research on the Inclusive Internet Index (2018) based on the scores of the Availability, Affordability, Relevance and Readiness categories, which assessed coverage in 86 countries in Asia, as of February 2018, Thailand has the overall rank of 31. For each category, Thailand ranked as follows:

Table 34.1:   Thailand's ICT indexes versus Asia & Pacific Region and Worldwide
(ITU, 2017)

Index	Thailand	Asia & Pacific	World
Mobile-cellular telephone subscriptions per 100 inhabitants	172.65	98.90	101.53
International internet bandwidth per Internet user (Bit/s)	49,243.82	48,000.00	74,464.00
Percentage of households with computer	28.41	37.80	46.61
Percentage of households with Internet access	59.84	45.50	51.46
Percentage of individuals using the Internet	47.50	41.50	45.91
Fixed (wired)-broadband subscriptions per 100 inhabitants	10.69	11.30	12.39
Active mobile-broadband subscriptions per 100 inhabitants	94.72	47.40	52.23
Mean years of schooling	7.90	8.15	8.52
Secondary gross enrollment ratio	129.00	83.06	84.00
Tertiary gross enrollment ratio	48.86	34.65	38.69

Availability was ranked 26; Affordability was ranked 27; Relevance was ranked 49; and Readiness was ranked 26. Therefore, Thailand was in the top 25% of all categories.

According to International Telecommunication Union's (ITU) ICT Development Index, Thailand was ranked number 78 in the world in 2017 (ITU, 2017). Thailand's key ICT development indexes compared to Asia and Pacific Region and Worldwide are shown in Table 34.1. According to data from Table 34.1, Thailand was better than the world's average in 6 key ICT development indexes and lower in 4 indexes.

In 2016, the Thai communications sector had a total value of THB 580 billions, or around 4% of the national GDP (Ninkitsaranont, 2018). Of the total communications sector, mobile communications had a combined value of THB 400 billions, or around 68.7% of the total (Ninkitsaranont, 2018). Thai consumers are the world's most social shoppers, with 51%, saying they bought products by interacting with merchants on social media compared with 32% in India and 27% in China; 50% of all online purchases in Thailand are made through mobile devices (Accenture, 2017). It is predicted that by 2021 smartphones will be used by 79.4% of the population, 63% are expected to be Internet-connected, and per capita digital spending will surge, nearly doubling by 2022 to reach US$470.40 (Accenture, 2017). These growth prospects are already attracting interest from global players.

From a governmental perspective, the Thai government started its ICT initiative in 1992 by setting up the National IT Committee, or NITC, chaired by the Prime Minister (Thuvasethakul and Koanantakool, 2002). In February 1996, NITC announced the first national IT policy known as "IT2000". Then, the second national IT policy "IT2010" was approved in March 2002 (National Information Technology Committee Secretariat, 2003). In May 2011, the latest national IT policy known as "IT2020 (2011–2020)" was released (National Information Technology Committee Secretariat, 2011). The vision of IT2020 states that "ICT is a key driving force in leading Thai people towards knowledge and wisdom and leading society towards equality and sustainable economy". The IT2020 policy framework set seven development strategies: (1) Universal and secure ICT and broadband infrastructure; (2) ICT human resources and ICT competent workforce; (3) ICT industry competitiveness and ASEAN integration; (4) Smart government: ICT for government service innovation and good governance; (5) ICT for Thailand competitiveness and vibrant economy; (6) ICT to enhance social equality; and (7) ICT and Environment: the Green ICT. This framework involves ICT development in five areas: (1) key concept of sustainable development in social, economic, and environment; (2) utilizing ICT to minimize inequality and offer equal opportunity for all Thai people; (3) compliance with sufficiency economy philosophy for economic development; (4) connecting and continuing the existing policies and plans; and (5) promoting the collaboration of private sectors (Wongwuttiwat and Lawanna, 2018). The focus on sustainability through digitization has been a core perspective that has also permeated across the non-governmental sector (Aoun *et al.*, 2011).

At present, Thailand is working to drive digital transformation by using new growth engines to build economic prosperity, social security and sustainability (Thailand Board of Investment, 2018). The Digital Economy is estimated to contribute 25% to the GDP by 2027 (We Are Social, 2017). The driving forces behind digital economic growth are high Internet penetration rate and large base of social media users. Concerning Thailand's startup ecosystem, the number of funded startups in Thailand increased from 3 in 2012 to 75 companies in 2016. Funding for startups has also increased by more than 120% between 2012–2016 with the top three startup sectors being e-commerce, fintech, and logistics (Thailand Board of Investment, 2018). In terms of IT human resources, the estimated number of IT graduates in 2016 is around 115,000 persons (Thailand Board of Investment, 2018).

34.4   Methodology

In order to assess current and emerging issues in the IT industry, a quantitative survey was adopted for data collection. Collecting data from IT professionals in Thailand was a challenge. The standard World IT Project (Palvia *et al.*, 2017; Palvia *et al.*, 2018) from the World IT Project was translated into Thai language by the local research assistant and then back-translated to English by one of the authors. To validate the accuracy of translation, both English versions were compared by the research team, and no issues were found.

One thousand paper-based surveys were distributed to IT professionals by the local research team in Thailand. The local team opted for a paper-based survey instead of an online survey. This was to increase the response rate in the Thai cultural context, where physical presence is preferred to email invitations for online surveys. The paper-based data collection strategy worked well. Seven hundred responses were received. After eliminating invalid responses and surveys with a high volume of missing data, there were 665 valid responses left. Participants were not given any gift or momentary incentive to complete the surveys. The respondents' identities were kept anonymous. Once the data were collected, a local research assistant did the primary data entry task. Then, one of the authors performed quality control and made sure that the data were coded in compliance with the requirements of the World IT Project. The process of instrument translation, validation and data collection lasted for a year.

Table 34.2 shows descriptive statistics for the Thai IT professionals who took part in the study. Note that some totals may be less than 665 due to missing data. In our sample, 73.1% of the respondents hold a Bachelor's degree and 26.6% hold a Master's degree. The majority of respondents have less than 9 years of work experience and 41.1% has less than 4 years of IT-related work experience. 98.3% have full-time positions. 65.9% work in IT departments, while the rest of the respondents' positions are external to the IT department. The majority of the respondents (73.3%) are in non-managerial positions.

34.5   Organizational IT Issues

Table 34.3 shows the organizational IT issues from the Thai respondents ranked in order of importance. The rating ranges from 1 to 5 on a Likert scale, with 1 as most important and 5 as least important. Eighteen

Table 34.2: Descriptive Statistics

Characteristics	N	%	Characteristics	N	%
Education:			**Years of work experience:**		
High school or less	0	0	0–4 years	247	37.1
Associate degree	2	0.3	5–9 years	253	38.0
Bachelor's degree	486	73.1	10–19 years	126	18.9
Master's degree	117	26.6	20–29 years	33	5.0
Ph.D.	0	0	30+ years	6	0.9
Years of IT experience:			**Organizational location:**		
0–4 years	275	41.4	IT department employee	438	65.9
5–9 years	245	36.8	IT worker in non-IT department	84	12.6
10–19 years	115	17.3	Contract employee	48	7.2
20–29 years	27	4.1	Consultant	43	6.5
30+ years	3	0.5	Vendor employee	51	7.7
Work as:			**Work position:**		
Mostly full time	654	98.3	Not part of management	490	73.3
Mostly part time	4	0.6	In lower management	129	19.4
Mostly over time	7	1.1	In middle management	36	5.4
Been laid off from IT job:			In senior management	10	1.5
Yes	14	2.1			
No	651	97.9			

Table 34.3: Organizational IT Issues in Thailand

Organizational IT Issues	Rank	Mean Rating*	Std. Deviation
IT reliability and efficiency	1	1.70	0.77
Security and privacy	2	1.79	0.78
Knowledge management	3	1.82	0.74
Project management	4	1.83	0.76
IT strategic planning	5	1.86	0.75
Business agility and speed to market	6	1.87	0.76
Continuity planning and disaster recovery	7	1.88	0.87
Attracting and retaining IT professionals	8	1.88	0.80
Alignment between IT and business	9	1.90	0.76
Revenue-generating IT innovations	10	1.98	0.81
Business productivity and cost reduction	11	2.04	0.75
IT service management (e.g., ITIL)	12	2.06	0.81
Business process reengineering	13	2.07	0.76
Enterprise architecture	14	2.21	0.80
Globalization	15	2.23	0.79
IT cost reduction	16	2.24	0.91
Outsourcing	17	2.76	0.88
Bring your own computing device (BYOD)	18	3.10	1.09

*Rating scale ranges from 1 to 5: 1 as most important and 5 as no importance.

organizational IT issues were ranked. The top five issues are IT reliability and efficiency, security and privacy, knowledge management, project management, and IT strategic planning.

The top five results could be categorized in two dimensions for analysis, the Technological (Tech) dimension, and the Governance Dimension.

When it comes to the Tech dimension, the Thai perspective emphasizes the importance of IT reliability and efficiency, which was viewed as critical. This is particularly relevant when it comes to having robust and functional infrastructure. It highlights the rise in automation and the digital economy in Thailand. IT reliability is central to achieving efficiency in organizational operations, and therefore central to organizational success. Consequently, security issues pose cyber threats to effective and efficient IT infrastructure, and therefore, are logically noted as a concern. Coupled with this is the concern to privacy posed by security breaches, which could harm the organization and its clients and constituents.

The Governance dimension encompasses IT strategic planning, knowledge management, and project management, as essential antecedents to having reliable and effective IT. The number of IT projects is increasing given the Thailand 4.0 national vision, with its aims for a knowledge-based digital economy, and for bridging the digital divide. Having a strategic focus, with proven knowledge and project management, is paramount to achieving such a vision.

The intertwine between these two dimensions, both Tech and Governance, demonstrates maturity among respondents, albeit, also raising concerns about such pivotal issues moving forward. While the respondents demonstrate good holistic foresight, could it also be the case that they are rating these issues highly due to a lack of relevant action, skills, and resources in their context when it comes to issues such as security and management? This could indeed be of value if further investigated in follow up studies.

34.6   Technology and Infrastructure Issues

Table 34.4 shows the Technology and Infrastructure issues from Thai respondents, ranked in order of importance. The rating ranges from 1 to 5 on a Likert scale, rating 1 as most important and 5 as least important. Sixteen technology and infrastructure issues were ranked. The top five issues are network/telecommunications, big data systems, collaborative and workflow tools, business process management systems, and customer relationship

Table 34.4: Technology and Infrastructure Issues in Thailand

IT-Related Issues	Rank	Mean Rating*	Std. Deviation
Networks/telecommunications	1	1.90	0.83
Big Data systems	2	1.93	0.81
Collaborative and workflow tools	3	1.94	0.78
Business process management systems	4	1.97	0.76
Customer relationship management (CRM) systems	5	1.99	0.83
Business intelligence/analytics	6	1.99	0.75
Mobile and wireless applications	7	2.00	0.86
Enterprise resource planning (ERP) systems	8	2.02	0.81
Enterprise application integration	9	2.02	0.76
Software as a service	10	2.11	0.89
Data mining	11	2.12	0.87
Mobile apps development	12	2.14	0.95
Social networking/media	13	2.18	0.90
Virtualization (desktop or server)	14	2.20	0.86
Service-oriented architecture (SOA)	15	2.28	0.84
Cloud computing	16	2.28	0.87

*Rating scale ranges from 1 to 5: 1 as most important and 5 as no importance.

management systems as well as business intelligence/analytics (tied for fifth place).

The technology and infrastructure issues can be thematically categorized into two components, the network and communication infrastructure component, and the data component.

When it comes to the network and communications infrastructure component, it is evident that integration between systems and services is viewed as pivotal along with developing effective and reliable infrastructure to address emergent challenges and opportunities such as business intelligence and analytics capabilities. This would require investment in technologies that support broad integration between systems, as well as storage capabilities, and advanced processing capabilities for analytical tools.

The data component builds upon the networking and communications infrastructure, by providing an architecture for seamless integration between systems, through uniform meta-data and indexing frameworks. This would enable collaborative and adaptive flow of information between processes and systems, such as data warehouses that could enable the implementation of effective modeling and analytics techniques and machine learning methods.

What is also interesting is the importance that the respondents placed on customer relationship management (CRM) systems, which highlights a shift towards digitization when it comes to dealing with customers as opposed to (or maybe complementing) traditional means that were focused on building personal relationships through high-context face-to-face communications.

Overall, the ranks point to a focus on integration between systems through networking and communication, with a future-focused perspective aimed at exploiting the emerging benefits of workflow digitization, big data analytics and business intelligence.

34.7   Individual IT Employee Issues

Table 34.5 shows the results for responses on IT employee issues in Thailand. A similar Likert scale was used as above with 1 being strongly

Table 34.5:   Individual IT Employee Issues in Thailand

Individual Issues	Mean Rating*	Std. Deviation
Job satisfaction		
In general, I like working here.	2.20	0.81
All in all, I am satisfied with my current job.	2.27	0.83
In general, I don't like my current job.	3.56	1.03
Work pressure		
I feel that the number of requests, problems or complaints that I deal with at work is more than expected.	2.86	0.98
I feel that the amount of work I do interferes with how well it is done.	2.39	0.86
I feel busy or rushed at work.	2.64	0.91
I feel pressured at work.	2.98	0.93
Work–life balance		
There is a blurring of boundaries between my job and my home life.	3.15	1.10
My work-related responsibilities create conflicts with my home responsibilities.	3.46	1.10
I do not get everything done at home because I find myself completing job-related work.	3.33	1.10
Workload and burnout		
I feel drained from activities at work.	2.98	1.07
I feel tired from my work activities.	2.95	1.04
Working all day is a strain for me.	2.85	1.00
I feel burned out from my work activities.	2.95	1.03

(Continued)

Table 34.5:   (*Continued*)

Individual Issues	Mean Rating*	Std. Deviation
Sense of accomplishment		
I feel I'm making an effective contribution to what this organization does.	2.31	0.72
In my opinion, I do a good job.	2.21	0.69
I have accomplished many worthwhile things in this job.	2.42	0.87
At my work, I feel confident that I am effective at getting things done.	2.17	0.73
Threats to one's job		
I am worried that future technology advancements may pose a threat to my job.	2.95	1.05
I believe that other people may be able to perform my work activities.	2.84	1.18
I am concerned that my job may be eliminated soon.	3.54	1.09
I am concerned that my job may be outsourced soon.	3.50	1.11
Career plans		
I will be with this organization 1 year from now.	3.39	1.33
I will take steps during the next year to secure a job at a different organization.	3.74	1.11
I will be with this organization 5 years from now.	3.20	1.14
I will be working in the IT field 1 year from now.	2.26	1.14
I will take steps during the next year to secure a job outside the IT field.	3.68	1.06
I will be working in the IT field 5 years from now.	2.46	1.08

*Rating scale ranges from 1 to 5: 1 as strongly agree, to 5 as strongly disagree.

agree and 5 being strongly disagree. Caution should be applied in interpreting the results as some items are stated in the negative (e.g., I don't like) and some are stated in the positive (e.g., I like).

Overall, the results of the table indicate that the IT employees are generally satisfied with their job, perceive that they have job security, but possibly due to high demand for skills and IT market growth opportunities, there is a high interest in seeking IT jobs in other organizations.

More specifically, IT employees in Thailand perceive a degree of job security and are happy with their jobs. They are not concerned about their job being outsourced or eliminated. They are mostly content when it comes to having the right work–life balance, as they complete work at office during work time, and maintain a clear delineation between work and home duties. They do not suffer from excessive workload or burnout issues. They also have a high sense of accomplishment at their work.

Looking ahead, the respondents were positive about staying in the IT industry in the short and long term, but plan to move to a different organization. This is interesting given that they reported satisfaction in their current roles. This could be due to a growing demand for IT skills and know-how, creating opportunities for them, and their possible perception that the 'grass is greener on the other side'. What is also interesting is the fact that, while they predominately believe that they will be leaving their current organization within a year, they are not actively seeking opportunities elsewhere. This could offer a glimpse at the work and recruitment culture in Thailand, where much happens through connections and word of mouth in a collectivist culture. The high demand may also mean that IT employees are being sought after and headhunted rather than having to seek and apply for open positions.

34.8 Conclusion

After assessing multiple factors affecting the IT Industry in Thailand, it seems highly likely that the national policy, Thailand 4.0, has successfully boosted growth and development of the IT industry. IT workers in Thailand are generally satisfied with their jobs, but seek mobility in a dynamic market. IT infrastructure, strategy, and governance are the corner-stones of a digital knowledge economy. While these issues are still of concern to Thai respondents, the fact that such issues were highlighted as significant demonstrates maturity and growing attention to such strategic issues in the IT industry. Nonetheless, major concerns in both Technical and Organizational dimensions exist, particularly in relation to IT infrastructure and the potential to benefit from big data and analytics infrastructure. There are also prominent concerns about security, privacy, as well as the importance of governance of knowledge and projects to address such concerns.

The Thai IT industry is indeed growing with a promise of a prosperous future, should it overcome the issues identified. A promising digital future and knowledge economy await, conditional on being able to adapt to emerging challenges.

References

Accenture (2017). Insights to Digital Commerce, Available at https://www. accenture.com/th-en/insight-digital-commerce-apac-perspective. Accessed October 2018.

Aoun, C., Vatanasakdakul, S., & Cecez-Kecmanovic, D. (2011). Can IS Save the World? Collaborative Technologies for Eco-Mobilisation. In *The proceedings of the 22nd Australasian Conference on Information Systems*, 29 November–2 December 2011, Sydney.

CountryWatch (2018). Thailand Country Review 2018. Houston: CountryWatch, Inc.

EU Gateway (2018). Information & Communication Technologies — Thailand Market Study. Available at https://www.eu-gateway.eu/sites/default/f iles/collections/document/file/eu-gateway-information-communication-th ailand.pdf. Accessed October 2018.

IBM (2013). IBM in Thailand — An Overview, Available at http://www-07.ibm. com/my/media/pdf/IBM-Brochure-Thailand.pdf. Accessed October 2018.

ITU (2017). ICT Development Index. Available at http://www.itu.int/net4/IT U-D/idi/2017/index.html. Accessed October 2018.

Kantabutra, V. (2001). Thailand's First Computers: The IBM 1620 and The IBM 1401, *Journal of the Thai Statistical Association*. Available at https://www.academia.edu/2819109/Thailand_s_First_Computers_The_ IBM_1620_and_The_IBM_1401. Accessed October 2018.

KPMG (2018). ASEAN Business Guide Thailand. Available at https://ho me.kpmg.com/content/dam/kpmg/sg/pdf/2018/07/ASEAN-GUIDE-Tha iland.pdf. Accessed October 2018.

National Information Technology Committee Secretariat (2003). Information Technology Policy Framework 2001–2010: Thailand Vision Towards a Knowledge-Based Economy. Bangkok: National Information Technology Committee Secretariat.

National Information Technology Committee Secretariat (2011). Executive Summary Thailand Information and Communication Technology Policy Framework (2011–2020). Available at http://www.mdes.go.th/view/10/All%20N ews/e-Publication/25. Accessed October 2018.

Ninkitsaranont, P. (2018). Krungsri Research. *Thailand Industry Outlook 2018–2020: Mobile Operator*. Available at https://www.krungsri.com/ bank/getmedia/e1267c95-bfa8-48eb-ba34-73024fac8f63/IO_Mobile_Opera tor_2017_EN.aspx. Accessed October 2018.

Palvia, P., Jacks, T., Ghosh, J., Licker, P., Romm-Livermore, C., Serenko, A., & Turan, A. H. (2017). The World IT Project: History, trials, tribulations, lessons, and recommendations. *Communications of the Association for Information Systems*, 41(18), 389–413.

Palvia, P., Ghosh, J., Jacks, T., Serenko, A., & Turan, A. (2018). Trekking the globe with the World IT Project. *Journal of Information Technology Case and Application Research*, 20(1), 3–8.

Palasri, S., Huter, S., & Wenzel, Z. (1998). *The History of the Internet in Thailand*. The Network Startup Resource Center, University of Oregon.

Thailand Board of Investment (2017). Thailand 4.0 Means Opportunity Thailand. Available at https://www.boi.go.th/upload/content/TIR_Jan_32824. pdf. Accessed October 2018.

Thailand Board of Investment (2018). Thailand's Digital Economy & Software Industry. https://www.boi.go.th/upload/content/digital_economy_5a4fa47 0adda5.pdf. Accessed October 2018.

The Economist Group (2018). The Inclusive Internet Index. Available at https:// theinclusiveinternet.eiu.com/. Accessed October 2018.

Thuvasethakul, C. & Koanantakool, T. (2002). National ICT Policy in Thailand. Available at http://www.nectec.or.th/users/htk/publish/20020302-Nation al-ICT-Policy-v16-word.pdf. Accessed October 2018.

Trading Economic (2018). Thailand GDP Annual Growth Rate. Available at https://tradingeconomics.com/thailand/gdp-growth-annual. Accessed October 2018.

We Are Social (2017). Digital in 2017: Global Overview. Available at https:// wearesocial.com/special-reports/digital-in-2017-global-overview. Accessed October 2018.

Wongwuttiwat, J. & Lawanna, A. (2018). The digital Thailand strategy and the ASEAN community. Available at https://onlinelibrary.wiley.com/doi /epdf/10.1002/isd2.12024. Accessed October 2018.

Worldometer (2018). Thailand Population. Available at http://www.worldomet ers.info/world-population/thailand-population/. Accessed October 2018.

Chapter 35

Information Technology Issues in Turkey

Aykut Hamit Turan[*,‡], Naciye Güliz Uğur[†,§],
and Prashant Palvia[‡,¶]

Sakarya University, Sakarya, Turkey
†*The University of North Carolina at Greensboro,
Grrensboro, NC, USA*
‡*ahturan@sakarya.edu.tr*
§*ngugur@sakarya.edu.tr*
¶*pcpalvia@uncg.edu*

Summary

In this chapter, we report organizational, technological and individual information technology (IT) issues of Turkish IT workers. The participants of our survey are mostly young IT professionals, working full time in non-managerial positions. IT reliability and efficiency, security and privacy, and outsourcing were among the most pressing organizational issues. Among technology issues, business intelligence and analytics, enterprise application integration, and networks and telecommunications were the top concerns. Cost reduction and globalization were among the least important organizational IT-related issues, while cloud computing and collaborative and workflow tools were the least important technology issues. Turkish IT employees seem to be satisfied with their work and perceived their workloads to be meaningful. They exhibited moderate levels of turnover intention and felt pretty secure in their jobs.

35.1   Introduction

Many studies in the literature investigate information technology (IT)-related issues in organizations in the US and in developed countries. However, countries in the process of rapid growth and industrialization do not have a clear understanding of IT related issues. One can argue that such countries differ in terms of market characteristics; culture, economics, and politics; educational background of the population, and technology perceptions. Turkey is one of the countries that have an emerging economy. The literature on IT employees and employee-related issues in an emerging market is scarce and, in particular, studies concerning IT in Turkey are not readily available. In this chapter, we aim to understand the perceptions of Turkish IT employees with regard to organizational IT, technological, and individual issues.

35.2   Country Background and History

Turkey, officially known as the Republic of Turkey, lies between Asia and Europe. Due to its geopolitical importance, Turkey has hosted many civilizations and as such has a very historic and cultural heritage. There are seventeen cultural and natural wonders located in Turkey on the UNESCO World Heritage List.

   Turkey has integrated with the Western world by becoming a member of the Council of Europe, NATO, OSCE, and the G-20. Turkey is also one of the founders of OECD countries. Since 1963, Turkey is a privileged partner of the European Economic Community and is a member of the Customs Union since 1995. Turkey began full membership negotiations with the European Union in 2005. Many economists, politicians, and scientists consider Turkey as a developing country. Turkey's gross national product is more than US\$1trillion and it is the world's 15th largest economy. In terms of GDP per capita, the World Bank has assessed Turkey in the middle-upper income layer (World Bank, 1993). According to Forbes magazine, as of March 2008, there were 35 billionaires in İstanbul, behind only London (36 people), New York (71 people), and Moscow (74 people).

35.3 Information Technology in Turkey

In recent years, great progress has been made in the field of IT in the country. Parallel to this, IT is quickly settling into the *sine qua non* of the people's everyday lives. In this sense, as in other countries, IT is seen as a critical element for progress.

The first computer in the country was an IBM computer in 1960 at the Highways General Directorate. In the 1970s, government institutions and universities slowly started to buy computers. In the 1980s, computers that are large and capable of making complex calculations started to appear at government institutions. The first personal computers were used in the 1990s, while Internet utilization started in 1993 (NTV, 2000a). The Turkish Microsoft operating system beta version was first published in 1983, and Turkish Windows 95 followed suit. The real success of Windows came with the XP version as the variety and the number of computers increased continuously beginning in 2011. Now personal computers can be found in every mid and higher income home (NTV, 2000b).

According to a study done by the Turkish Statistic Institution, as of 2017, the computer usage rate of people aged between 16 and 74 is 56.6% and the Internet adoption rate is 66.8%. In 2017, the ratio of households with broadband Internet access reached 78.3% and one out of four people does Internet shopping (TUIK, 2017).

Another study by Deloitte, a consultancy company, was carried out in 33 countries with more than 53 thousand participants called "Global Mobile User Study" and according to this study, Turkish mobile phone users glance at their screen in average 78 times every day, meaning in every 13 minutes. The same figures were 70 times and every 15 minutes in 2015. Turkish people are the most dependent people to their mobile phones and the majority check their phones one last time just before going to sleep (Haberturk, 2008).

E-government services are picking up in the country and offer hundreds of services online. About 42.4% of Turkish citizens use these (TUIK, 2017). Business trends that typically emerge in Europe and the USA eventually affect Turkish IT users as well. Turkish companies outsource many of their processes and operations, especially call centers. The call center industry has reached a market value of 5.1 Billion Turkish Liras (over US$1billiion), and the total outsourcing business is expected to reach a volume of US$20 billion in a very short time (Global, 2018).

35.4   Methodology

The standard instrument from the World IT Project (Palvia *et al.*, 2017; Palvia *et al.*, 2018) was translated into Turkish using the back-translation method. First, one of the authors translated the instrument into Turkish and then another author translated it back into English. By involving a third another academician, the back-translation was compared with the original copy and discrepancies were resolved.

The completion of the survey by each participant took about 30 minutes. The survey was posted online and IT workers were invited to participate. In total, 600 emails were sent out. After two weeks, a reminder was sent. Later, authors personally visited IT workers in the city of Sakarya and requested them to participate in the survey in a face-to-face setting. Companies located in the Special Industrial Zones were also visited. There are three such zones in the Sakarya province. In some instances, IT managers or CIOs of companies were visited and invited to distribute the survey to their IT employees. In the face-to-face survey administration, respondents could ask questions and request clarifications. Online surveys were directly imported to Excel and surveys administered by personal visits were manually entered into Excel.

There was no specific incentive or bonus offered to participants. An ethical report was obtained from the Sakarya University Ethical Commission; it was presented during the face-to-face implementation and was made available if requested by online participants. Communication between the investigators and the core World IT Project team were mainly via emails and phone.

In our sample, Turkish IT employees are mostly in early ages (39.7% are between 21 and 29 years of age). The second largest group is between 30 and 39 years of age (27.9%), followed by the 18–20 years group (17.1%). The most common job roles were programming (25.8%), system administrator and training (11.5%), and system analysis and design (7.7%). The least common roles were integration, security, financial, and email and messaging systems. All respondents had Turkish nationality, but only one respondent was born in Germany. The sample is largely composed of males (80.1%). About 60% of our sample is made up of non-management employees; the rest are in some management position. More descriptive statistics are provided in Table 35.1.

Table 35.1: Descriptive Statistics

Characteristics	N	%	Characteristics	N	%
Education:			Years of work experience:		
High school or less	60	20.9	0–4 years	100	34.8
Associate degree	71	24.7	5–9 years	62	21.6
Bachelor's degree	107	37.3	10–19 years	88	30.7
Master's degree	38	13.2	20–29 years	28	9.8
Ph.D.	11	3.8	30+ years	9	3.1
Years of IT experience:			Organizational location:		
0–4 years	97	33.8	IT department employee	198	69.0
5–9 years	90	31.4	IT worker in non-IT department	32	11.1
10–19 years	76	26.5	Contract employee	28	9.8
20–29 years	20	7.0	Consultant	11	3.8
30+ years	4	1.4	Vendor employee	18	6.3
Work as:			Work position:		
Mostly full time	211	73.5	Not part of management	172	59.9
Mostly part time	36	12.5	In lower management	46	16.0
Mostly over time	40	13.9	In middle management	50	17.4
Been laid off from IT job:			In senior management	19	6.6
Yes	33	11.5			
No	254	88.5			

35.5 Organizational IT Issues

The survey participants were asked to rate 18 organizational IT related issues in terms of their importance level based on a 5-point Likert scale, with 1 being the most important and 5 being of no importance. The average rating for each issue was computed. The 18 issues are ranked from 1 to 18 based on their average ratings and are shown in Table 35.2.

According to our sample of IT employees and managers, IT reliability and efficiency is the most important organizational IT issue. Security and privacy came up as the second important organizational issue, followed by outsourcing, and continuity planning and disaster recovery, which tied for the third and fourth places, respectively. The next two issues were tied as well: business productivity and cost reduction, and BYOD. The least important organizational issues listed were globalization, IT cost reduction, enterprise architecture, business agility and speed to market, project management, and alignment between IT and business. Note that some of the least important issues are very close to each other in terms of their average ratings.

Table 35.2:   Organizational IT Issues

Organizational IT Issue	Rank	Mean Rating*	Std. Deviation
IT reliability and efficiency	1	1.82	0.724
Security and privacy	2	1.89	0.870
Outsourcing	3	1.94	0.755
Continuity planning and disaster recovery	4	1.94	0.751
Business productivity and cost reduction	5	2.08	0.843
Bring your own computing device (BYOD)	6	2.08	0.957
Knowledge management	7	2.12	0.819
IT strategic planning	8	2.15	0.881
IT service management (e.g., ITIL)	9	2.19	0.766
Revenue-generating IT innovations	10	2.21	0.907
Attracting and retaining IT professionals	11	2.23	0.820
Business process reengineering	12	2.25	0.843
Alignment between IT and business	13	2.27	0.812
Project management	14	2.27	0.933
Business agility and speed to market	15	2.28	0.872
Enterprise architecture	16	2.28	0.746
IT cost reduction	17	2.31	0.907
Globalization	18	2.35	0.826

*Rating scale ranges from 1 to 5: 1 as most important and 5 as no importance.

Reliable and efficient information technologies play an important role in Turkish organizations' success and failure. Working on reliable information, organizations can make better decisions and achieve a competitive advantage by improving their work performance. Turkish people are usually skeptical about technology and outputs of information systems and tend to validate outputs and information in a variety of ways. Hence, the reliability of systems coming up at the top was not very surprising for the Turkish sample.

Privacy and security have become very important in the global economy. Users have become more aware and concerned with security and privacy issues, especially in the developed economies, given the culture and the lax business practices. In Turkey in 2016, 50 million residents' data was leaked from the government's databases. This event and similar vulnerabilities in Turkish data security have heightened users' worries about data security and privacy. Besides, the ISO/EC Information Systems Management Standards have become mandatory in Turkish enterprises — leading to greater awareness of security and privacy.

Another important organizational issue is outsourcing. Outsourcing has become an important trend in the last couple of decades in Europe and the USA. It seems that Turkish firms are also realizing the importance of outsourcing their IT activities. China and Eastern European countries are the preferred destinations for outsourcing in Turkish firms. Today, many Turkish firms have started to outsource their operations abroad and this trend has become an important one. The most important reason for IT outsourcing is the significant savings in costs. Turkish organizations do not want to bear the high costs of IT investments and outsourcing provides an attractive alternative.

Recovering from natural disasters and assuring continuity of operations are emerging concerns in Turkish firms. Globally increasing terrorism, natural disasters due to global warming and climate change, fast-changing business plans, and merging of organizations all require firms to be prepared for continuity planning and disaster recovery. Turkey lies on major earthquake fault lines. During the winter, Northern parts of Turkey face the risks of floods. Thanks to investments in infrastructure in the last few years, power shortages have been largely eliminated. However, natural disasters and terrorism pose significant risks and organizations need to be always on the guard.

Regarding the least important issues, globalization came at the bottom. Perhaps the reason is that globalization has now become a fact and a way of life, and as such IT professionals do not perceive it as a new trend that needs to be managed. Globalization has likely been already integrated into organizational goals and policies and as such does not require any special efforts. IT cost reduction was also not a major concern as the cost of IT products is continuously decreasing even as the technology continues to advance. Other issues which came out low in rank include enterprise architecture, business agility, and project management. Turkey, like many other developing nations, does not face intense competitive pressures and environmental turbulence, which require more attention to such issues.

35.6 Technology Issues

The respondents were also asked to rate several IT issues in terms of their importance based on a 5-point Likert scale, with 1 being of most importance and 5 being of no importance. Sixteen technology-related issues were rated on this scale and are shown in Table 35.3.

The top five most important issues rated by respondents were business intelligence/analytics, enterprise application integration, networks/telecommunications, mobile applications development, and business process management systems.

Table 35.3:   Technology Issues

IT-Related Issues	Rank	Mean*	Std. Deviation
Business intelligence/analytics	1	1.85	0.86
Enterprise application integration	2	1.91	0.88
Networks/telecommunications	3	1.93	1.06
Mobile apps development	4	1.94	0.89
Business process management systems	5	1.97	0.69
Mobile and wireless applications	6	1.99	0.94
Big data systems	7	2.04	0.76
Software as a Service	8	2.07	0.86
Service-oriented architecture (SOA)	9	2.08	1.01
Customer relationship management (CRM) systems	10	2.10	0.90
Social networking/media	11	2.10	1.16
Data mining	12	2.15	0.97
Virtualization (Desktop Or Server)	13	2.20	0.85
Enterprise resource planning (ERP) systems	14	2.21	0.90
Collaborative and workflow tools	15	2.24	0.85
Cloud computing	16	2.30	1.03

*Rating scale ranges from 1 to 5: 1 as most important and 5 as no importance.

As data of all variety grow exponentially, business and society are becoming increasingly data-driven. This trend is seen worldwide, especially in the advanced nations, but was also observed in Turkey — becoming the number one issue. Turkish companies are aware that proper analyses of big data could enhance their competitive advantage. Due to business intelligence and analytics, organizations can arrive at better strategies and utilize resources in more appropriate ways. As a result, Turkish organizations are making a significant investment in analysis tools and adopting policies to build data warehouses and large data repositories.

Enterprise application integration was the second most important issue. Turkish companies have long been investing in information technologies, and their unchecked proliferation, especially in large firms, has become a significant problem. Enterprise application integration deals with the integration of existing applications in the IT infrastructure. Furthermore, Turkish enterprises have used a variety of applications, both standalone and ERP-based. Furthermore, as new technologies, systems, and applications enter the Turkish market, their integration with existing systems has become an important issue.

Networks and telecommunications were rated the third most important issue. Enhancing the power of the Internet and creating widespread business networks have created both technical and business challenges. In Turkey, as the number of Internet-based services has increased, organizations rely increasingly on networks and telecommunications equipment to enhance their importance. In the network age, the integration of various business networks and processes has become a crucial issue for organizations. Therefore, Turkish organizations are rapidly moving their operations online in order to achieve integration based on Internet technologies.

The fourth most important issue was mobile applications development. Mobile devices, especially mobile phones, are fast becoming the norm for communication among individuals in today's mobile age. As a result, Turkish companies are striving to develop mobile applications and mobile websites for their customers, suppliers, and other trade partners. Business process management systems were rated as the fifth most important issue. As efficiency and flexibility have become important for organizations, process management activities have gained significance. Process management systems offer organizations the ability to manage their processes from end to end, with little human intervention and greater agility and flexibility.

Somewhat surprisingly, the least important technology issue in Turkey was cloud computing. While cloud computing is an emerging phenomenon and there are many companies providing these services in Turkey, it was not perceived as an important technology by the IT staff. The underlying reason may be related to "trust" issues. Turkish companies may not find it secure to store their information on a platform that they do not have any control over. Instead, they make investments to create their own secure server rooms and use own software to access the data. Thus security issues may override any cost and expertise advantages of cloud computing.

The second and third least important issues were collaborative and workflow tools and ERP systems. While collaboration has become a standard practice in today's corporate life, it may be that Turkish IT employees may be collaborating without the use of software tools and may not see the need for such tools. As to ERP systems, while ERP are becoming increasingly popular in developed countries, they require a significant upfront investment and are not always implemented successfully. Thus rather than starting from ground zero as required in ERP systems, the emphasis in Turkey seems to be the integration of existing applications and new applications in the software portfolio — as it was rated the second most important issue.

35.7   Individual IT Employee Issues

Individual IT employee issues were investigated under seven broad themes. The detailed analysis of each theme is provided below in Table 35.4.

Table 35.4:   Individual IT Related Issues

Items	Mean*	Std. Deviation
Job satisfaction		
In general, I like working here.	2.03	0.91
All in all, I am satisfied with my current job.	2.22	0.84
In general, I don't like my current job.	3.41	1.31
Work pressure		
I feel that the number of requests, problems or complaints that I deal with at work is more than expected.	2.77	1.15
I feel that the amount of work I do interferes with how well it is done.	3.05	1.04
I feel busy or rushed at work.	2.95	0.99
I feel pressured at work.	3.25	1.17
Work–life balance		
There is a blurring of boundaries between my job and my home life.	2.77	1.21
My work-related responsibilities create conflicts with my home responsibilities.	3.05	1.13
I do not get everything done at home because I find myself completing job-related work.	3.14	1.15
Workload and burnout		
I feel drained from activities at work.	3.06	1.09
I feel tired from my work activities.	2.90	1.06
Working all day is a strain for me.	3.04	1.05
I feel burned out from my work activities.	2.97	1.13
Sense of accomplishment		
I feel I'm making an effective contribution to what this organization does.	2.14	0.90
In my opinion, I do a good job.	2.18	0.84
I have accomplished many worthwhile things in this job.	2.13	0.84
At my work, I feel confident that I am effective at getting things done.	2.10	0.89
Threats to one's Job		
I am worried that future technology advancements may pose a threat to my job.	3.36	1.19
I believe that other people may be able to perform my work activities.	2.62	0.99
I am concerned that my job may be eliminated soon.	3.60	1.26
I am concerned that my job may be outsourced soon.	3.47	1.16

(Continued)

Table 35.4: (*Continued*)

Items	Mean*	Std. Deviation
Career plans		
I will be with this organization 1 year from now.	2.29	1.07
I will take steps during the next year to secure a job at a different organization.	3.24	1.14
I will be with this organization 5 years from now.	2.81	1.14
I will be working in the IT field 1 year from now.	2.37	1.15
I will take steps during the next year to secure a job outside the IT field.	3.43	1.19
I will be working in the IT field 5 years from now.	2.40	1.23

*Rating scale ranges from 1 to 5: 1 as strongly agree and 5 as strongly disagree.

In general, the IT industry in Turkey is vibrant and alive. Finance organizations lead the industry with applications using cutting-edge technologies and tools. Many young people and especially new graduates are hired by Turkish companies as IT personnel. As the Turkish economy transitions from manufacturing to the service industry, the demand for IT staff is expected to continue to rise.

As per our scale, anything below 3 shows an agreement with the stated item and above 3 shows disagreement. In terms of job satisfaction, respondents seemed to be satisfied with their jobs. The standard deviation is also low, pointing to general agreement on this issue. The IT sector in Turkey pays quite well and young IT professionals usually make good money compared to their peers from other industries. In addition, many young people in the IT industry are single and do not seem to feel the pressure of their married peers in terms of living expenses.

The participants seem to perceive high workloads in their IT jobs. However, the standard deviation in the results is somewhat high. Therefore, depending on the organization, there could be varying degrees of workloads among IT employees. There is a general tendency in Turkish organizations for managers to give more work and put more pressure on the younger employees. But given their age and enthusiasm, they usually take this load on willingly and do not complain.

In terms of work–life balance and work–home conflict factor, the average scores turned out to be very close to the mid-point. Thus the respondents did not seem to have significant work–home conflicts. Once again, the Turkish IT workers in our sample are mostly young and many are single; thus they may not experience much work–home conflict. Yet, the standard deviation is high, meaning there could be some exceptions to the overall findings. Older and more experienced IT employees working for SMEs may find difficulty in balancing work and life needs, as Turkish SMEs prefer to hire fewer but experienced IT staff and have them do all IT-related work.

Similar results were observed for work exhaustion, with mean values of the four items close to the mid-point. The finding is related to the observation that there were no major concerns about workload or work pressure. Once again, the standard deviation is high, indicating that exceptions exist. Some Turkish organizations would not even hire the minimum number of IT professionals and expect young and new graduates to do all the work. As was observed earlier, many SMEs prefer to hire very few IT personnel to carry out the tasks. On the other hand, IT employees working in public organizations may not have exhausting workloads.

The IT employees, who rated themselves, seem quite content in terms of their self-efficacy and accomplishments. Remarkably, there was considerable agreement as the standard deviation was low. These employees are young, well-educated and know that there is a high demand for their IT skills in the job market. They also regard their professional IT skills valuable to their organization. As an example, two of the authors of this chapter are employed by the Management Information Systems department of Sakarya University and, on the average, the department's IS graduates are offered full-time jobs within one month after graduation.

Despite the above observations, Turkish IT workers do not seem to feel secure in their jobs. They see a persistent threat to their jobs due to advancements in technology and the ever-increasing number of IT graduates. Turkish universities offer new undergraduate and graduate programs in IT every year and the Turkish Higher Education Council continues to increase the quota of the number of such students. Also, while the IT workers believe that others cannot perform their jobs, they are also aware that if they do not enhance their skills continuously, they could be replaced. Furthermore, in the Turkish IT industry, outsourcing of IT work is a recent trend and could pose further risks to job security in the longer term.

Looking at the career and turnover data, Turkish IT employees do not anticipate changing their jobs or the profession in the short term. In the longer 5-year term, the propensity to leave the organization or the profession increases some, but only slightly. Current gloomy economic conditions in Turkey and the relatively high salaries in the IT industry help explain these results. Basically, once you are unemployed, it is very difficult to find another job and more so a higher paying job. The unemployment rate is about 10% in Turkey and unemployment is even higher among young people. With the comparatively high salaries in the IT field, employees generally do not like to change their companies or desert the profession.

35.8 Conclusion

This chapter provided an understanding of the organizational IT issues, technology issues and individual concerns of IT employees in Turkey. The most important IT-related organizational issues in Turkey include IT reliability and efficiency, security and privacy, and outsourcing. Among technology issues, business intelligence and analytics, enterprise application integration, and networks and telecommunications were among the top concerns. In general, the Turkish IT employees seem to be satisfied with their jobs and found their workloads to be reasonable, having little desire to change jobs or the profession itself. Turkey has different business and environmental dynamics and culture compared to many of the advanced nations. Thus understanding these issues in the Turkish context provides new insights. To the best of our knowledge, these results have been provided for the first time in an organized manner and should provide much value to both academics and practitioners.

References

Global (2018). "Türkiye'de outsourcing 20 milyar TL büyüklüğe ulaşacak". Retrieved from http://www.globaltechmagazine.com/turkiyede-outsourcing-20-milyar-tl-buyukluge-ulasacak/ [Accessed June 6, 2018].

Haberturk (2018). "Telefonsuz 13 Dakika Dayanamıyoruz". Retrieved from http://www.haberturk.com/turkiye-de-akilli-telefon-kullanim-orani-nedir-1793115-ekonomi [Accessed June 6, 2018].

NTV (2000a). "Türkiye'de Bilişimin Öyküsü -1". Retrieved from http://arsiv.ntv.com.tr/news/28088.asp [Accessed June 6, 2018].

NTV (2000b). "Türkiye'de Bilişimin Öyküsü -3". Retrieved from http://arsiv.ntv.com.tr/news/28113.asp [Accessed June 6, 2018].

Palvia, P., Jacks, T., Ghosh, J., Licker, P., Romm-Livermore, C., Serenko, A., & Turan, A. H. (2017). The World IT Project: History, trials, tribulations, lessons, and recommendations. *Communications of the Association for Information Systems*, 41(18), 389–413.

Palvia, P., Ghosh, J., Jacks, T., Serenko, A., & Turan, A. (2018). Trekking the globe with the World IT Project. *Journal of Information Technology Case and Application Research*, 20(1), 3–8.

TUIK (2017). "HanehalkıBilişim Teknolojileri Kullanım Araştırması, 2017"
    Retrieved from http://www.tuik.gov.tr/HbPrint.do?id=24862 [Accessed
    June 6, 2018].
World Bank (1993). *Turkey: Informatics and Economic Modernization.*
    Washington, DC: World Bank.

Chapter 36

Information Technology Issues in the UK

Anne Powell[*,‡], Penny Hart[†,§], and Tim Jacks[*,¶]

*Southern Illinois University Edwardsville,
Edwardsville, IL, USA
† University of Portsmouth,
Portsmouth, UK
‡ apowell@siue.edu
§ penny.hart@port.ac.uk
¶ tjacks@siue.edu

Summary

This chapter addresses the top issues in the United Kingdom (UK) for IT professionals. The information technology (IT) profession is a growth industry in the UK and the demand for IT employees continues to rise. While London is still the largest tech hub in the UK, several other areas have seen a rise in technology innovation in the recent past. Because of technical innovations in the country, the UK has been ranked as the most entrepreneurial country in Europe, and the fourth most entrepreneurial country in the world. Reported in this chapter are the results of a survey which was completed by 95 IT employees. The typical response profile was a full-time, managerial IT employee with at least a bachelor's degree and 10 or more years of IT experience. The top three organizational IT issues were (1) IT reliability and efficiency, (2) security and privacy, and (3) alignment between IT and business. The top three technology issues were (1) mobile and wireless application, (2) business intelligence/analytics, and (3) software as a service (SaaS). IT employees in the UK are largely satisfied and believe they contribute value to their company. The participants are not concerned about their jobs being outsourced and most expect to stay within the IT industry in the near future.

36.1 Introduction

The UK has one of the strongest economies in Europe and a well-deserved status of thought leadership in the history of computing. While the UK is currently facing challenges with a still potential? exit from the European Union, it nevertheless has a dominant tech industry and a business environment conducive to entrepreneurship. This chapter will sketch out the dominant themes for information technology (IT) in the UK which provide rich results for comparison with other countries, both developed and underdeveloped. It will be seen that the UK has a unique constellation of issues that differs in important ways from other countries in the World IT Project. These differences are driven in part by the UK's economic dominance as well as historical factors. A better understanding of the issues facing IT workers in the UK will foster insight into what is similar about the IT occupation around the world and what is unique to the specific context of the UK.

36.2 Country Background and History

The UK consists of four countries: England, Scotland, Wales, and Northern Ireland. The population of the UK is 66,000,000 and is expected to rise to over 75 million by 2050. It is a unitary parliamentary democracy and constitutional monarchy. Theresa May was the Prime Minister between 2016 and 2019. Queen Elizabeth II has reigned since 1952. The United Kingdom has the fifth largest free market economy in the world with a Gross Domestic Product (GDP) of 2.6 million in 2017 (World Bank, 2017). Germany is the only country in the European Union with a larger GDP than the UK. The official language is English and approximately 95% of those living in the UK speak only English.

Historically, the UK has been a democracy and a major power in the world. The influence of the UK in language and legal systems can be seen in its many former colonies including the USA, India, and South Africa. In 1945, after World War II, the UK was one of four super powers in the world along with the USA, China, and the Soviet Union.

The top industries in the UK include finance/banking, aerospace, construction, and oil/gas. The IT industry is considered a fast growth industry (Nig, 2018). Reviewing the top imports and exports in the UK results in two very similar lists. Top exports include machinery including computers, cars, and mineral fuels. The top imports include machinery, including computers, cars, and electrical machinery/equipment. Over half of the UK's total imports and exports are purchased from/sold to other European countries (Workman, 2019), although the effect of the withdrawal of the UK from the European Union (a.k.a. "Brexit") is as yet unknown.

36.3   Information Technology in the UK

In the 1930s and 1940s, the UK pioneered the development of Computing, most notably through the efforts of Alan Turing. Famous for his code-breaking work during World War II, Turing was a founder of computer technology and machine intelligence. Computer research took place at Manchester, Cambridge, and the NPL in the 1950s and the first business data processing machine, LEO, was in use in 1953. Leadership in digital computing innovation passed to the United States, a country that, in the 1950s and 1960s, had government support of scientific research and welcomed talented immigrants, policies the United Kingdom did not have as such, or as stated in one article "UK policies, on the other hand, killed its technology industry, quite literally in the case of Alan Turing" (Satell, 2015, p. 4).

Throughout the second half of the 20th century, computer manufacturing and software development increased in the UK, notably with the advent of the personal computer, improved business applications, and the Internet. During the 2000s, the UK tech industry made their mark in start-up tech businesses. During the 2000s, London became a tech hotspot with numerous international tech investment projects (Heath, 2015) and both global technology companies have made their home in London as well as numerous tech start-ups (Ismail, 2018). The IT sector has a strong growth forecast and the tech sector is growing at a faster pace than the overall economy in the UK (Ismail, 2018). According to the UK's Office for National Statistics (EMP13), 1,309,000 workers were employed in the Information and Communications sector in 2018, which accounted for 4% of the UK's working population.

Tech growth in the UK is not limited to London. Several tech hubs have been identified outside of London. For example, the fastest growing sector in Scotland is digital technology. Cambridge is home to over 1500 tech businesses with hundreds of tech start-ups created every year. With several other tech hubs throughout the UK, IT continues to be a growth industry. Firms are engaged in all current IT industry trends, including artificial intelligence and machine learning, Cloud Computing, Internet of Things, augmented and virtual reality, distributed ledgers and data analytics (Deloitte, 2018).

36.4   Methodology

The World IT Project (Palvia *et al.*, 2017; Palvia *et al.*, 2018) was used as the data collection method. The survey was promulgated by email, followed up by phone calls. The sampling strategy was broadly purposive — the intention was to contact IT professionals across a range of organizations and industries throughout the UK. In practical terms, convenience and snowball sampling were used. The researchers' contacts in business were approached, and further respondents were sought using business directories for London and the Southeast. Contacts in some organizations acted as gatekeepers, forwarding the survey on to their colleagues.

Respondents included a number of large organizations in the Southeast of the UK, Information Services departments in Universities in the South, the Midlands and the Northeast, and smaller organizations with whom the researchers had working relationships or contacts through alumni from the university. There was a preponderance of male middle managers in the early returns, and efforts were made to gain a more distributed set of data. No incentives were offered to the participants.

Table 36.1: Descriptive Statistics

Characteristics	N	%	Characteristics	N	%
Education:			Years of work experience:		
High school or less	6	6.3	0–4 years	6	6.3
Associate degree	12	12.5	5–9 years	15	15.6
Bachelor's degree	36	36.5	10–19 years	17	17.7
Master's degree	32	33.3	20–29 years	22	22.9
Ph.D.	11	11.5	30+ years	36	37.5
Years of IT experience:			Organizational location:		
0–4 years	15	15.6	IT department employee	70	72.9
5–9 years	15	15.6	IT worker in non-IT department	8	8.3
10–19 years	25	26.0	Contract employee	8	8.3
20–29 years	22	22.9	Consultant	9	9.4
30+ years	18	18.8	Work position:		
Work as:			Not part of management	33	34.4
Mostly full time	82	85.4	In lower management	16	16.7
Mostly part time	8	8.3	In middle management	25	26.0
Mostly over time	5	5.2	In senior management	21	21.9
Been laid off from IT job:					
Yes	23	24.0			
No	72	75			

Table 36.1 provides descriptive statistics of the IT professionals who took part in the study from the UK. Of the 97 responses in the UK dataset, the typical response profile was a full-time IT worker with at least a bachelor's degree and 10 or more years of IT experience. The majority of the respondents were in some level of management (lower, mid, or senior). Additional details on demographics can be seen in Table 36.1.

36.5   Organizational IT Issues

Table 36.2 shows the organizational IT issues. UK respondents identified (1) IT reliability and efficiency, (2) security and privacy, and (3) alignment between IT and business as the three most important organizational issues in the UK. Cybersecurity and information security continue to grow in importance for organizations. A survey of 3000 CIOs (Garfinkel, 2018) found that they gave high priority to addressing security vulnerabilities that threaten their organizations, and that there is a move towards dedicated CISO (Chief Information Security Officer) roles at the Executive Board level.

Table 36.2:   Organizational IT Issues in the U.K.

Organizational IT Issues	Rank	Mean Rating*	Std. Deviation
IT reliability and efficiency	1	1.48	0.713
Security and privacy	2	1.58	0.833
Alignment between IT and business	3	1.90	0.946
IT strategic planning	4	2.01	0.869
Project management	5	2.11	0.893
Continuity planning and disaster recovery	6	2.15	1.151
Attracting and retaining IT professionals	7	2.16	1.089
Knowledge management	8	2.16	0.915
Business productivity and cost reduction	9	2.37	0.803
Business agility and speed to market	10	2.42	0.925
IT service management (e.g., ITIL)	11	2.45	1.025
IT cost reduction	12	2.55	0.819
Enterprise architecture	13	2.59	1.005
Business process reengineering	14	2.68	0.877
Revenue-generating IT innovations	15	2.70	1.048
Bring your own computing device (BYOD)	16	3.22	1.339
Globalization	17	3.23	1.128
Outsourcing	18	3.48	1.095

*Rating scale ranges from 1 to 5: 1 as most important and 5 as no importance.

It is interesting to note that IT reliability and efficiency and security and privacy were also the top two issues in the USA dataset from the World IT Project. But where attracting and retaining IT professionals was ranked third in the USA dataset, this was only seventh in the UK, which emphasizes some of the differences between the two countries.

36.6   Technology and Infrastructure Issues

As can be seen in Table 36.3, the top three technology and infrastructure issues in the UK were (1) mobile and wireless applications, (2) business intelligence/analytics, and (3) software as a service (SaaS).

In the UK, 91% of those in the 18–44 age group owned smartphones in 2016. The mobile app developer market in the UK is strong and growing. The UK has some of the best app developers and is known as a powerhouse of the global app development market (Dogtiev, 2018). With so many app development companies in the UK, particularly London and Manchester, it is not surprising that mobile and wireless applications would be a top IT issue in the country. Mobile growth throughout the world has a big impact

Table 36.3:   Technology and Infrastructure Issues in the U.K.

IT-Related Issues	Rank	Mean Rating*	Std. Deviation
Mobile and wireless applications	1	2.20	1.053
Business intelligence/analytics	2	2.29	0.972
Software as a aervice	3	2.36	0.975
Cloud computing	4	2.40	1.174
Networks/telecommunications	5	2.40	1.275
Collaborative and workflow tools	6	2.41	0.878
Virtualization (desktop or server)	7	2.46	1.196
Enterprise application integration	8	2.51	1.016
Big data systems	9	2.74	1.207
Data mining	10	2.76	1.122
Customer relationship management (CRM) systems	11	2.77	1.162
Business process management systems	12	2.83	0.948
Service-oriented architecture (SOA)	13	2.92	1.012
Social networking/media	14	2.92	1.043
Enterprise resource planning (ERP) systems	15	2.98	1.095
Mobile apps development	16	3.00	1.196

*Rating scale ranges from 1 to 5: 1 as most important and 5 as no importance.

on SaaS. As more developing countries adopt mobile technologies, the use of mobile SaaS and the number of SaaS buyers will escalate (Salkar, 2017).

Business analytics and intelligence has been identified as a key IT issue in many studies recently and shows up as a top issue for many of the countries in the World IT Project. The SIM IT issues and trends list showed Data Analytics as the third most important IT management issue in 2017 and ranked second as the most difficult skill to find (Kappelman *et al.*, 2018). With cost reduction, faster and better decision-making, and greater ability to gauge customer needs and wants as all known benefits of data analytics, its growing importance and need for skilled employees keeps it in the forefront of UK tech issues.

As is the case in many countries, it is obvious that organizational issues in the IT industry are seen as more pressing issues than technical issues. The top eight organizational issues in the UK all received higher scores than the top technical issue.

36.7   Individual IT Employee Issues

Table 36.4 provides information into both individual issues and the IT occupation as a whole. To some extent, the responses reflect the seniority and

Table 36.4:   Individual IT Employee Issues in the U.K.

Individual Issues	Mean Rating*	Std. Deviation
Job satisfaction		
In general, I like working here.	1.86	0.858
All in all, I am satisfied with my current job.	2.04	0.910
In general, I don't like my current job.	4.04	0.944
Perceived work overload		
I feel that the number of requests, problems or complaints that I deal with at work is more than expected.	3.07	1.034
I feel that the amount of work I do interferes with how well it is done.	2.91	1.022
I feel busy or rushed at work.	2.84	1.188
I feel pressured at work.	2.74	1.103
Work–home conflict		
There is a blurring of boundaries between my job and my home life.	3.07	1.240
My work-related responsibilities create conflicts with my home responsibilities.	3.40	1.161
I do not get everything done at home because I find myself completing job-related work.	3.21	1.383
Work exhaustion/strain		
I feel drained from activities at work.	2.86	1.199
I feel tired from my work activities.	2.58	1.135
Working all day is a strain for me.	3.12	1.193
I feel burned out from my work activities.	3.29	1.237
Professional self-efficacy		
I feel I'm making an effective contribution to what this organization does.	1.98	0.799
In my opinion, I do a good job.	1.72	0.559
I have accomplished many worthwhile things in this job.	1.78	0.732
At my work, I feel confident that I am effective at getting things done.	1.92	0.663
Job insecurity		
I am worried that future technology advancements may pose a threat to my job.	3.75	1.211
I believe that other people may be able to perform my work activities.	2.68	1.024
I am concerned that my job may be eliminated soon.	3.98	1.000
I am concerned that my job may be outsourced soon.	4.18	0.922
Turnover intention		
I will be with this organization 1 year from now.	2.28	1.155
I will take steps during the next year to secure a job at a different organization.	3.52	1.080

(*Continued*)

Table 36.4: (*Continued*)

Individual Issues	Mean Rating*	Std. Deviation
I will be with this organization 5 years from now.	3.13	1.142
I will be working in the IT field 1 year from now.	1.80	0.870
I will take steps during the next year to secure a job outside the IT field.	4.08	1.048
I will be working in the IT field 5 years from now.	2.21	1.091

*Rating scale ranges from 1 to 5: 1 as strongly agree, to 5 as strongly disagree.

work history of those responding to the survey. As in the last tables, a lower average score means higher agreement with each statement. Responses indicate that, in general, IT employees in the UK have a high level of job satisfaction and feel they are making significant and worthwhile contributions to their organizations and their own jobs. Balancing work life with home life is not seen as an issue but work can impinge on their home time. While UK IT workers indicate a moderate amount of work overload and work exhaustion, these feelings are not uncommon in the IT (or any) industry.

While the UK has not been at the forefront of new technology, IT workers in the UK appear very optimistic about the field. They are highly satisfied, confident that their work is valued by their organizations, and have little job insecurity. The low job insecurity correlates with the result in Table 36.2 that lists Outsourcing as the least important organizational IT issue in the UK. The respondents' concerns are more about globalization than the effect of future technology on their organizations, reflecting their confidence in their ability to innovate.

The low job insecurity and high intent to remain in the IT field may be attributable to the UK's recent rise in IT start-ups. London is the largest and fastest growing tech hub in the UK. In fact, London ranks only behind the Silicon Valley of the USA in global tech connections (Ismail, 2018). The international development index ranks the UK fourth in its readiness to compete in the digital economy (Heath, 2015).

36.8 Conclusion

The UK had a strong start in technical innovation that sputtered out after WWII, but is now roaring back. While no one can predict the effect Brexit may have, the IT industry is expected to continue to grow and is seen as the most entrepreneurial country in Europe. The top three technology issues reported from the UK include mobile and wireless applications, business intelligence, and software as a service. These technology issues align with the UK's strength in mobile application development. IT employees in the UK report being very satisfied in their work life and feel their contributions are valuable. They have little concern about outsourcing or that their jobs will be eliminated and plan to remain in the IT field. IT in the UK is a high-growth field and it is expected that demand for IT employees will remain high.

References

Davenport, T., & Dyche, J. (2013). Big Data in Big Companies. Retrieved from https://www.sas.com/content/dam/SAS/en_us/doc/whitepaper2/bigdata-bigcompanies-106461.pdf.

Deloitte (2018). UK Technology, Media and Telecommunications Predictions 2019. Retrieved from http://www.deloitte.co.uk/tmtpredictions/.

Dogtiev, A. (2018). Top UK App Developers. Retrieved from http://www.businessofapps.com/guide/uk-app-developers/.

Garfinkel, J. (2018). CIO Agenda Survey 2018. *Gartner Symposium/ITxpo 2018*, Orlando. Retrieved from https://www.gartner.com/en/newsroom/press-releases/2018-10-16-gartner-survey-of-more-than-3000-cios-reveals-that-enterprises-are-entering-the-third-era-of-it.

Heath, A. (2015). It's taken years — but the UK is finally building a great technology industry, *The Telegraph*. Retrieved from https://www.telegraph.co.uk/finance/comment/11676983/Its-taken-years-but-the-UK-is-finally-building-a-great-technology-industry.html.

Ismail, N. (2018). What are the biggest tech hubs in the UK, *Information Age*. Retrieved from https://www.information-age.com/biggest-tech-hubs-uk-right-business-123472568/.

Ismail, N. (2018). UK Tech Expanding Faster Than the Rest of the Economy, *Information Age*. Retrieved from https://www.information-age.com/tech-nation-2018-report-uk-tech-faster-economy-123471982/.

Kappelman, L., Johnson, V., McLean, E., & Maurer, C. (2018). The 2017 SIM IT Issues and Trends Study, *MISQ Executive*, 17(1), 53–88.

Nig (2018). The UK's fastest growth industries. Retrieved from https://www.nig.com/trading-support/news/the-uk-s-fastest-growth-industries.

Office for National Statistics UK (2018). Dataset EMP13: Employment by Industry. Retrieved from https://www.ons.gov.uk/employmentandlabourmarket/peopleinwork/employmentandemployeetypes/datasets/employmentbyindustryemp13.

Palvia, P., Jacks, T., Ghosh, J., Licker, P., Romm-Livermore, C., Serenko, A., & Turan, A. H. (2017). The World IT Project: History, trials, tribulations, lessons, and recommendations. *Communications of the Association for Information Systems*, 41(18), 389–413.

Palvia, P., Ghosh, J., Jacks, T., Serenko, A., & Turan, A. (2018). Trekking the globe with the World IT Project. *Journal of Information Technology Case and Application Research*, 20(1), 3–8.

Salkar, A. (2017). Emerging nations to dominate SaaS market by 2020. Retrieved from https://www.crrux.com/i/ARTICLE/Emerging_nations_to_dominate _SaaS_market_by_2020/c76c0be22a06920a408393d7168cd151.

Satell, G. (2015). Here's What Killed The British Technology Industry, *Forbes*. Retrieved from https://www.forbes.com/sites/gregsatell/2015/01/24/here s-what-killed-the-british-technology-industry/#646be04d7f16.

Wauters, R. (2019). From tech city to tech nation. Retrieved from https://tech. eu/features/3905/from-tech-city-to-tech-nation-report/.

Workman, D. (2018). United Kingdom's Top 10 Imports. Retrieved from http:// www.worldstopexports.com/united-kingdoms-top-10-imports/.

Workman, D. (2018). United Kingdom's Top 10 Exports, Retrieved from http:// www.worldstopexports.com/united-kingdoms-top-exports/.

Chapter 37

Information Technology Issues in the US

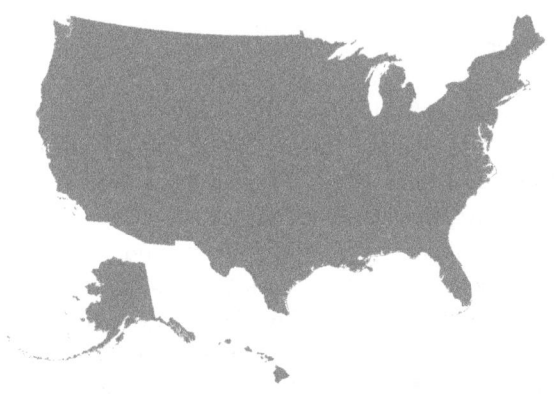

Tim Jacks[*,‡] and Prashant Palvia[†,§]

*Southern Illinois University Edwardsville,
Edwardsville, IL, USA
†University of North Carolina at Greensboro,
Greensboro, NC, USA
‡tjacks@siue.edu
§pcpalvia@uncg.edu

Summary

This chapter addresses the top issues in the United States of America (US) for information technology (IT) professionals. The IT profession continues to grow in the US and employment demand remains high.

The World IT Project gathered 308 responses in the Midwest and
Southeast of the US The typical response profile was a full-time,
non-managerial IT worker with a bachelor's degree and 10 or more
years of IT experience. The top three organizational IT issues were:
IT reliability and efficiency, Security and privacy, and Attracting
and retaining IT professionals. The top three technology issues were:
Networks/telecommunications, enterprise application integration, and
collaborative and workflow tools. IT professionals in the US show a high
level of job satisfaction and sense of accomplishment at work, a moderate amount of work pressure, lower concerns for work–life balance and
burnout, low concerns about job security, and a fairly strong desire to
stay both with their jobs and their profession in the coming years.

37.1   Introduction

The World Information Technology (IT) Project was officially launched in
2013 with a charter to examine the important issues confronting IT employees around the world. The label "IT worker" captures a broad range of
jobs from systems architecture (hardware) to application development (software) to business analysis (services). Primary importance of the project was
to take stock of the context of IT workers globally and move the research
agenda beyond simply looking at IT workers in the US and projecting those
issues on the rest of the world. We needed to create a global view that
focuses on understanding the major IS issues in the world in the context
of each country's or region's unique cultural, economic, political, religious,
and societal environments. Such a comparative examination across countries would facilitate more global understanding, cooperation, and knowledge transfer among many nationalities, both in academia as well as in
business.

An understanding of the technological, organizational, and individual
issues facing IT professionals in the US can be regarded as important by
itself. However, what is especially compelling is the ability to directly compare the ratings and rankings of US responses to the exact same questions
by their counterparts in other countries, especially non-Western countries.
This chapter will draw a brief outline of the context of the US and its special
role in the history of the IT profession. The results will provide a baseline
for comparison with other countries. While the US is not the largest country
in the world by population (China and India are) or the largest by landmass
(Russia is), it remains an important touchstone for the rest of the world
due to its economic power.

37.2   Country Background and History

The US has the largest free market economy in the world, with a Gross Domestic Product (GDP) of over \$19 trillion in 2017 (World Bank, 2017). By comparison, China is second at over \$12 trillion and the EU accounts for \$18 trillion. The US GDP continues to grow steadily. For example, in 2013 when the World IT Project first began, it was \$16.6 trillion. The US population is on a similar growth trend, from 316 million in 2013 to about 325 million in 2017.

Governmental regulation in the US is considered low compared to other countries. For example, there is no equivalent digital privacy protection for US citizens like the General Data Protection Regulation (GDPR) in the European Union as of 2018, and weaker environmental regulation (i.e., air quality, water quality, recycling, natural preservation) compared to other industrialized countries.

The US tends to be conservative on social issues such as healthcare, gun control, human rights, and immigration. Religion is traditionally separated from government activities in contrast to countries like Iran, where politics and religion are closely aligned. English is the dominant language. Most citizens speak only one language and tend to not travel outside of the US-based on its geographic isolation. Due to its economic prosperity since World War II, the US has developed a culture of consumerism, materialism, and American exceptionalism.

The top industries in the US include real estate, finance, manufacturing, retail, and IT. The top exports in the US are aircraft, cars, and integrated circuits. Its top imports are cars, crude petroleum, and computers. One of the US's most influential exports historically is the IT profession itself.

37.3   Information Technology in the US

The US can be considered the birthplace of IT, beginning with the early history of Silicon Valley in the 1960s. Most of today's most powerful IT companies were founded in the US, including Apple, Google, IBM, HP, Intel, Oracle, AT&T, Verizon, Microsoft, Amazon, Cisco, and Facebook. The Internet itself began as a US government-sponsored military project called ARPANET in 1967. Most of the typical IT certification organizations such as COMP-TIA, SANS, ISACA, and CIS are based in the US. Professional IT organizations like AITP, SIM, and IEEE, all began in the US.

Despite massive layoffs in the IT sector when the dot-com bubble burst in 2000, the need for IT jobs has remained high. IT jobs are expected to grow at 13% from 2016 to 2026 (BLS, 2018). Typical jobs include Network Architects, Programmers, Support Specialists, Systems Analysts, Database Administrators, Security Analysts, and Web Developers (BLS, 2018; White, 2018). While software application developers are most likely to experience off-shore outsourcing (Statista, 2017), cybersecurity jobs in IT are experiencing 0% unemployment (Morgen, 2016).

37.4 Methodology

Because of the size of the US, it was important to gather data as broadly as possible in order to avoid any regional bias. To that end, data were collected by the two co-authors in different geographic regions: one in the Midwest (with Tim Jacks as the lead CI) and one in the Southeast of the US (with Prashant Palvia as the lead CI). The data were collected in 2016 by using Palvia *et al.* (2017) instrument. Solicitations for participation in the World IT Project (Palvia *et al.*, 2018) were emailed to networks of contacts. Professional contacts came from industry partners affiliated with both universities. In most cases, the CIO or equivalent executive was solicited to encourage staff's participation.

The total number of responses gathered was 378, with a final total of 308 usable responses after incomplete and unusable responses were removed. Responses that indicated they did not work primarily in an IT-related field were not used in the analysis. Unusable responses also included ones where the respondent got tired of answering truthfully and began to "straight-line" answers (choosing the same answer for every single question) or leaving questions blank. If more than half the responses showed such behaviors, then the entire response was discarded. Finally, the US data collection efforts experienced the same challenges as other countries in terms of getting participants to complete a long 160-question survey. Two factors which helped overcome this obstacle were to (1) explain the importance of this global research to the executives at each organization, and (2) set expectations for the time commitment slightly higher than the typical time (i.e., "it will take about 30 minutes to complete this survey" when the average was closer to 20 minutes). The online format of the survey enabled participants to complete it relatively quickly, but they were also allowed to stop and return to the survey later.

All surveys were conducted online via Qualtrics and both data collection sites used the same survey. Because the survey was created in English, no translation was necessary as it was for some countries. Firms representing a wide variety of industry and organizational types and sizes were targeted in order to be as representative of the overall population as possible.

Items from the survey came from a variety of sources in the IS literature. IT organizational issues were pulled from annual key issues studies such as Luftman and Ben-Zvi (2010) and Luftman *et al.* (2012). See Table 37.2 for the complete list. For the IT technology issues, including an exhaustive list of all possible technologies was not feasible, therefore specific ones had to be selected based on their frequency in the IS literature and the industry experience of the core team members. These are shown in Table 37.3. Individual issues for IT employees address common themes in the IS literature such as job satisfaction (Moore, 1997), perceived work overload (Kirmeyer and Dougherty, 1988), perceived work–home conflict (Kreiner, 2006), strain (Moore, 2000), professional self-efficacy (Moore, 1997), job insecurity (Ashford *et al.*, 1989), turnover intention in the organization (Moore, 1997), and turnover intention in the IS profession (i.e., turnaway) (Moore, 1997). The complete list of individual issues is in Table 37.4. For the full survey, instrument and additional background on the challenges and lessons learned in conducting a global research project, see Palvia *et al.* (2017).

Out of 308 responses in the US dataset, the typical response profile was a full-time non-managerial IT worker with a bachelor's degree and 10 or more years of IT experience. 70% of the respondents were male, which is typical of the IT workforce in the US. Additional demographics are shown in Table 37.1.

37.5 Organizational IT Issues

The top three organizational IT issues were (1) IT reliability and efficiency, (2) Security and privacy, and (3) Attracting and retaining IT professionals. The high ranking of IT reliability and efficiency is most likely attributable to this being the core responsibility of IT workers. IT processes and technology must be reliable and efficient in order to provide value to the business. Reliability of IT is a necessary, but not sufficient, condition for organizational success. It is likely that IT reliability was ranked higher than, for example, IT Strategic Planning, due to the fact that most respondents were not in management. Security and Privacy in the number two slot are further evidence of the enormous importance of cybersecurity in the 21st century. Between privacy regulation increasing, cybercrime on the

Table. 37.1:   Descriptive Statistics

Characteristics	N	%	Characteristics	N	%
Education:			Years of work experience:		
High school or less	10	3.2	0–4 years	13	4.2
Associate degree	45	14.6	5–9 years	54	17.5
Bachelor's degree	177	57.3	10–19 years	67	21.7
Master's degree	75	24.3	20–29 years	74	23.9
Ph.D.	1	0.3	30+ years	99	32.0
Years of IT experience:			Organizational location:		
0–4 years	67	21.7	IT department employee	252	81.6
5–9 years	46	14.9	IT worker in non-IT department	16	5.2
10–19 years	68	22.0	Contract employee	20	6.5
20–29 years	75	24.3	Consultant	18	5.8
30+ years	53	17.2	Work position:		
Work as:			Not part of management	194	62.8
Mostly full time	283	91.6	In lower management	45	14.6
Mostly part time	8	2.6	In middle management	42	13.6
Mostly over time	17	5.5	In senior management	27	8.7
Been laid off from IT job:					
Yes	71	23.0			
No	237	76.7			

rise, and the inherent vulnerabilities in internet of things devices, this category getting high marks should be no surprise. Attracting and Retaining IT Professionals in the number three slot may be indicative of a global trend in needing more IT people, but it may also be more of a US-specific issue where IT job-hopping in the 1990s became a cultural norm for the IT occupation.

The fourth-ranked issue of Alignment between IT and Business had a mean rating (1.88) very close to the third (1.86). The theme of friction between IT and business is almost as old as the occupation itself and dates back to the 1980s. Since that time, IT has become a part of most firms' core business strategy as opposed to merely serving a back-office function. Yet IT-business strategic (mis)alignment has been an ongoing concern for senior managers (Kappelman *et al.*, 2018) due to different priorities, attitudes and occupational cultures (Jacks *et al.*, 2018) and the World IT Project supports this view as well.

The bottom three organizational IT issues were outsourcing (18), BYOD (17), and globalization (16). The low ranking of Outsourcing is an indication that IT professionals in the US feel simultaneously a strong sense of job

Table 37.2: Organizational IT Issues in the US

Organizational IT Issues	Rank	Mean Rating*	Std. Deviation
IT reliability and efficiency	1	1.53	0.688
Security and privacy	2	1.56	0.755
Attracting and retaining IT professionals	3	1.86	0.741
Alignment between IT and business	4	1.88	0.799
IT strategic planning	5	1.94	0.833
Knowledge management	6	2.03	0.765
Continuity planning and disaster recovery	7	2.06	1.007
Project management	8	2.08	0.841
IT service management (e.g., ITIL)	9	2.20	0.874
Enterprise architecture	10	2.28	0.917
Business productivity and cost reduction	11	2.37	0.809
Business process reengineering	12	2.38	0.917
Business agility and speed to market	13	2.46	0.905
IT cost reduction	14	2.49	0.899
Revenue-generating IT innovations	15	2.58	1.026
Globalization	16	2.91	1.081
Bring your own device (BYOD)	17	3.58	1.152
Outsourcing	18	3.69	1.030

*Rating scale ranges from 1 to 5: 1 as most important and 5 as no importance.

security and a strong desire to keep as many jobs in-house (and even in the US) as possible. The low ranking of BYOD seems at odds with the high ranking of security and privacy. It may simply be that a majority of respondents are not allowed by their organization's security policies to attach their home devices to the network. And the low ranking of globalization in the US is not surprising given the US's historical lack of focus on events (both economic and political) outside their borders. The full results are displayed in Table 37.2.

37.6 Technology and Infrastructure Issues

The top three technology issues were (1) networks/telecommunications, (2) enterprise application integration, and (3) collaborative and workflow tools. The top-rated issue of networks/telecommunications indicates a focus on speed and connectivity above all other concerns. This result would be easier to interpret in a developing country where the telecommunications infrastructure may be unreliable, but this would not be likely in many parts of the US (and especially not in the metropolitan areas where our data were collected). It therefore must be an indication that enterprise architecture is completely reliant on networked applications such that any network outage

Table 37.3:   Technology and Infrastructure Issues in the US

IT-Related Issues	Rank	Mean Rating*	Std. Deviation
Networks/telecommunications	1	2.10	1.024
Enterprise application integration	2	2.19	0.927
Collaborative and workflow tools	3	2.27	0.899
Business intelligence/analytics	4	2.28	0.920
Virtualization (desktop or server)	5	2.31	1.079
Mobile and wireless applications	6	2.38	1.049
Business process management systems	7	2.46	0.973
Software as a service	8	2.50	1.049
Cloud computing	9	2.58	0.993
Enterprise resource planning (ERP) systems	10	2.61	1.080
Big data systems	11	2.62	1.055
Data mining	12	2.63	1.087
Customer relationship management (CRM) systems	13	2.65	1.080
Service-oriented architecture (SOA)	14	2.65	1.056
Mobile apps development	15	2.77	1.123
Social networking/media	16	3.19	1.154

*Rating scale ranges from 1 to 5: 1 as most important and 5 as no importance.

is a mission-critical concern for the organization. This interpretation is further supported by the next two highest rated issues: Enterprise application integration and collaborative and workflow tools.

It is especially important to note that eight of the IT organizational issues had mean ratings higher than *any* of the Technology issues. Today's IT professionals seem to agree that understanding business issues is more important than understanding technology. Another surprising finding was that none of the elements from today's popular SMAC acronym (social media, mobile, analytics, cloud; see Choudhuri, 2017) appeared in the top three and two of them were at the very bottom of the list. Due to the rapid rate of change in the IT field, it seems very likely that the rankings would change from year to year and this data only captures a snapshot in time. The full list of issues is shown in Table 37.3.

37.7   Individual IT Employee Issues

The individual issues' ratings (Table 37.4) provide insights into both individual concerns and the IT occupation as a whole. In general, IT professionals in the US show a high level of job satisfaction and sense of

Table 37.4: Individual IT Employee Issues in the US

Individual Issues	Mean Rating*	Std. Deviation
Job satisfaction		
In general, I like working here.	1.67	0.75
All in all, I am satisfied with my current job.	1.92	0.89
In general, I don't like my current job.	4.13	0.93
Perceived work overload		
I feel that the number of requests, problems or complaints that I deal with at work is more than expected.	3.31	1.19
I feel that the amount of work I do interferes with how well it is done.	3.27	1.17
I feel busy or rushed at work.	2.99	1.16
I feel pressured at work.	3.16	1.17
Work–home conflict		
There is a blurring of boundaries between my job and my home life.	3.38	1.25
My work-related responsibilities create conflicts with my home responsibilities.	3.59	1.19
I do not get everything done at home because I find myself completing job-related work.	3.69	1.16
Work exhaustion/strain		
I feel drained from activities at work.	3.03	1.20
I feel tired from my work activities.	2.94	1.21
Working all day is a strain for me.	3.57	1.11
I feel burned out from my work activities.	3.47	1.17
Professional self-efficacy		
I feel I'm making an effective contribution to what this organization does.	1.87	0.80
In my opinion, I do a good job.	1.65	0.60
I have accomplished many worthwhile things in this job.	1.82	0.80
At my work, I feel confident that I am effective at getting things done.	1.78	0.70
Job insecurity		
I am worried that future technology advancements may pose a threat to my job.	3.62	1.15
I believe that other people may be able to perform my work activities.	2.59	1.04
I am concerned that my job may be eliminated soon.	3.86	1.09
I am concerned that my job may be outsourced soon.	3.95	1.13
Turnover intention		
I will be with this organization 1 year from now.	2.00	1.00
I will take steps during the next year to secure a job at a different organization.	3.64	1.10
I will be with this organization 5 years from now.	2.66	1.06

(*Continued*)

Table 37.4: (Continued)

Individual Issues	Mean Rating*	Std. Deviation
Turnover intention — IT profession		
I will be working in the IT field 1 year from now.	1.57	0.79
I will take steps during the next year to secure a job outside the IT field.	4.21	0.98
I will be working in the IT field 5 years from now.	1.91	1.02

*Rating scale ranges from 1 to 5: 1 as strongly agree, to 5 as strongly disagree.

accomplishment at work, a moderate but not high amount of work pressure, lower concerns for work–life balance and burnout, low concern about job security, and a fairly strong desire to stay both with their job and their occupation in the coming years. This shows a fairly positive outlook, more so than in some other countries. Part of this optimism may be due to the maturity of the IT occupation and the economic strength of the US compared to the rest of the world.

While stress, burnout, and turnover are common within the IT profession (Joseph et al., 2007), the World IT dataset puts these concerns in a larger context. Work exhaustion/strain and work–home conflict, for example, had some of the lowest scores in the entire survey and were rated lower than all of the technology issues in Table 37.3. The low rating (3.95 on a reversed scale) of concern about jobs being outsourced is consistent with outsourcing being at the bottom of the list of organizational concerns for IT workers.

IT workers in the US tend to have very high job satisfaction, a high regard for their own performance at work, low fears about job security, and high intention of staying with their organization and staying in the IT field. This is consistent with the high economic outlook for IT jobs in the US and should make IT a very attractive field for recent college graduates.

Despite this rosy outlook, it should be noted that individual issues tend to be extremely context-sensitive, so it is likely that issues like professional self-efficacy may be tied to age and level of education, while issues of job security may be affected by whether an IT worker has been laid off in the past or not and whether they are old enough to remember the dot-com bust. These concerns and more will be pursued in future research based on the World IT Project dataset. Table 37.4 shows the complete list of individual issues.

37.8   Conclusion

The World IT Project dataset for the US provides a baseline for close comparison with the results of other countries in this book. A total of 308 responses in the Midwest and Southeast of the US were collected and the typical response profile was a full-time non-managerial IT worker with a bachelor's degree and 10 or more years of IT experience. The top three organizational IT issues were (1) IT reliability and efficiency, (2) security and privacy, and (3) attracting and retaining IT professionals. The top three technology issues were (1) networks/telecommunications, (2) enterprise application integration, and (3) collaborative and workflow tools. Demand for IT employees remains high in a strong US economy and should continue to grow. IT worker responses reflected this optimism and showed high levels of job satisfaction, professional self-efficacy, and intention to remain in their jobs and the IT field, moderate amounts of work pressure, and low levels of concern for work–life balance, job burnout, and job security.

References

Ashford, S. J., Lee, C., & Bobko, P. (1989). Content, cause, and consequences of job insecurity: A theorybased measure and substantive test. *Academy of Management Journal*, 32(4), 803–829.

BLS (2018). Computer and Information Technology Occupations. *Bureau of Labor Statistics*. Accessed at https://www.bls.gov/ooh/computer-and-information-technology/home.htm.

Choudhuri, I. (2017). Social, Mobile, Analytics and Cloud: The Rising SMAC Trend. *Linked-In*. Accessed at https://www.linkedin.com/pulse/social-mobile-analytics-cloud-rising-smac-trend-roy-choudhuri/.

Jacks, T., Palvia, P., Iyer, L., Sarala, R., & Daynes, S. (2018). An ideology of IT occupational culture: The ASPIRE values. *ACM SIGMIS DATABASE: The Database for Advances in Information Systems*, 49(1), 93–117.

Kappelman, L., Johnson, V., McLean, E., & Maurer, C. (2018). The 2017 SIM IT Issues and Trends Study. *MIS Quarterly Executive*, 17(1), 53–88.

Kirmeyer, S. L. & Dougherty, T. W. (1988). Work load, tension, and coping: Moderating effects of supervisor support. *Personnel Psychology*, 41(1), 125–139.

Kreiner, G. E. (2006). Consequences of work-home segmentation or integration: A person-environment fit perspective. *Journal of Organizational Behavior*, 27(4), 485–507.

Luftman, J. & Ben-Zvi, T. (2010). Key issues for IT executives 2009: Difficult economy's impact on IT. *MIS Quarterly Executive*, 9(1), 203–213.

Luftman, J., Zadeh, H. S., Derksen, B., Santana, M., Rigoni, E. H., & Huang, Z. D. (2012). Key information technology and management issues 2011–2012: An international study. *Journal of Information Technology*, 27(3), 198–212.

Moore, J. E. (2000). One road to turnover: An examination of work exhaustion in technology professionals. *MIS Quarterly*, 24(1), 141–168.

Morgan, S. (2016). Zero-percent cybersecurity unemployment, 1 million jobs unfilled. *CSO Online*. Accessed at https://www.csoonline.com/article/3120998/techology-business/zero-percent-cybersecurity-unemployment-1-million-jobs-unfilled.html.

Palvia, P., Jacks, T., Ghosh, J., Licker, P., Romm-Livermore, C., Serenko, A., & Turan, A. H. (2017). The World IT Project: History, trials, tribulations, lessons, and recommendations. *Communications of the Association for Information Systems*, 41(18), 389–413.

Palvia, P., Ghosh, J., Jacks, T., Serenko, A., & Turan, A. (2018). Trekking the globe with the World IT Project. *Journal of Information Technology Case and Application Research*, 20(1), 3–8.

Statista (2017). Outsourced or off-shored IT functions worldwide as of 2017. *Statista*. Accessed at https://www.statista.com/statistics/662991/worldwide-cio-survey-outsourced-it-functions/.

White, S. (2018). The 7 most in-demand tech jobs for 2018 — and how to hire for them. *CIO*. Accessed at https://www.cio.com/article/3235944/hiring-and-staffing/hiring-the-most-in-demand-tech-jobs-for-2018.html.

Chapter 38

Information Technology Issues in Vietnam

Prashant Palvia[*,‡], Hao Wu[*,§], and Tim Jacks[†,¶]

[*]University of North Carolina at Greensboro,
Greensboro, NC, USA
[†]Southern Illinois University Edwardsville,
Edwardsville, IL, USA
[‡]pcpalvia@uncg.edu
[§]h_wu@uncg.edu
[¶]tjacks@siue.edu

Summary

In this chapter, we analyze and report information technology (IT) issues
for Vietnamese IT workers. The participants of our survey answered

questions about organizational, technological and individual IT issues. They are mostly young IT professionals, holding a Bachelor's degree, working full time, and in non-managerial positions. Security and privacy, revenue-generating IT innovations, business agility and speed to market, and knowledge management were among the most pressing organizational IT issues. The top three technology issues are mobile apps development, networks/telecommunications, and business intelligence/analytics. Vietnamese IT employees seem to feel satisfied with their IT jobs. As to the responses about work pressure, work–life balance, and workload, they do not feel under too much pressure. They also feel a sense of accomplishment in general about doing their jobs. Although Vietnamese IT workers do not think future technology advancements may threaten their job, nor do they have a concern about their jobs being eliminated or outsourced, they do feel that other people may be able to perform their work activities such that their position could be replaced.

38.1 Introduction

Vietnam has a population of about 96 million, which ranked 15th in the world. Not only is there cheap labor and a large market size but also strong foreign investment, education emphasizing on IT, adoption of international IT standards, a government-controlled IT development policy, etc. All these factors have helped contribute to Vietnam's growth in its (still young) IT industry. The literature related to the IT industry in Vietnam has emphasized trade policy (Adhikari *et al.*, 1992), business guides (Vierra and Vierra, 2011), rural development (Frohlich *et al.*, 2013), labor market (Gallup, 2002), economic development (Van, 2002), private entrepreneurship (Hoang and Dung, 2009), and market growth (Buiter and Rahbari, 2011; Pincus, 2015). Interestingly, there are few studies emphasizing on IT issues systematically on different levels. This chapter aims to fill this gap and obtain a clear understanding of the IT industry in Vietnam by analyzing the organizational, individual, and technological issues in the World IT Project.

38.2 Country Background and History

Vietnam, known as the Socialist Republic of Vietnam, is a socialist country in Asia. It is located in the eastern part of the Indo-China Peninsula in Southeast Asia, with an area of about 330,000 square kilometers. Vietnam is a multi-ethnic country, with the Jing people as the main body. Historically,

Vietnam's central and northern regions have long been Chinese territory. In 968, it officially separated from China. After that, Vietnam experienced many feudal dynasties and continued to expand southward, but all the dynasties maintained a vassal relationship with China. After the middle of the 19th century, it gradually became a French colony. After the August 1945 revolution, Ho Chi Minh announced the establishment of the Democratic Republic of Vietnam. In 1976, he changed the country's name to the Socialist Republic of Vietnam. The Communist Party of Vietnam is the only legal ruling party in the country (Hong Lien and Sharrock, 2014).

Vietnam is a member of the Association of Southeast Asian Nations. In 2014, Vietnam's population was about 90 million, ranking 13th in the world. Men accounted for 50.2%, and women accounted for 49.8%. The urban population accounts for 33%, and the rural population accounts for 67%. Vietnam is a multi-ethnic country, and the government has officially identified 54 different ethnicities. Every Vietnamese ethnicity has its own language, a way of life, and cultural heritage (VIR, 2014).

Vietnam is a developing country. In 1986, reforms and opening-up policies began to be implemented by the government. In 1996, the Eighth National Congress of the Communist Party of Vietnam proposed to vigorously promote national industrialization and modernization. In 2001, the Ninth National Congress of the Communist Party of Vietnam decided to establish a socialist-oriented market economic system and identified three major economic, strategic priorities, namely, focusing on industrialization and modernization, developing various economic components, giving play to the dominant position of the state-owned economy, and establishing a supporting market economy. In the past 20 years of reform and opening-up, the economy has maintained a relatively fast growth rate. From 1990 to 2006, the gross domestic product (GDP) grew at an average annual rate of 7.7%, the economic aggregate continued to expand, the structure of the three industries became more coordinated, and the level of opening up to the outside world continued to increase. Vietnam has established a development pattern which is dominated by the state-owned economy and supported by multiple other economic components. In 2006, Vietnam officially joined the World Trade Organization (WTO) and successfully held an informal meeting of leaders from the Asia-Pacific Economic Cooperation (APEC) (Vierra and Vierra, 2011).

The main industrial products in Vietnam are coal, crude oil, natural gas, liquefied petroleum gas, aquatic products, etc. Vietnam is also a traditional agricultural country with an agricultural population accounting for about

75% of the total population. Cultivated land and forest land account for 60% of the total area. Food crops include rice, corn, potatoes, sweet potatoes, and cassava. The main economic crops are coffee, rubber, cashew nuts, tea, peanuts, silk and so on. Vietnam has trade relations with more than 150 countries and regions in the world. Since 2013, Vietnam's foreign trade has maintained rapid growth. Import and export of goods, service trade, tourism, and transportation have played an essential role in stimulating economic development (Vierra and Vierra, 2011).

Vietnam has formed an education system that includes early childhood education, primary education, secondary education, higher education, teacher education, vocational education, and adult education. The general education system is 12 years and is divided into three stages. The first stage is a 5-year primary school, the second stage is a 4-year junior high school, and the third stage is a 3-year high school. In 2000, Vietnam announced that it had achieved the goal of universal compulsory primary education. In 2001, 9 years of compulsory education became popular. There are nearly 400 colleges and universities across the country. Famous universities include Hanoi National University, Ho Chi Minh City National University, Hue University, Taiyuan University, and Da Nang University (Tran, 2014).

38.3 Information Technology in Vietnam

With the continuous influx of new foreign investment, Vietnam's IT industry has entered a period of prosperity. The country now has more than 600 software companies. Recently, Intel, Canon, and other internationally renowned companies have settled in Vietnam. In addition, companies such as Alcatel, Fujitsu, and Siemens are also interested in investing in Vietnam. It is anticipated that Vietnam will become a leader of low-cost IT products and services and compete with countries such as Thailand and the Philippines.

Intel's trends often represent the trend of the global IT industry. The presence of Intel means that it will bring a larger supply chain network to Vietnam. The entire IT industry will pay attention to these developments, and other companies are likely to follow. In addition to professional suppliers such as NEC, TDK, Kelly services, DaewOn Semiconductor Packaging, Dainippon Screen, Mnters and Shinco Jinpeng, Intel will also give priority to local suppliers and technology companies in Vietnam. The growing number of suppliers will make Vietnam even more attractive to global IT manufacturers.

It was not until 2000 that Vietnam began to open to foreign manufacturing investors. However, it immediately defeated countries such as Thailand, Malaysia, and the Philippines that have developed well in the IT industry in the region and won the favor of Intel. After an extensive investigation, Intel chose to enter the Saigon Industrial Park because it provides good infrastructure and sufficient labor resources. The rapid growth of the local IT industry proves that Vietnam's IT industry is profitable. Vietnam's domestic IT industry-related infrastructure construction and educational achievements are rapidly catching up with Thailand and are expected to catch up with the Philippines in the near future. However, Malaysia still has advantages in its existing infrastructure construction, and its domestic manufacturing equipment. In addition, there are more laborers in Malaysia who are fluent in English. Although Vietnam is only a small competitor in Southeast Asia, its neighbor China has a huge market, a large number of suppliers, and a solid scientific research foundation. For big Japanese investors, investing in Vietnam does not cause any concerns about cultural and political conflicts. In addition, Vietnam's population is younger, and its desire for knowledge is strong.

Of course, Vietnam's cheap labor is one of the biggest reasons for attracting IT companies to invest in Vietnam. The labor cost in Vietnam is lower than in the Philippines, Malaysia and other countries, including even China. Education is another important reason for the rapid development of the IT industry in Vietnam. In Thailand, more than 40% of students receive college education after graduating from high school, compared with only 10% of colleges and universities in Vietnam. However, high school in Vietnam values the subject of mathematics, which provides a distinct advantage for people who choose to pursue a career in IT.

Vietnamese legislators are advocating more government subsidies for their IT industry, and the government is gradually increasing its focus on the country's Internet and communications technologies as a basis for broad-based development of the national economy. There is no doubt that Vietnamese people value the development of their IT industry. Just as the development of electricity was part of Russia's modernization in the era of Lenin, Internet construction is similarly part of the Vietnamese government's modernization plan. Part of that plan is to promote the use of mobile phones that are transmitted over the airwaves to make up for the shortage of traditional land-line telephones. So far, there are six mobile phone companies in Vietnam competing for consumers, and several of them are ready to build a third-generation network. In Vietnam, the number of mobile phone users has doubled every 2 years. With the support of Intel, Vietnam's new project, the latest W1HAV technology (wireless access, for global microwave access and interoperability) is under construction. Even in developed countries, W1HAV technology for commercial use is rare. The W1HAV project aims to connect rural areas and public access to the Internet through remote computing centers. Through these means, the Vietnamese government hopes to provide better and more effective public services in a wide variety of forms, from healthcare consultations to farming news.

These ambitious projects illustrate the increasing emphasis placed by Vietnamese officials on the use of IT and wireless communication technologies as tools to eradicate poverty and to provide better job opportunities in the knowledge industry, beyond traditional low-cost manufacturing jobs. It is premature to say that the rise of the IT industry in Vietnam has led international IT companies to ignore the IT industry already established in Thailand, the Philippines, or other countries and regions. However, it is without a doubt that the Vietnamese government is concentrating on developing its own IT industry to overtake other leading countries in Asia (VAST, 2017).

38.4   Methodology

Data were collected from Vietnamese organizations representing a broad variety of industries, and the companies were recruited to participate using purposive sampling. The original survey instrument of World IT Project (see Palvia *et al.*, 2017; Palvia *et al.*, 2018) has been modified. No incentives were provided to the respondents. The responses resulted in 298 usable surveys after incomplete surveys were discarded. Most of the respondents (about 76.9%) hold a bachelor's degree. The first and second organizational location groups are IT department employees (56.4%) and contract employees (26.5%). About half (52.4%) of them were not part of management, 16.1% are in lower management level, 25.2% are in middle management level, 6.4% are in senior management level and above. More descriptive statistics are provided in Table 38.1.

38.5   Organizational IT Issues

Table 38.2 shows the organizational IT issues from the investigators' report ranked in order of importance. The participants from Vietnam were asked to rate 18 organizational IT-related issues in terms of their level of importance.

Table 38.1:   Descriptive Statistics

Characteristics	N	%	Characteristics	N	%
Education:			Organizational location:		
High school or less	4	1.34	IT department employee	168	56.38
Associate degree	28	9.40	IT worker in non-IT department	29	9.73
Bachelor's degree	229	76.85	Contract employee	79	26.51
Master's degree	33	11.07	Consultant	4	1.34
Ph.D.	4	1.34	Vendor employee	18	6.04
Years of work experience:			Work as:		
0–4 years	84	28.19	Mostly full time	272	91.28
5–9 years	144	48.32	Mostly part time	3	1.01
10–19 years	67	22.48	Mostly over time	23	7.72
20–29 years	2	0.67	Been laid off from IT job:		
30+ years	1	0.34	Yes	12	4.03
Years of IT experience:			No	286	95.97
0–4 years	82	27.52	Part of management:		
5–9 years	148	49.66	Not part of management	156	52.35
10–19 years	64	21.48	In lower management	48	16.11
20–29 years	3	1.01	In middle management	75	25.17
30+ years	1	0.34	In senior management	19	6.38

The rating scale ranges from 1 to 5, with 1 as most important and 5 meaning not important. The 18 issues, the average rating for each issue, the standard deviation of the response, and the rank of the 18 issues are presented below.

According to Table 38.2, security and privacy is the most important organizational IT issue. Revenue-generating IT innovations came up as the second important organizational issue, followed by business agility and speed to market and knowledge management which are the third and fourth, respectively. The next three issues are IT reliability and efficiency, alignment between IT and business, and attracting and retaining IT professionals. The least important organizational issues listed are globalization, business process reengineering, outsourcing, and BYOD.

As technology advances and the use of technology increases, companies become more and more dependent on it. This dependence, however, makes companies more vulnerable to security and privacy threats. Thus, it is not surprising that the highest-ranked issue in our survey is security and privacy. Vietnam is in the stage of fast development. More foreign investors and government-controlled projects require security and privacy to be a higher concern, and new cybersecurity legislation was recently passed.

In order to stay relevant and keep growing in business, innovation plays a crucial role in Vietnamese IT companies. Thus, it is reasonable that

Table 38.2: Organizational IT Issues in Vietnam

Organizational IT Issues	Rank	Mean Rating*	Std. Deviation
Security and privacy	1	2.21	0.78
Revenue-generating IT innovations	2	2.26	0.80
Business agility and speed to market	3	2.29	0.83
Knowledge management	4	2.30	0.76
IT reliability and efficiency	5	2.31	0.74
Alignment between IT and business	6	2.35	0.74
Attracting and retaining IT professionals	7	2.37	0.75
Project management	8	2.37	0.70
Continuity planning and disaster recovery	9	2.42	0.85
Enterprise architecture	10	2.47	0.80
IT strategic planning	11	2.47	0.67
IT cost reduction	12	2.51	0.80
IT service management (e.g., ITIL)	13	2.53	0.73
Business productivity and cost reduction	14	2.56	0.74
Globalization	15	2.62	0.81
Business process reengineering	16	2.68	0.69
Outsourcing	17	2.76	0.84
Bring your own computing device (BYOD)	18	2.88	1.04

*Rating scale ranges from 1 to 5: 1 as most important and 5 as no importance.

revenue-generating IT innovations comes up as the second most important issue on the list. IT companies in developing countries like Vietnam are facing challenges during their evolution. They're focusing on generating revenue rather than cutting costs. They are also emphasizing improving existing projects, instead of finishing them up as soon as possible and initiating the next one. Seeking external resources rather than internal resources is another trend. This evolution requires IT cultural transformation which pays attention to speed, innovation, and revenue generation.

Business agility and speed to market ranks number three on the organizational issues list. In the phase of rapid development, IT companies in Vietnam are facing intense competition and constant change. In an environment like this, it is very important that businesses pay attention and respond to change proactively. Companies will gain a tremendous advantage if they can drive value through IT to enable better, faster, and more efficient business processes. Increasing speed to market enables businesses to reach customers more quickly, develop innovation faster, and stay one step ahead of their competitors.

Knowledge management is another important issue. A successful knowledge management system allows an organization to understand the flow of knowledge, clearly define the organizational goals and assessment measures, generate new knowledge, and apply that knowledge to new technology. Thus, this issue is closely related to the IT reliability and efficiency issue, which ranks right below it. Knowledge management boosts efficiency and improves decision-making. In order to perform better in knowledge management, companies should also improve IT reliability by reducing unwanted, unanticipated, and unexplained variance in organizational processes (Hollnagel, 1993).

Regarding the least important organizational issues, BYOD (bring your own computing device) appears at the bottom. "BYOD is making significant inroads in the business world, with about 75% of employees in high growth markets such as Brazil and Russia and 44% in developed markets already using their own technology at work" (Ian, 2013). Perhaps the reason why the Vietnamese IT industry thinks BYOD is not that important is that, in a fast-developing environment, companies do not generally have any restrictions or policies that prohibit this particular activity. Business process reengineering and outsourcing are also on the bottom due to the fact that most businesses in the IT industry in Vietnam are government-controlled or foreign investments. They do not need to worry about these because they are under "good" policy or they already have matured experience from global-leading companies. Globalization is not a major concern for IT professionals in Vietnam, not because it is unimportant, but because it is not new anymore and has already been embedded deeply in all organizational contexts throughout the IT industry.

38.6 Technology and Infrastructure Issues

The participants from Vietnam were also asked to rate 16 IT issues in terms of their level of importance. The rating scale ranges from 1 to 5, with 1 as most important and 5 as of no importance. Based on the investigators' report, Table 38.3 reveals the 16 issues, the average rating for each issue, the standard deviation of the response, and the rank of each issue.

According to the report, the top three technology issues are mobile apps development, networks/telecommunications, and business intelligence/ analytics. The least important technology issues are cloud computing, virtualization (desktop or server), enterprise resource planning (ERP) systems, mobile and wireless applications, and software as a service. Interestingly, mobile and wireless applications appear in the last position while mobile application development appears in the top place. Mobile communication

Table 38.3: Technology and Infrastructure Issues in the US

IT-Related Issues	Rank	Mean Rating*	Std. Deviation
Mobile apps development	1	2.43	0.81
Networks/telecommunications	2	2.44	0.81
Business intelligence/analytics	3	2.45	0.86
Customer relationship management (CRM) systems	4	2.50	0.81
Big data systems	5	2.53	0.87
Social networking/media	6	2.58	0.75
Business process management systems	7	2.63	0.81
Service-oriented architecture (SOA)	8	2.63	0.81
Data mining	9	2.65	0.85
Enterprise application integration	10	2.67	0.76
Collaborative and workflow tools	11	2.68	0.71
Cloud computing	12	2.70	0.75
Virtualization (desktop or server)	13	2.70	0.83
Enterprise resource planning (ERP) systems	14	2.70	0.89
Software as a Service	15	2.73	0.94
Mobile and wireless applications	16	2.73	0.78

*Rating scale ranges from 1 to 5: 1 as most important and 5 as no importance.

devices have changed the way companies communicate. Making progress in mobile applications development represents a more efficient, responsive, and proactive business. Technology creates mobile and wireless applications for computing devices such as smartphones and tablets, which changes the business landscape as it allows companies to create customized, secured applications. It provides an integrated experience for customers, suppliers, and other business partners. Perhaps the reason why these two issues are ranked so differently is that IT professionals in Vietnam think the development side of mobile applications is more important than the mobile application itself.

Networks and telecommunications is rated as the second most important technology issue. In developing countries like Vietnam, the power of exchanging information over the Internet over a significant distance needs to be enhanced in order to provide quality business services to the world. Hence, the Vietnamese government is gradually increasing its focus on the country's Internet and communications technologies as a basis for broad-based development of the national economy. In Vietnam, companies rely increasingly more on networks and telecommunications to ensure their business performance. Six Internet service providers create a healthy competitive environment.

Business intelligence and analytics is another important part of the IT field. In today's era of big data, companies all over the world use business intelligence and analytics to detect significant events, identify, and monitor business trends in order to adapt quickly to the changing environment. By emphasizing business intelligence and analytics, organizations can gain insights about consumer behavior, turn data into actionable information, improve efficiency, and enhance the ability to identify suitable business opportunities. In Vietnam, job openings for business intelligence and analytics have been on the rise over the past few years representing an important trend.

It seems that customer relationship management (CRM) systems and enterprise resource planning (ERP) systems are not among the most important technology issues. Although these systems contribute tremendously to IT services, in developing countries like Vietnam, less mature companies do not have enough investment ambition for such systems, nor do they have the experience necessary for implementing. In contrast, companies are more willing to invest in more practical areas such as mobile applications development, networks and telecommunications, and business intelligence and analytics.

38.7   Individual IT Employee Issues

Seven themes of individual IT employee issues were investigated from the participants. The rating scale ranges from 1 to 5, with 1 as strongly agree and 5 as strongly disagree. The detailed results of each theme are provided below in Table 38.4.

For job satisfaction, participants seem to feel satisfied with their IT jobs. The low standard deviation means there is consensus on this issue. The number one factor for job satisfaction is salary. IT is one of the highest paid industries, which makes IT work very attractive. In addition, the emphasis on mathematics throughout Vietnamese students' education makes young people more suitable to and, more importantly, motivated to pursue a career in IT.

As for the responses about work pressure, work–life balance, and workload, it seems that because work time is not long, and workload is not large, Vietnamese IT employees do not feel under too much pressure. The IT workers in our sample are mostly young professionals or new employees, and it is normal that they are the group of people who will be assigned more tasks, in order to get trained and become familiar with the industry.

Table 38.4:   Individual IT Employee Issues in Vietnam

Individual Issues	Mean Rating*	Std. Deviation
Job satisfaction		
In general, I like working here.	2.17	0.71
All in all, I am satisfied with my current job.	2.29	0.71
In general, I don't like my current job.	3.50	0.90
Work pressure		
I feel that the number of requests, problems or complaints that Ideal with at work is more than expected.	3.26	0.91
I feel that the amount of work I do interferes with how well it is done.	3.14	0.94
I feel busy or rushed at work.	2.73	0.87
I feel pressured at work.	3.06	0.87
Work–life balance		
There is a blurring of boundaries between my job and my home life.	3.54	1.03
My work-related responsibilities create conflicts with my home responsibilities.	3.68	0.99
I do not get everything done at home because I find myself completing job-related work.	3.13	1.01
Workload and burnout		
I feel drained from activities at work.	3.46	0.97
I feel tired from my work activities.	3.47	0.94
Working all day is a strain for me.	3.49	1.03
I feel burned out from my work activities.	3.41	0.99
Sense of accomplishment		
I feel I'm making effective contribution to what this organization does.	2.31	0.69
In my opinion, I do a good job.	2.24	0.66
I have accomplished many worthwhile things in this job.	2.35	0.70
At my work, I feel confident that I am effective at getting things done.	2.47	0.75
Threats to one's job		
I am worried that future technology advancements may pose a threat to my job.	3.30	0.98
I believe that other people may be able to perform my work activities.	2.57	0.69
I am concerned that my job may be eliminated soon.	3.37	0.86
I am concerned that my job may be outsourced soon.	3,39	0.83
Career plans		
I will be with this organization 1 year from now.	3.12	1.00
I will take steps during the next year to secure a job at a different organization.	3.18	0.92

(Continued)

Table 38.4: (*Continued*)

Individual Issues	Mean Rating*	Std. Deviation
I will be with this organization 5 years from now.	2.87	0.96
I will be working in the IT field 1 year from now.	3.22	1.01
I will take steps during the next year to secure a job outside the IT field.	3.29	0.92
I will be working in the IT field 5 years from now.	2.74	0.95

*Rating scale ranges from 1 to 5: 1 as strongly agree, and 5 as strongly disagree.

Interestingly, the participants disagree that they have a large workload, feel their everyday life is affected by work, or that they are strongly pressured.

The IT employees in Vietnam feel a sense of accomplishment in general about doing their jobs. The low standard deviation confirms their sense of fulfillment. The IT skills of these young professionals are in high demand in the job market in Vietnam. IT employees typically have a 10% higher education rate than the rest of society. Hence, their sense of accomplishment comes from their actual capability as well as confidence.

Although Vietnamese IT workers do not think future technology advancements may threaten their jobs or have concerns about their jobs being eliminated or outsourced (because they are in a developing country, and their IT industry is in a fast growth period), they do feel threats that other people may be able to perform their work activities such that their position may be replaced. With more and more IT major graduates, the competition for IT jobs will indeed be getting more intense.

Looking at the career plan data, Vietnamese IT employees incline slightly to change their jobs or the profession in the short term. In the longer 5-year term, the propensity to leave the organization or the profession decreases. Under the developing economic environment in Vietnam, IT jobs are always in demand and growing. The technology world is in constant change, which requires the IT professionals to keep their skill sets up to date and discover new strengths. This is both valuable and attractive to young people. High salaries in the IT field is another reason why people want to stay in.

38.8 Conclusion

This chapter provided an analysis of the organizational IT issues, technology issues, and individual concerns of IT employees in Vietnam.

Security and privacy, revenue-generating IT innovations, business agility and speed to market, and knowledge management were ranked the most critical organizational IT issues. The top three technology issues are mobile apps development, networks/telecommunications, and business intelligence/analytics. Vietnamese IT employees seem to feel satisfied with their IT jobs. As for the responses about work pressure, work–life balance, and workload, it seems that they do not feel too much pressure. They also, in general, feel a sense of accomplishment about doing their jobs. Although Vietnamese IT workers do not think future technology advancements may threaten their jobs or have concerns about their jobs being eliminated or outsourced, they do feel other people could perform their work activities so that their position could be replaced. Vietnam's IT industry is growing substantially. It is without a doubt that Vietnam and its government are concentrating on developing the IT industry to overtake other leading countries. Moreover, Vietnam is sparing no effort to attract more IT companies that want to invest in the region. Thus, while the IT industry is thriving, understanding the top issues in the Vietnamese context provides new insights, and the results will provide value to both academics and practitioners.

References

Adhikari, R., Kirkpatrick, C., & Weiss, J. (1992). *Industrial and Trade Policy Reform in Developing Countries.* Manchester University Press.

Buiter, W., & Rahbari, E. (2011). Global growth generators: Moving beyond emerging markets and BRICs. *Centre for Economic Policy Research* (11).

Cook, I. (2013). BYOD — Research findings. *Logicalis White Paper.* Retrieved from https://cxounplugged.com/2012/11/ovum_byod_research-findings-released.

Frohlich, H., Schreinemachers, P., Stahr, K., & Clemens, G. (Eds.) (2013). *Sustainable Land Use and Rural Development in Southeast Asia: Innovations and Policies for Mountainous Areas.* Springer Science & Business Media.

Gallup, J. (2002). The wage labor market and inequality in Viet Nam in the 1990s. *Policy Research Working Paper Series*, World Bank.

Hoang V., & Dung Tran, T. (2009). The Cultural Dimensions of the Vietnamese Private Entrepreneurship. *The IUP Journal of Entrepreneurship and Development*, VI(3&4), 54–78.

Hollnagel, E. (1993). The phenotype of erroneous actions. *International Journal of Manmachine Studies*, 39, 1–32.

Lien, V., & Sharrock, P. (2014). *Descending Dragon, Rising Tiger: A History of Vietnam.* Reaktion Books.

Palvia, P., Jacks, T., Ghosh, J., Licker, P., Romm-Livermore, C., Serenko, A., & Turan, A. H. (2017). The World IT Project: History, trials, tribulations, lessons, and recommendations. *Communications of the Association for Information Systems*, 41(18), 389–413.

Palvia, P., Ghosh, J., Jacks, T., Serenko, A., & Turan, A. (2018). Trekking the globe with the World IT Project. *Journal of Information Technology Case and Application Research*, 20(1), 3–8.

Pincus, J. (2015). Why Doesn't Vietnam Grow Faster?: State Fragmentation and the Limits of Vent for Surplus Growth. *Journal of Southeast Asian Economies*, 32(1), 26–51.

Tran, L. (2014). *Higher Education in Vietnam: Flexibility, Mobility and Practicality in the Global Knowledge Economy (Palgrave studies in global higher education)*. Basingstoke: Palgrave Macmillan.

Van Tho, T. (2003). Economic development in Vietnam during the second half of the 20th century: How to avoid the danger of lagging behind. *The Vietnamese Economy: Awakening the Dormant Dragon*. Taylor & Francis.

Vierra, K.and Vierra, B. (2011). *Vietnam Business Guide: Getting Started in Tomorrow's Market Today*. John Wiley & Sons.

VAST (2017). Recent development and implementation plan 2017–2022 of Vietnam Space Center Project. *Vietnam Academy of Science and Technology*. Retrieved from http://www.vast.ac.vn/en/news/activities/1755-recent-development-and-implementation-plan-2017-2022-of-vietnam-space-center-project.

VIR (2014). Conquering the Fansipan. *Vietnam Investment Review*. Retrieved from https://www.vir.com.vn/conquering-the-fansipan-30707.html.

Index

World Scientific–Now Publishers Series in Business

(Continuation of series card page)